CLIMATE CHANGE SCIENCE PROGRAM and
SUBCOMMITTEE ON GLOBAL CHANGE RESEARCH

James R. Mahoney
Department of Commerce
National Oceanic and Atmospheric
Administration
Director, Climate Change Science Program
Chair, Subcommittee on Global Change
Research

Ghassem Asrar, Vice Chair
National Aeronautics and Space
Administration

Margaret S. Leinen, Vice Chair
National Science Foundation

James Andrews
Department of Defense

Mary Glackin
National Oceanic and Atmospheric
Administration

Charles (Chip) Groat
U.S. Geological Survey

William Hohenstein
Department of Agriculture

Linda Lawson
Department of Transportation

Melinda Moore
Department of Health and Human Services

Patrick Neale
Smithsonian Institution

Aristides Patrinos
Department of Energy

Jacqueline Schafer
U.S. Agency for International Development

Michael Slimak
Environmental Protection Agency

Harlan Watson
Department of State

EXECUTIVE OFFICE AND OTHER LIAISONS

Kathie L. Olsen
Office of Science and Technology Policy
Co-Chair, Committee on Environment and
Natural Resources

David Conover
Department of Energy
Director, Climate Change Technology Program

Philip Cooney
Council on Environmental Quality

David Halpern
Office of Science and Technology Policy

Margaret R. McCalla
Office of the Federal Coordinator
for Meteorology

Erin Wuchte
Office of Management and Budget

This document was prepared in compliance with Section 515 of the Treasury and General Government Appropriations Act for Fiscal Year 2001 (Public Law 106-554) and information quality guidelines issued by the Department of Commerce and NOAA pursuant to Section 515 (<http://www.noaanews.noaa.gov/stories/iq.htm>). For purposes of compliance with Section 515, this climate change science research program is an "interpreted product," as that term is used in NOAA guidelines.

This document describes a strategic plan for developing scientific knowledge regarding climate change. It does not express any regulatory policies of the United States or any of its agencies, or make any findings of fact that could serve as predicates for regulatory action. Agencies must comply with required statutory and regulatory processes before they could rely on any statements in this document or by the CCSP as a basis for regulatory action.

Strategic Plan
for the U.S. Climate Change
Science Program

A Report by the Climate Change Science Program and
the Subcommittee on Global Change Research

Members of Congress:

Transmitted herewith is a copy of the *Strategic Plan for the Climate Change Science Program*. This document describes the Climate Change Science Program (CCSP) approach to enhancing scientific understanding of global climate change. This plan is summarized in a companion document, *The U.S. Climate Change Science Program:Vision for the Program and Highlights of the Scientific Strategic Plan*.

This strategic plan responds to the President's direction that climate change research activities be accelerated to provide the best possible scientific information to support public discussion and decisionmaking on climate-related issues. The plan also responds to Section 104 of the Global Change Research Act of 1990, which mandates the development and periodic updating of a long-term national global change research plan coordinated through the National Science and Technology Council. This is the first comprehensive update of a strategic plan for U.S. global change and climate change research since the original plan for the U.S. Global Change Research Program was adopted at the inception of the program in 1989.

The President established the U.S. Climate Change Science Program in 2002 as part of a new cabinet-level management structure to oversee public investments in climate change science and technology. The new management structure also includes the Climate Change Technology Program, which is responsible for accelerating climate-related technology research and development. The CCSP incorporates the U.S. Global Change Research Program, established by the Global Change Research Act, and the Climate Change Research Initiative, established by the President in 2001. The Program coordinates and integrates scientific research on global change and climate change sponsored by 13 participating departments and agencies of the U.S. Government.

The CCSP, under the direction of the Assistant Secretary of Commerce for Oceans and Atmosphere, reports through the Interagency Working Group on Climate Change Science and Technology to the cabinet-level Committee on Climate Change Science and Technology Integration. The chairmanship of these coordinating bodies rotates annually between the Departments of Commerce and Energy, with the Director of the Office of Science and Technology Policy serving as the Executive Director of the cabinet-level committee.

The CCSP strategic plan, though disseminated by the Department of Commerce, was developed through a multi-agency collaboration and has benefited substantially from external review of an earlier discussion draft by a special committee of the National Academy of Sciences – National Research Council, as well as extensive public review by hundreds of scientists and stakeholders. The strategic plan document contains a detailed discussion of the goals and priorities for the program and how climate and global change research activities will be integrated.

The CCSP strategic plan reflects a commitment to high-quality research that advances the frontiers of science and outlines an integrated approach for developing an improved understanding of climate change and its potential impacts. The program described in this strategic plan will meet the highest standards of credibility and transparency to support public evaluation of climate change issues.

We thank the participating departments and agencies of the CCSP for their close cooperation and support and look forward to working with Congress in the continued development of these important programs.

Spencer Abraham
Secretary of Energy
Chair, Committee on Climate Change
Science and Technology Integration

Donald L. Evans
Secretary of Commerce
Vice Chair, Committee on Climate Change
Science and Technology Integration

John H. Marburger III, Ph.D.
Director, Office of Science and Technology Policy
Executive Director, Committee
on Climate Change
Science and Technology Integration

Strategic Plan for the Climate Change Science Program

FOREWORD

In February 2002, President George W. Bush announced the formation of a new management structure, the Climate Change Science Program (CCSP), to coordinate and direct the U.S. research efforts in the areas of climate and global change. These research efforts include the U.S. Global Change Research Program (USGCRP) authorized by the Global Change Research Act of 1990, and the Climate Change Research Initiative (CCRI), launched by the President in June 2001 to reduce significant uncertainties in climate science, improve global observing systems, develop science-based information resources to support policymaking and resource management, and communicate findings broadly among the international scientific and user communities. The CCSP aims to balance the near-term (2- to 4-year) focus of the CCRI with the breadth of the USGCRP.

This strategic plan has been prepared by the 13 federal agencies participating in the CCSP, with coordination by the CCSP staff under the leadership of Dr. Richard H. Moss. The development of this plan has benefited from the contributions of an extraordinarily large number of climate scientists and users of climate information. More than 250 federal government scientists participated in drafting both the *Discussion Draft Strategic Plan* published by CCSP in November 2002, and the revised plan presented in this document. The CCSP Climate Science Workshop, held in Washington, DC, in December 2002, provided critique and suggestions from 1,300 climate scientists and information users from throughout the United States and 35 other nations. In January 2003, CCSP received 270 sets of comments from participants in the workshop and other interested specialists. At the request of CCSP, the National Academy of Sciences – National Research Council (NRC) convened an expert committee to review the *Discussion Draft Strategic Plan* and the workshop discussions. The NRC committee provided extensive advice in its review published in February 2003.

We thank all of those individuals and groups who have contributed substantially to the completion of this strategic plan, and we hope that the plan will provide useful guidance for the continued national investment in climate change science. While acknowledging the useful contributions of many specialists, the CCSP interagency team takes responsibility for the plan, and invites continued correspondence regarding improvements that should be considered over the months and years ahead. A prominent web link for receiving and displaying comments will be maintained at the web site <www.climatescience.gov>. In view of the evolving nature of science, and in view of the continuing emergence of key climate change issues, we expect that periodic updates of selected elements of the strategic plan will be published when warranted.

James R. Mahoney, Ph.D.
Assistant Secretary of Commerce for Oceans and Atmosphere, and Director, Climate Change Science Program

July 2003

CHAPTER

About the Strategic Plan

This Strategic Plan for the Climate Change Science Program describes a strategy for developing knowledge of variability and change in climate and related environmental and human systems, and for encouraging the application of this knowledge. The strategy seeks to optimize the benefits of research that is conducted, sponsored, or applied by 13 agencies and departments of the U.S. government. These agencies coordinate their research through the Climate Change Science Program (CCSP), which incorporates the U.S. Global Change Research Program (USGCRP) and the Climate Change Research Initiative (CCRI).

Scientists and research program managers from the 13 participating agencies and the Climate Change Science Program Office drafted the Strategic Plan. It reflects a commitment by its authors to high-quality science, which requires openness to review and criticism by the wider scientific community. The process by which the plan was drafted proceeded with the transparency essential for scientific credibility.

External comments have played a key role in revising the initial draft of the plan. Significant input was received during the CCSP workshop held in December 2002, in Washington, D.C., which was attended by 1,300 scientists and stakeholders. Written comments on the *Discussion Draft Strategic Plan* were submitted during a public comment period. When collated, these comments amounted to nearly 900 pages of input from scientists and stakeholders. In addition, a special committee of the National Research Council (NRC) reviewed the draft plan at the request of the CCSP. The NRC committee also will submit a final report on both the content of the plan and the process used to produce it.

This document is not a detailed blueprint for conducting specific research projects or applying research results. Research strategies and implementation plans for specific areas of science are the focus of detailed planning documents that have already been prepared for some research elements, and are under preparation for others. These implementation plans spell out agency roles and provide added detail on the prioritization and sequencing of research activities. Annual budget information will continue to be provided through the program's annual report to Congress, *Our Changing Planet*. As a baseline reference for this Strategic Plan, the overall program budget is assumed to continue at the level of the FY2004 President's Budget Request. Actual budget commitments will be made on an annual basis.

While this document provides overviews of research, it does not report in detail on past or expected scientific conclusions.

Summaries of the current state of scientific knowledge are available in a series of reports, evaluations, and assessments conducted by international and national scientific bodies, as well as in the open scientific literature.

Brief definitions of the terms used throughout this chapter are listed in Box 1-1. For a more complete listing of the definitions used throughout the plan, refer to Annex D.

Vision and Mission of the Climate Change Science Program

Climate and climate variability play important roles in shaping the environment, natural resources, infrastructure, economy, and other aspects of life in all countries of the world. Potential human-induced changes in climate and related environmental systems, and the

BOX 1-1

DEFINITION OF KEY TERMS

Adaptation
Adjustment in natural or human systems to a new or changing environment that exploits beneficial opportunities or moderates negative effects.

Climate
Climate can be defined as the statistical description in terms of the mean and variability of relevant measures of the atmosphere-ocean system over periods of time ranging from weeks to thousands or millions of years.

Climate Change
A statistically significant variation in either the mean state of the climate or in its variability, persisting for an extended period (typically decades or longer). Climate change may be due to natural internal processes or to external forcing, including changes in solar radiation and volcanic eruptions, or to persistent human-induced changes in atmospheric composition or in land use.

Climate Feedback
An interaction among processes in the climate system in which a change in one process triggers a secondary process that influences the first one. A positive feedback intensifies the change in the original process, and a negative feedback reduces it.

Climate System
The highly complex system consisting of five major components: the atmosphere, the hydrosphere, the cryosphere, the land surface, and the biosphere, and the interactions among them. The climate system evolves in time under the influence of its own internal dynamics and because

of external forcings such as volcanic eruptions, solar variations, and human-induced forcings such as the changing composition of the atmosphere and land-use change.

Climate Variability
Variations in the mean state and other statistics of climatic features on temporal and spatial scales beyond those of individual weather events. These often are due to internal processes within the climate system. Examples of cyclical forms of climate variability include El Niño-Southern Oscillation (ENSO), the North Atlantic Oscillation (NAO), and the Pacific Decadal Variability (PDV).

Decision Support Resources
The set of observations, analyses, interdisciplinary research products, communication mechanisms, and operational services that provide timely and useful information to address questions confronting policymakers, resource managers, and other users.

Global Change
Changes in the global environment (including alterations in climate, land productivity, oceans or other water resources, atmospheric chemistry, and ecological systems) that may alter the capacity of the Earth to sustain life (from the Global Change Research Act of 1990, PL 101-606).

Mitigation (Climate Change)
An intervention to reduce the causes of change in climate. This could include approaches devised to reduce emissions of greenhouse gases to the atmosphere; to enhance their removal from the

atmosphere through storage in geological formations, soils, biomass, or the ocean; or to alter incoming solar radiation through several "geo-engineering" options.

Observations
Standardized measurements (either continuing or episodic) of variables in climate and related systems.

Prediction (Climate)
A probabilistic description or forecast of a future climate outcome based on observations of past and current climatological conditions and quantitative models of climate processes (e.g., a prediction of an El Niño event).

Projection (Climate)
A description of the response of the climate system to an assumed level of future radiative forcing. Changes in radiative forcing may be due to either natural sources (e.g., volcanic emissions) or human-induced causes (e.g., emissions of greenhouse gases and aerosols, or changes in land use and land cover). Climate "projections" are distinguished from climate "predictions" in order to emphasize that climate projections depend on scenarios of future socioeconomic, technological, and policy developments that may or may not be realized.

Weather
The specific condition of the atmosphere at a particular place and time. It is measured in terms of parameters such as wind, temperature, humidity, atmospheric pressure, cloudiness, and precipitation.

options proposed to adapt to or mitigate these changes, may also have substantial environmental, economic, and societal consequences. Because of the pervasiveness of the effects of climate variability and the potential consequences of human-induced climate change and response options, citizens and decisionmakers in public and private sector organizations need reliable and readily understood information, including a clear understanding of the reliability limits of such information, to make informed judgments and decisions. Over the past 15 years, the United States has invested heavily in scientific research, monitoring, data management, and assessment for climate change analyses to build a foundation of knowledge for decisionmaking. The seriousness of the issues and the unique role that science can play in helping to inform society's course give rise to CCSP's guiding vision:

> *A nation and the global community empowered with the science-based knowledge to manage the risks and opportunities of change in the climate and related environmental systems.*

The core precept that motivates CCSP is that the best possible scientific knowledge should be the foundation for the information required to manage climate variability and change and related aspects of global change. Thus the CCSP mission is to:

> *Facilitate the creation and application of knowledge of the Earth's global environment through research, observations, decision support, and communication.*

CCSP will add significant integrative value to the individual Earth and climate science missions of its 13 participating agencies and departments, and their national and international partners. A critical role of the interagency program is to coordinate research and integrate and synthesize information to achieve results that no single agency, or small group of agencies, could attain.

CCSP Goals

CCSP will develop knowledge that addresses the following basic questions:

> *How will variability and potential change in climate and related systems affect natural environments and our way of life? How can we use and improve this knowledge to protect the global environment and to provide a better living standard for all?*

Five CCSP goals have been identified to focus and orient research in the program to ensure that knowledge developed by the participating agencies and research elements can be integrated and synthesized to address these broad questions.

CCSP Goal 1: Improve knowledge of the Earth's past and present climate and environment, including its natural variability, and improve understanding of the causes of observed variability and change

Climate conditions change significantly over the span of weeks, seasons, years, decades, and even longer time scales. CCSP research will improve understanding of natural oscillations in climate on time scales from weeks to centuries, including improving and

---

> # GUIDING VISION
> # FOR THE CCSP
>
> A nation and the global community empowered with the science-based knowledge to manage the risks and opportunities of change in the climate and related environmental systems.

harnessing ENSO forecasts, a large-scale climate oscillation with implications for resource and disaster management. Research will sharpen qualitative and quantitative understanding of climate extremes, and whether any changes in their frequency or intensity lie outside the range of natural variability, through improved observations, analysis, and modeling. The program also will expand observations, monitoring, and data/information system capabilities and increase confidence in our understanding of how and why climate is changing. Fostering improved interactions and connectivity between research and ongoing operational measurements and activities will be another important aspect of the program's work.

CCSP Goal 2: Improve quantification of the forces bringing about changes in the Earth's climate and related systems

Combustion of fossil fuels, changes in land cover and land use, and industrial activities produce greenhouse gases (GHGs) and aerosols and alter the composition of the atmosphere and physical and biological properties of the Earth's surface. These changes have several important climatic effects, some of which can be quantified only poorly at present. Research conducted through CCSP will address reducing uncertainty in the sources and sinks of GHGs; aerosols and their precursors; the long range atmospheric transport of GHGs and aerosols and their precursors; and the interactions of GHGs and aerosols with global climate, ozone in the upper and lower layers of the atmosphere, and regional-scale air quality. It will improve quantification of the interactions among the carbon cycle, other biological/ecological processes, and land cover and land use to better project atmospheric concentrations of key greenhouse gases and to support improved decisionmaking. The program will also improve capabilities for developing and applying emissions scenarios in research and analysis, in cooperation with the Climate Change Technology Program (CCTP).

CCSP Goal 3: Reduce uncertainty in projections of how the Earth's climate and related systems may change in the future

While a great deal is known about the mechanisms that affect the response of the climate system to changes in natural and human influences, many questions remain to be addressed. There is also a high level of uncertainty regarding precisely how much climate will change overall and in specific regions. A primary objective of CCSP is to develop information and scientific capacity needed to sharpen qualitative and quantitative understanding through interconnected observations, data assimilation, and modeling activities. CCSP-supported research will address basic climate system properties, and a number of "feedbacks" or secondary changes that can either reinforce or dampen the initial effects of greenhouse gas and aerosol emissions and changes in land use and land cover. The program will also address the potential for future changes in extreme events and uncertainty regarding potential rapid or discontinuous changes in climate. CCSP will also build on existing U.S. strengths in climate research and modeling and enhance capacity for development of high-end coupled climate and Earth system models.

CCSP Goal 4: Understand the sensitivity and adaptability of different natural and managed ecosystems and human systems to climate and related global changes

Seasonal to annual variability in climate has been connected to impacts on ecosystems and many aspects of human life. Longer time scale natural climate cycles and human-induced changes in climate could have additional effects. Improving our ability to assess potential implications on ecosystems and human systems of variations and future changes in climate and environmental conditions could enable governments, businesses, and communities to reduce damages and seize opportunities by adapting infrastructure, activities, and plans. CCSP research will examine the interactions of multiple interacting changes and effects (e.g., the carbon dioxide "fertilization effect," deposition of nitrogen and other nutrients, changes in landscapes that affect water resources and habitats, changes in frequency of fires or pests) to improve knowledge of sensitivity and adaptability to climate variability and change. CCSP research will also improve methods to integrate our understanding of potential effects of different atmospheric concentrations of greenhouse gases and to develop methods for aggregating and comparing potential impacts across different sectors and settings.

CCSP Goal 5: Explore the uses and identify the limits of evolving knowledge to manage risks and opportunities related to climate variability and change

Over the last decade, the scientific and technical community has developed a variety of products to support management of risks and opportunities related to climate variability and change. CCSP will foster additional studies and encourage evaluation and learning from these experiences in order to develop decision support processes and products that use knowledge to the best effect, while communicating levels of uncertainty. CCSP will develop resources (e.g., observations, databases, data and model products, scenarios, visualization products, scientific syntheses, assessments, and approaches to conduct ongoing consultative mechanisms) to support policymaking, planning, and adaptive management.

Core Approaches

CCSP will employ four core approaches in working toward its goals, including:

1) *Scientific Research*. Plan, sponsor, and conduct research on changes in climate and related systems

2) *Observations*. Enhance observations and data management systems to generate a comprehensive set of variables needed for climate-related research

3) *Decision Support*. Develop improved science-based resources to aid decisionmaking

4) *Communications*. Communicate results to domestic and international scientific and stakeholder communities, stressing openness and transparency.

Each of these approaches is essential for achievement of the CCSP's goals. The first two of these approaches will rely heavily on existing programmatic strengths and mechanisms, while the latter two approaches will require development of new capabilities and initiatives over the coming years.

Approach 1: Plan, sponsor, and conduct research on changes in climate and related systems

Fundamental, long-term research on a broad range of global change issues.
Over the past 15 years, the USGCRP element of CCSP has provided planning and sponsorship of the world's most extensive program of scientific research, monitoring, data management, and assessment for climate change analyses.

Results of this program include the first ever global characterization of many aspects of the Earth's environment; the development of decadal-scale global observations of a limited number of environmentally important variables; detailed knowledge of a variety of processes important in the functioning of the Earth system; the development of ENSO forecasts and derived products used in management, planning, and emergency preparedness; and significant improvement in the capability of models used to project the future evolution of the Earth system, as evidenced by improvements in their ability to simulate variability in the present and recent past. USGCRP accomplishments are evidenced by large numbers of peer-reviewed scientific papers and other reports; unique data archives; contributions of U.S.-based scientists to the body of work produced by the Intergovernmental Panel on Climate Change (IPCC) and other assessment activities; and increased public awareness of issues associated with climate variability and change.

A substantial percentage of future CCSP budgets will be devoted to continuing this essential investment in scientific knowledge, facilitating the discovery of the unexpected and advancing the frontiers of research. CCSP agencies will coordinate their work through seven interdisciplinary research elements that have evolved from the framework for research presented in *Global Environmental Change: Research Pathways for the Next Decade*, a report from the National Research Council that lays out advances in knowledge needed to improve predictive capability in Earth system science (NRC, 1999a).

The following "snapshot" of CCSP research elements depicts their foci at the time of the preparation of this Strategic Plan. The program will encourage evolution of these research elements over the coming decade in response to new knowledge and societal needs. Over time, a greater degree of integration across the research elements and a greater degree of involvement with users are expected.

Atmospheric Composition. The atmosphere creates a protective envelope for life on Earth, providing key ingredients for respiration and photosynthesis and shielding the planet from harmful incoming radiation. It transports materials globally on rapid time scales, yet it can retain some pollutants for centuries or longer. Changes have been observed in relative quantities of key constituents that affect climate and in processes that affect the composition of the atmosphere itself (e.g., self-cleansing of pollutants). CCSP-supported research focuses on how the composition of the global atmosphere is altered by human activities and natural phenomena, and how such changes influence climate, ozone, ultraviolet radiation, pollutant exposure, ecosystems, and human health. Specific objectives address processes affecting the recovery of the stratospheric ozone layer from reduced ozone levels observed in recent decades; the properties and distributions of greenhouse gases and aerosols; long-range transport of pollutants and implications for regional air quality; and integrated assessments of the effects of these changes. Atmospheric composition issues involving interactions with climate variability and change—such as interactions between the climate system and stratospheric water vapor and ozone, or the potential effects of global climate change on regional air quality—are of particular interest at present.

Climate Variability and Change. Climate is a crucial aspect of the physical environment. The historical record of climate shows evidence of variability on multiple time scales, as well as rapid change. CCSP-supported research on climate variability and change focuses on how climate elements that are particularly important to human and natural systems—especially temperature, precipitation, clouds, winds, and storminess—are affected by changes in the Earth system. Specific objectives include improved predictions of seasonal to decadal climate variations (e.g., ENSO); improved detection, attribution, and projections of longer term changes in climate; the potential for changes in extreme events at regional to local scales; the possibility of abrupt climate change; and development of approaches (including characterization of uncertainty) to inform national dialogue and support public and private sector decisionmaking.

Global Water Cycle. Water is crucial to life on Earth. Water changes phase from solid to liquid to gas through a natural cycle that also transports and converts energy. Water in its different phases affects the Earth's radiative balance (e.g., changes in water vapor, clouds, and high-latitude ice formations are important climate feedbacks). Humans depend on predictability in the water cycle (e.g., water works are engineered to operate within certain tolerances of precipitation, evaporation, flow, and storage), and changes beyond expectations can have serious implications. CCSP-supported research on the global water cycle focuses on how natural processes and human activities influence the distribution and quality of water within the Earth system, whether changes are predictable, and on the effects of variability and change in the water cycle on human systems. Specific areas include identifying trends in the intensity of the water cycle and determining the causes of these changes (including feedback effects of clouds on the global water and energy budgets as well as the global climate system); predicting precipitation and evaporation on time scales of months to years and longer; and modeling physical/biological and socioeconomic processes to facilitate efficient water resources management.

Land-Use/Land-Cover Change. Land cover and use influence climate and weather at local to global scales. Land surface (cover) characteristics affect the exchange of greenhouse gases, including water vapor,

between the land surface and the atmosphere, the radiation balance of the continents, the exchange of sensible heat between continents and the atmosphere, and the uptake of momentum from the atmosphere. Land-cover characteristics are key inputs to climate models. Land cover and use also affect water runoff, infiltration, and quality; biogeochemistry (including the carbon and nitrogen cycles); the distribution of microorganisms, plants, and animals; and other factors. Understanding and projecting observed and future states of land cover and land use will require close integration of the natural and social sciences. Research within this program element will focus on the interactions among changes in land use and land cover, global change, and socioeconomic factors, including the predictability of land-use and land-cover change. Specific foci will identify and quantify the human drivers of land-use and land-cover change; improve monitoring, measuring, and mapping of land use and land cover, and the management of these data; and develop projections of land-cover and land-use change under various scenarios of climate, demographic, economic, and technological trends.

Global Carbon Cycle. Although water vapor is the most significant greenhouse gas, increased atmospheric concentration of carbon dioxide in recent decades is the largest single forcing agent of climate change. Methane is also a significant contributor. Evidence of increases in the atmospheric concentrations of these gases since pre-industrial times is unequivocal. The natural carbon cycle involves several reservoirs (the oceans, the biosphere, and the atmosphere) and is in approximate balance. Relatively small human perturbations can have major impacts, however, and our knowledge of these and their implications for environmental change is insufficient to manage carbon effectively. CCSP-supported research on the global carbon cycle focuses on identifying the size, variability, and potential future changes to reservoirs and fluxes of carbon within the Earth system, and providing the scientific underpinning for evaluating options to manage carbon sources and sinks. Specific programs and projects focus on North American and oceanic carbon sources and sinks; the impact of land-use change and resource management practices on carbon sources and sinks; projecting future atmospheric carbon dioxide and methane concentrations and changes in land-based and marine carbon sinks; and the global distribution of carbon sources and sinks and how they are changing.

Ecosystems. Ecosystems provide a variety of environmental goods and services that are necessary to sustain life. For the purposes of this research plan, ecosystems include agricultural lands, commercial forests, and other ecosystems that are essential to human survival and a desirable quality of life. Provision of essential resources depends on a variety of physical and chemical inputs and is affected by climate variability and change, and by human influences such as introduction of nutrients and pollutants and fragmentation of landscapes. Improving projections of future climate and global changes depends on developing improved understanding of ecosystem processes under multiple natural and human influences. CCSP-supported research on ecosystems focuses on: (1) how natural and human-induced changes in the environment interact to affect the structure and functioning of ecosystems (and the goods and services they provide) at a range of spatial and temporal scales, including those ecosystem processes that in turn influence regional and global environmental changes; and (2) what options society may have to ensure that desirable ecosystem goods and services will be sustained or enhanced in the context of still uncertain regional and global environmental changes. Among the specific focus areas are the cycling of nutrients, such as nitrogen, and how these nutrients interact with the carbon cycle; key processes that link ecosystems with climate; and options for managing agricultural lands, forests, and other ecosystems to sustain goods and services essential to societies.

Human Contributions and Responses. Human activities are an important influence on the global environment. Human responses to change, through adaptation and mitigation, will strongly influence whether environmental changes have positive or negative effects on society. CCSP-supported research on human contributions and responses to global change focuses on the interactions of changes in the global environment and human activities. The current focus of this research is on the potential effects of climate variability and change on human health and welfare; human influences on the climate system, land use, and other global environmental changes; analyses of societal vulnerability and resilience to global environmental change; decisionmaking under conditions of significant complexity and uncertainty; and integrated assessment methods.

Enhanced short-term focus on reducing key scientific uncertainties to support informed public review of adaptation and mitigation strategies. President Bush created the CCRI in June 2001, and directed that it focus on short-term (i.e., within 5 years) actions to reduce high-priority scientific uncertainties about global climate change where possible, and to synthesize the available scientific information to support public discussion of global climate change response strategies. CCSP manages CCRI activities jointly with its management of the long-term USGCRP studies, using the same interagency management and scientific working group structures.

Enhanced modeling capacity to accelerate incorporation of new knowledge into comprehensive climate models and to develop model products for decision support. Models are an essential tool for synthesizing observations, theory, and experimental results to investigate how the Earth system works and how it may be affected by human activities. Comprehensive climate models represent the current imperfect state-of-the-science understanding of the major components of the climate system and the transfer of water, energy, chemicals, and mass among them. CCSP will foster two complementary streams of climate modeling activities. The first is a research activity that will support continued model experiments and accelerate incorporation of new knowledge into comprehensive climate and Earth system models. Closely associated with the research activity will be the sustained and timely delivery of predictive model products for assessments and other decision support resources.

The challenge is that of balancing fundamental long-term Earth system research with an enhanced short-term focus on climate change uncertainties. CCSP will focus attention on key climate change issues that are important for public discussion, while maintaining sufficient breadth to facilitate scientific discovery on a broad range of global environmental changes. Establishing a careful balance between focus and breadth is essential and will require input from both decisionmakers and the science community. The NRC has already played a significant role in shaping the program through a series of reports and evaluations of the program and is expected to help establish this balance through future interactions and evaluations.

Approach 2: Enhance observations and data management systems to generate a comprehensive set of variables needed for climate-related research

Since the early years of the USGCRP, an expanded program of global observations has been developed to characterize climate variability and change on a global and regional basis. These observations have come from a variety of sources, including paleoclimate studies interpreting climate parameters over thousands of years, satellite remote-sensing systems, and numerous *in situ* systems at the terrestrial surface (including the polar regions), in the atmosphere, and in the surface and deep oceans. The suite of available observations includes long-term observations associated with the National Oceanic and Atmospheric Administration's satellite monitoring program and global weather observations, which have not historically been considered as part of the USGCRP; several long-term surface-based measurement networks operated by the National Aeronautics and Space Administration (NASA), National Oceanic and Atmospheric Administration, Department of Energy, and other agencies; and several long-running NASA research satellite series, as well as a large number of limited-duration measurements obtained during research programs. Satellite observations made under the USGCRP have provided wide-scale, synoptic measurements and data sets of the global distribution of important environmental parameters and their spatial and temporal variability.

Prior and current investments in new observations will significantly enhance our knowledge of a number of environmental parameters in the coming years as the promise of these investments is realized. But two considerations have created a need for enhanced attention to global and regional observation and data management systems: (1) The large quantity and diverse format of available observations requires a major expansion of the capacity to prioritize, quality assure, archive, disseminate (in useful format), and assimilate the elements of this extensive record; and (2) the importance of integrated evaluation of climate and ecosystem parameters calls for the development of new requirements for integrated observation systems, followed by system design and implementation. This will enable the research community to address additional research issues including ecosystem and land-use/land-cover forcing and feedback relationships with other climate parameters, and impacts of climate variability and change on terrestrial and marine ecosystem dynamics, productivity, and biodiversity.

The United States has taken a leading role in fostering the development of a more broadly defined and integrated global observing system for all Earth parameters (e.g., including geological as well as climate information). The United States is hosting a ministerial-level Earth Observation Summit in July 2003, with participation by many developed and developing nations as well as many intergovernmental and international non-governmental organizations. This summit will initiate a 10-year commitment to design, implement, and operate an expanded global observing system that builds on the major observational programs currently operated by the United States and many other governments and international organizations. CCSP agencies have provided the leadership, definition, and support for the Earth Observation Summit, and CCSP will closely integrate the U.S. observation and data management programs with the international programs launched at the summit.

CCSP relies on both research and operational systems. The former are designed principally to address research questions, both those posed by CCSP as well as those posed by other federal environmental research efforts—for example, those of other subcommittees of the Committee on Environment and Natural Resources (see the CCSP Background and Management Overview section later in this chapter). A set of fundamental geophysical parameters must be measured to describe Earth system processes. A subset of these parameters is measured on an operational basis to serve other needs such as weather forecasting. CCSP uses measurements from the best available sources. CCSP working groups and participating agencies must explicitly include both research and operational systems in implementation planning. As many CCSP scientific objectives require long-term data records, facilitating the transition of responsibility for mature research observations to operational systems is a key element of the CCSP observing strategy. CCSP will also adhere to NRC climate monitoring principles, as well as the Global Climate Observing System (GCOS) climate monitoring principles for satellites.

Approach 3: Develop improved science-based resources to aid decisionmaking

Since the earliest years of the USGCRP and its counterparts around the world, the use of available scientific information to address key questions about changes in climate and related systems has continuously grown in importance. The available scientific record has been used for many years to improve understanding of a range of questions, from detecting climate change and attributing it to particular causes, to applying satellite and ground-based observations and related analyses in resource management applications. CCSP will build on this record and respond to significant new demands for additional information resources to support adaptive management of natural resources, planning, and policymaking. The program will improve approaches for sustained interactions with stakeholders that consider needs for information from a "user perspective." It will encourage development of new methods, models, and other resources that facilitate economic analysis, decisionmaking under conditions of complexity and uncertainty, and integration and application of information from the natural and social sciences in specific decision contexts.

Critical to the success of CCSP's contributions to decision support are partnerships with a variety of federal and non-federal entities that rely on the outcomes and products of global change research. These include the Climate Change Technology Program, subcommittees of the Committee on Environment and Natural Resources, and agencies with charters to provide essential public services, such as weather forecasting, disaster preparedness and response, management of resources, and enhancement of agricultural efficiency.

Evaluation and communication of uncertainty and levels of confidence is a crucial issue for the development of credible decision support resources. Uncertainties can arise from lack of knowledge, from problems with data, models, terminology, or assumptions, and from other sources, creating room for considerable misunderstanding. CCSP research will contribute to reduction of uncertainty, although research can also unexpectedly increase uncertainty. Because uncertainty can never be completely eliminated, CCSP will develop

systematic approaches for assessing and updating levels of confidence and uncertainty and communicating this information in ways that are appropriate to the particular decision at hand. This will enable decisionmakers to understand the uses and limits of the information they are seeking to apply. CCSP will develop and employ transparent and systematic approaches for decision support under conditions of uncertainty, and for evaluating and reporting levels of confidence and uncertainty, including when uncertainties expand unexpectedly as the result of research. In addition, CCSP will strive to improve clarity in understanding how and when uncertainty is likely to be reduced by different research initiatives.

Approach 4: Communicate results to domestic and international scientific and stakeholder communities, stressing openness and transparency

The domestic and international communities addressing global climate change are already well developed. This is evident in publications in the scientific literature, IPCC collaborations, and many other scientific forums; in policy discussions in Washington and other world capitals; and in the media throughout the world. CCSP has a major responsibility to communicate with interested partners in the United States and throughout the world, and to learn from these partners on a continuing basis. Because of the large commitment of public resources to CCSP activities, CCSP also has a responsibility to report its findings in the form of educational materials suitable for use at various educational and public information levels, so that the dissemination of its findings will be effective.

Global climate change is complex and often subject to disputed interpretations even among scientists. Further, the economic and policy dimensions of the issue often give rise to even greater debates among individuals with different policy views. Controversy about climate change can sometimes be characterized as "public debate by headlines"—selective citations from the scientific literature, advocacy-oriented quotations from interested persons (often citing their own expertise), and other debating tactics. As an essential part of its mission, CCSP undertakes the significant responsibility of enhancing the quality of discussion by stressing openness and transparency in its findings and reports.

CCSP will employ four methods to ensure the credibility of its reported findings: (1) use of structured analyses (usually question-based) for CCSP scientific synthesis, assessment, and projection reports; (2) use of transparent methodologies that openly report all key assumptions, methods, data, and uncertainties; (3) continuous use of web-based and other forms of information dissemination so that CCSP information is freely available to all interested users; and (4) frequent use of "draft for comment" methods to seek external review before completion of each key document. CCSP will also continue to urge all of its sponsored researchers to seek publication of their findings in the peer-reviewed scientific literature.

Some of these credibility-enhancing steps have already been introduced through the public dissemination and review of the CCSP *Discussion Draft Strategic Plan* in November 2002, the open public workshop with 1,300 participants in December 2002, the subsequent public comment period, and the invited review by a committee of the NRC.

CCSP Prioritization

Research priorities of USGRP and now CCSP have progressed over time as the fundamental scientific issues have evolved. Prioritization is an ongoing process in the program and reflects changes in needs and scientific progress and opportunity. The CCSP priorities will be reviewed on an annual cycle through the budget process.

Initial CCSP priorities have developed in response to a report requested by the Administration of a National Academies' NRC committee. The NRC report, *Climate Change Science: An Analysis of Some Key Questions*, characterized areas of uncertainty in scientific knowledge concerning climate change, and identified research areas that will advance the understanding of climate change. In particular, the report concluded that "predictions of global climate change will require major advances in understanding and modeling of (1) the factors that determine atmospheric concentrations of greenhouse gases and aerosols, and (2) the so-called "feedbacks" that determine the sensitivity of the climate system to a prescribed increase in greenhouse gases." The report also noted the limitations of current observing systems as well as the inadequacy of computational resources. Finally, the report called for an enhancement of the research enterprise dealing with environmental change and environment-society interactions in order to address the consequences of climate change and better serve the nation's decisionmakers. This includes "support of (a) interdisciplinary research that couples physical, chemical, biological, and human systems; (b) improved capability to integrate scientific knowledge, including its uncertainty, into effective decision support systems; and (c) an ability to conduct research at the regional or sectoral level that promotes analysis of the response of human and natural systems to multiple stresses."

In response to this NRC report, the Administration established the CCRI "to study areas of uncertainty [about global climate change] and identify priority areas where investments can make a difference." CCRI represents a focusing of resources and enhanced interagency coordination of ongoing and planned research on those elements of USGCRP that can best address major gaps in the understanding of climate change.

Initially, CCRI prioritizes research on three sets of uncertainties highlighted by the NRC: (1) atmospheric concentrations and effects of aerosols; (2) climate feedbacks and sensitivity, initially focusing on polar feedbacks; and (3) carbon sources and sinks, focusing particularly on North America in the immediate term. These priorities are discussed more completely in CCRI text boxes found in Chapter 3 (aerosols), Chapter 4 (climate feedbacks and sensitivity), and Chapter 7 (carbon cycle).

In addition, CCRI will focus on climate observing systems including efforts to: (a) document historical records; (b) improve observations for model development and applications; (c) enhance biological and ecological observing systems; and (d) improve data archiving and information system architectures. These activities involve substantial collaboration with the international climate science community and with several ongoing international monitoring development programs. Details on these efforts are highlighted in Chapters 12 and 13.

Development of state-of-the-art climate modeling that will help us better understand the causes and impacts of climate change is also a CCSP priority. Based on recommendations in several NRC reports on U.S. climate modeling (NRC 1999b, 2001d) and USGCRP evaluations (see, e.g., USGCRP, 2000), CCSP agencies are prioritizing new activities to strengthen the national climate modeling infrastructure. Details on these efforts are included in Chapter 10.

Finally, in the area of decision support resources, CCSP has identified over 20 synthesis and assessment products that will focus on key uncertainties and decisionmaking issues. These are described in Chapter 2. In addition, through CCRI, the program will develop additional resources to support national discussion and planning, adaptive management, and policymaking. Chapter 9 includes a description of research to develop new methods for use of scientific information under conditions of complexity and uncertainty, and Chapter 11 describes CCSP's decision support activities more broadly.

CCSP Criteria for Prioritization

As the program evolves in response to emerging needs and scientific opportunities, CCSP will employ the following overall criteria in establishing priorities for work elements selected for support:

1) Scientific or technical quality
   - The proposed work must be scientifically rigorous as determined by peer review.
   - Implementation plans will include periodic review by external advisory groups (both researchers and users).
2) Relevance to reducing uncertainties and improving decision support tools in priority areas
   - Programs must substantially address one or more CCSP goals.
   - Programs must respond to needs for scientific information and enhance informed discussion by all relevant stakeholders.
3) Track record of consistently good past performance and identified metrics for evaluating future progress
   - Programs addressing priorities with good track records of past performance will be favored for continued investment to the extent that time tables and metrics for evaluating future progress are provided.
   - Proposed programs that identify clear milestones for periodic assessment and documentation of progress will be favorably considered for new investment.
4) Cost and value
   - Research should address CCSP goals in a cost-effective way.
   - Research should be coordinated with and leverage other national and international efforts.
   - Programs that provide value-added products to improve decision support resources will be favored.

CCSP Background and Management Overview

CCSP Background

CCSP was created by the President in February 2002, as part of a new cabinet-level management structure (see Figure 1-1) to oversee public investments in climate change science and technology. The new structure also includes CCTP, which is responsible for climate change-related technology research and development. Joint oversight of CCSP and CCTP is intended to increase the degree of coordination and integration, and to apply the knowledge created by CCSP to technology development decisionmaking. CCSP and CCTP report through the Interagency Working Group on Climate Change Science and Technology (IWGCCST) to the cabinet-level Committee on Climate Change Science and Technology Integration (CCCSTI). The IWGCCST membership includes deputy secretary and deputy administrator level representatives of the relevant cabinet departments and agencies, and representatives of the Executive Office of the President including the Office of Management and Budget (OMB), the Council on Environmental Quality (CEQ), and the Office of Science and Technology Policy (OSTP). IWGCCST meets regularly, supplemented with ad hoc meetings as needed. This structure oversees a combined annual budget exceeding $3 billion.

CCSP integrates federal research on global change and climate change, as sponsored by 13 federal departments and agencies (the Departments of Agriculture, Commerce, Defense, Energy, Health and Human Services, the Interior, State, and Transportation; together with

the Environmental Protection Agency, the National Aeronautics and Space Administration, the National Science Foundation, the Agency for International Development, and the Smithsonian Institution). OSTP, CEQ, OMB, and the National Economic Council (NEC) provide oversight. By leveraging the complementary strengths of the 13 agencies, CCSP integrates the planning of research and applications that are implemented by the participating agencies.

In addition to the new management structure instituted by the President, the National Science and Technology Council and its subsidiary bodies—the Committee on Environment and Natural Resources and the Subcommittee on Global Change Research (SGCR)—continue to coordinate climate and global change research. Agency representation on the SGCR and CCSP is identical to ensure coordination.

CCSP integrates USGCRP and CCRI. The USGCRP was established in 1989, and codified in the Global Change Research Act of 1990 as a high-priority national research program to address key uncertainties about natural and human-induced changes in the Earth's global

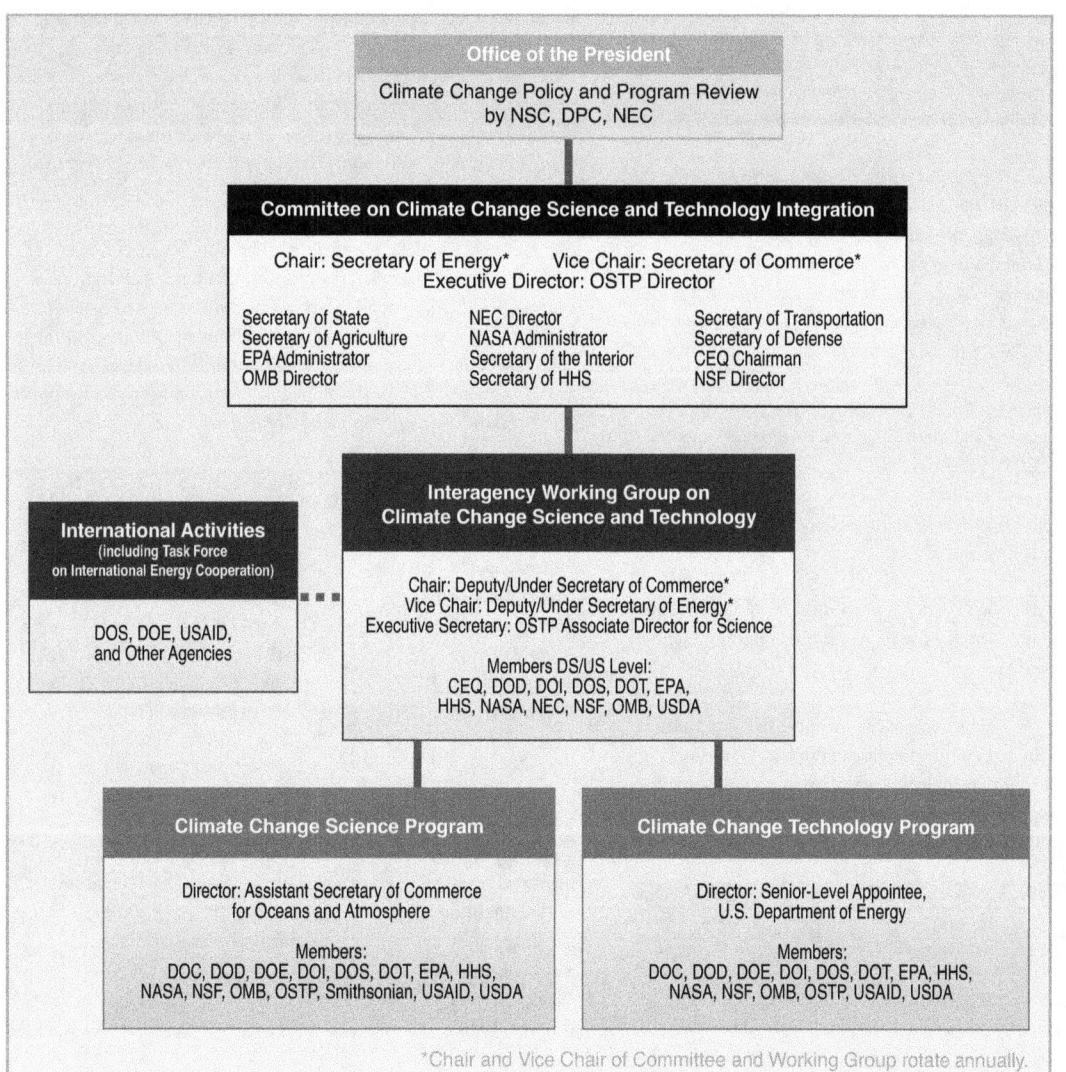

Figure 1-1: The Climate Change Science and Technology Programs are overseen by a cabinet-level management structure of the U.S. government.

environmental system; to monitor, understand, and predict global change; and to provide a sound scientific basis for national and international decisionmaking. The rationale for establishing the program was that the issues of global change are so complex and wide-ranging that they extend beyond the mission, resources, and expertise of any single agency, requiring instead the integrated efforts of several agencies. USGCRP is organized into a set of linked research program elements, which together support scientific research across a range of interconnected issues of climate and global change. As previously described, CCRI was created in June 2001, by the President in response to a report prepared by the National Academy of Sciences at the request of the Administration. The goal of CCRI is to reduce key uncertainties in climate science and measurably improve the integration of scientific knowledge, including measures of uncertainty, into effective decision support systems and resources. Specific examples of research proposed as part of CCRI in the FY04 President's Budget Request are highlighted throughout relevant chapters of this plan.

Management Overview

The CCSP approach to management integrates the planning and implementation of individual climate and global change research programs of the participating federal agencies and departments to reduce overlaps, identify and fill programmatic gaps, and synthesize products and deliverables generated under the auspices of CCSP.

Five mechanisms are used to achieve this management approach:

- *Executive Direction—IWGCCST/CCSP/SGCR Representatives*
  Overall priority-setting, program direction, management review, and accountability to deliver program goals.
- *Agency Implementation—CCSP-Participating Departments and Agencies*
  Conducting research, developing and operating observing systems, and producing CCSP-required products, often in collaboration and often as defined or refined in interagency working groups.
- *Interagency Planning and Implementation—Interagency Working Groups*
  Coordinated planning and implementation to align agency programs with CCSP priorities.
- *External Guidance and Interactions—Advisory Groups*
  External guidance, oversight, and interactions to ensure scientific excellence, credibility, and utility.
- *Program Support—CCSP Office*
  Value-added staffing and day-to-day coordination of CCSP-wide program integration, strategic planning, product development, and communications.

Interaction among the groups responsible for these five mechanisms is essential, especially to prioritize research efforts and decision support activities. A more complete description of each of these five mechanisms and the CCSP management approach can be found in Chapter 16, Program Management and Review.

Roadmap for the Strategic Plan

The strategy describes the goals of the CCSP and its component programs and elements, the products that are expected to result, and the approaches and criteria that will be adopted to implement

the program. The products include data sets, geographic information systems and other approaches for visualization of data, model studies, scientific publications, state-of-science reports, assessments, comparative evaluations of response options (including "If…, then…" scenario analyses), and other decision support resources. The plan also identifies enhancements needed in observing systems, data and information capabilities, modeling, and decision support resources to meet the program's goals.

Following this introduction, Chapter 2 describes the approach of CCSP for planning and sponsoring research on variability and change in climate and related systems. This approach is based on five overall CCSP goals and a set of interdisciplinary and interagency research elements. Integration of research from agency programs and research elements is an essential component of the development of synthesis products and assessments that address CCSP goals. The chapter introduces the cross-cutting goals, as well as related CCSP deliverables.

Chapters 3-9 provide a more detailed description of the research elements of the plan. These research elements focus on crucial components and interactions within the Earth system and have evolved from recommendations of the research community and the NRC. Each of these chapters describes research questions; provides an overview of the current state of knowledge; outlines milestones, products, and benefits from the research; and identifies needed inputs from and linkages with other national and international programs.

Chapters 10 and 12-13 describe the cross-cutting issues of modeling, observations, and data management. These are all areas where CCSP has substantial existing capabilities, but in which additional capacity will be developed to achieve the goals that have been set. Chapter 11 focuses on decision support. This chapter lays out the goals and strategy for participating in state-of-the-science syntheses and assessments, and for developing additional resources to support policymaking, planning, and adaptive resource management.

Chapters 14-16 describe communications, international cooperation, and management issues that cut across all areas of the program.

This plan was drafted by scientists and research program managers from the participating CCSP agencies and the Climate Change Science Program Office.

CHAPTER 1 AUTHORS

Lead Authors
James R. Mahoney, DOC
Richard H. Moss, CCSPO
Ghassem Asrar, NASA
Margaret S. Leinen, NSF
James Andrews, DOD
Mary Glackin, NOAA
Charles (Chip) Groat, USGS
William Hohenstein, USDA
Linda Lawson, DOT
Melinda Moore, HHS
Patrick Neale, Smithsonian Institution
Aristides Patrinos, DOE
Michael Slimak, USEPA
Harlan Watson, DOS

Integrating Climate and Global Change Research

CHAPTER CONTENTS

CCSP Goals and Products

"Critical Dependencies" and Approaches to Integration

Synthesis and Assessment Products

Chapter 2 describes the Climate Change Science Program's (CCSP) overall approach to climate research on different components of the Earth system and the integration of this research to achieve program objectives. This chapter expands on CCSP's five goals (see Chapter 1 and Box 2-1), including a brief overview of current scientific understanding, and illustrates the relationship between these goals and the program's interdisciplinary research elements and cross-cutting activities. Tables listing topics to be covered by CCSP synthesis and assessment products are included and indicate the target time frame for completion of each product. These products will fulfill the requirements for updated synthesis and assessment contained in Section 106 ("Scientific Assessment") of the 1990 Global Change Research Act.

This chapter of the plan also includes research products, milestones, and activities, which are grouped under research focus areas for each goal. These are given here as examples and are not meant to constitute exhaustive lists; completion dates are not provided in this chapter, but are in related discussions in Chapters 3-9, and range from 2004 to beyond 2007 (greater than 4 years). Other sections of this chapter discuss integration of research to give a comprehensive picture of the cumulative effects of natural processes and human activities on the future evolution of climate and related systems.

CCSP Goals and Products

CCSP's goals span the chain of climate-related issues including natural climate conditions and variability; forces that influence climate; cycles and processes that affect atmospheric concentrations of greenhouse gases and aerosols; climate responses; consequences for ecosystems, society, and our nation's economy; and application of knowledge to decisionmaking. This comprehensive characterization provides a useful organizing scheme for examining key climate change issues. CCSP research must focus on the full range of these complex issues if it is to lay the basis for informed discussion and decisionmaking.

The following text explains the five CCSP goals. The discussion of each goal includes a set of topics to be addressed by synthesis and assessment products related to the goal during the next 4 years, and examples of key research activities to be carried out in moving toward the goal, with links to the discussion of research activities in subsequent chapters. Near-term synthesis and assessment products will draw on a large body of research either already completed or currently underway throughout the program, in addition to new research results as they become available. Over time, CCSP research activities and the development of new synthesis and assessment products will evolve in partnership, as scientific research progresses and as new questions related to policymaking, planning, and adaptive management arise.

BOX 2-1

LIST OF CCSP GOALS
INCLUDING EXAMPLES OF KEY EXPECTED OUTCOMES

CCSP Goal 1: Improve knowledge of the Earth's past and present climate and environment, including its natural variability, and improve understanding of the causes of observed variability and change

- Better understand the natural long-term cycles in climate (e.g., Pacific Decadal Variability, North Atlantic Oscillation).
- Improve and harness the capability to forecast El Niño-La Niña events and other seasonal-to-interannual cycles of variability.
- Sharpen understanding of climate extremes through improved observations, analyses, and modeling, and determine whether any changes in their frequency or intensity lie outside the range of natural variability.
- Increase confidence in understanding of how and why climate has changed.
- Expand observations and data/information system capabilities.

CCSP Goal 2: Improve quantification of the forces bringing about changes in the Earth's climate and related systems

- Reduce uncertainty about the sources and sinks of greenhouse gases, emissions of aerosols and their precursors, and their climate effects.
- Monitor recovery of the ozone layer and improve understanding of the interactions among climate change, ozone depletion, and other atmospheric processes.
- Increase knowledge of the interactions among pollutant emissions, long-range atmospheric transport, climate change, and air quality management.
- Develop information on the carbon cycle, land cover and use, and biological/ecological processes by helping to quantify net emissions of carbon dioxide, methane, and other greenhouse gases, thereby improving the evaluation of carbon sequestration strategies and alternative response options.
- Improve capabilities to develop and apply emissions and related scenarios for conducting "If…, then…" analyses in cooperation with the Climate Change Technology Program.

CCSP Goal 3: Reduce uncertainty in projections of how the Earth's climate and related systems may change in the future

- Improve characterization of the circulation of the atmosphere and oceans and their interactions through fluxes of energy and materials.
- Improve understanding of key "feedbacks" including changes in the amount and distribution of water vapor, extent of ice and the Earth's reflectivity, cloud properties, and biological and ecological systems.
- Increase understanding of the conditions that could give rise to events such as rapid changes in ocean circulation owing to changes in temperature and salinity gradients.
- Accelerate incorporation of improved knowledge of climate processes and feedbacks into climate models to reduce uncertainty about climate sensitivity (i.e., response to radiative forcing), projected climate changes, and other related conditions.
- Improve national capacity to develop and apply climate models.

CCSP Goal 4: Understand the sensitivity and adaptability of different natural and managed ecosystems and human systems to climate and related global changes

- Improve knowledge of the sensitivity of ecosystems and economic sectors to global climate variability and change.
- Identify and provide scientific inputs for evaluating adaptation options, in cooperation with mission-oriented agencies and other resource managers.
- Improve understanding of how changes in ecosystems (including managed ecosystems such as croplands) and human infrastructure interact over long periods of time.

CCSP Goal 5: Explore the uses and identify the limits of evolving knowledge to manage risks and opportunities related to climate variability and change

- Support informed public discussion of issues of particular importance to U.S. decisions by conducting research and providing scientific synthesis and assessment reports.
- Support adaptive management and planning for resources and physical infrastructure affected by climate variability and change; build new partnerships with public and private sector entities that can benefit both research and decisions.
- Support policymaking by conducting comparative analyses and evaluations of the socioeconomic and environmental consequences of response options.

CCSP Goal 1: Improve knowledge of the Earth's past and present climate and environment, including its natural variability, and improve understanding of the causes of observed variability and changes

The climate system is highly variable, with average conditions changing significantly over the span of seasons, years, decades, and even longer time scales. Examining the period since the industrial revolution, when human activities started to significantly alter the Earth's land surface and increase the atmospheric concentrations of greenhouse gases (GHGs) and aerosols, an increasing body of evidence gives a picture of a warming world and other related changes. A key issue is whether there is a cause and effect relationship between

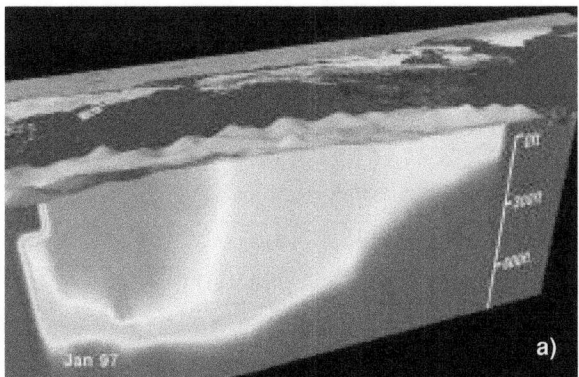

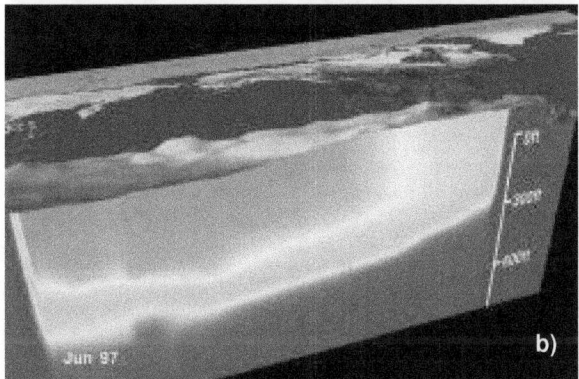

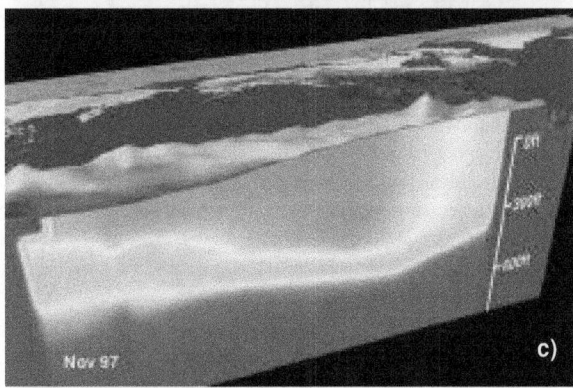

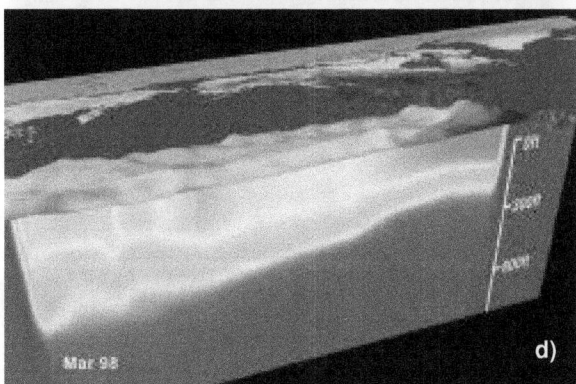

Figure 2-1: Evolution of the 1997-1998 El Niño event. This figure depicts water temperature and sea surface topography in the equatorial Pacific Ocean: (a) January 1997; (b) June 1997; (c) November 1997; (d) March 1998. Source: NASA Goddard Space Flight Center, Scientific Visualization Studio; NOAA Pacific Marine Environmental Laboratory.

these developments. The Intergovernmental Panel on Climate Change (IPCC) and the National Research Council (NRC) have independently concluded (in the words of the NRC) that "changes observed over the last several decades are likely mostly due to human activities," but that "a causal linkage between the buildup of greenhouse gases in the atmosphere and the observed climate changes during the 20th century cannot be unequivocally established" (NRC, 2001a). Uncertainty about natural oscillations in climate on scales of several decades or longer and inconsistencies in the temperature profiles of different data sets are critical scientific questions that must be addressed to improve confidence in the understanding of how and why climate has changed.

El Niño-Southern Oscillation (ENSO) is a widely-recognized, large-scale climate oscillation in the equatorial Pacific Ocean that changes phase every few years, with significant implications for resource and disaster management. There are other climate cycles that last decades, centuries, and even millennia, as revealed in studies of the Earth's climate history. Recently improved (but still imperfect) capacity to forecast ENSO has yielded significant benefits in the United States and other countries, enabling decisionmakers to prepare for impacts.

Improved data and information on the climate system's natural variability and recent changes are needed to increase confidence in studies that seek to detect and attribute changes in climate to particular causes. Expanded observations, modeling, and data/ information system capabilities will also improve society's ability to respond to the fluctuations and changes in the environment— whether natural or human-induced—that the nation and the world will face in the next 20 to 30 years and beyond. Substantial amounts of data are already collected through both an extensive observations system developed under the U.S. Global Change Research Program (USGCRP) and a set of operational monitoring systems used for weather forecasting and other uses (that historically were not identified as part of the USGCRP). New technologies can extend existing observation and monitoring networks and improve accessibility to these data, with benefits for both scientific understanding and resource management.

Examples of Key Research Activities

Focus 1.1: Better understand natural long-term cycles in climate [e.g., Pacific Decadal Variability (PDV), North Atlantic Oscillation (NAO)]

- An assessment of potential predictability beyond ENSO (e.g., associated with PDV, NAO) and improvements in the representation of major modes of climate variability in climate change projection models (Chapter 4.2)[1]
- Identification of impacts of natural oscillations in climate on marine fisheries and marine ecosystems (Chapter 8.2)
- Identification of adaptation strategies effective for managing the impacts of seasonal and year-to-year climate variability (Chapter 9.2)

---

[1] Parenthetical references indicate the chapter and question where additional information on this research can be found.

	TOPICS FOR PRIORITY CCSP SYNTHESIS PRODUCTS	SIGNIFICANCE	COMPLETION
CCSP GOAL 1	Temperature trends in the lower atmosphere—steps for understanding and reconciling differences.	Inconsistencies in the temperature profiles of different data sets reduce confidence in understanding of how and why climate has changed.	within 2 years
	Past climate variability and change in the Arctic and at high latitudes.	High latitudes are especially sensitive and may provide early indications of climate change; new paleoclimate data will provide long-term context for recent observed temperature increases.	within 2 years
	Reanalyses of historical climate data for key atmospheric features. Implications for attribution of causes of observed change.	Understanding the magnitude of past climate variations is key to increasing confidence in the understanding of how and why climate has changed and why it may change in the future.	2-4 years

Focus 1.2: Improve and harness the capability to forecast El Niño-La Niña and other seasonal-to-interannual cycles of variability

- Improved predictions of El Niño-La Niña, particularly the onset and decay phases, and improved probability forecasts of regional manifestations of seasonal climate anomalies resulting from ENSO (Chapter 4.2)
- An assessment of potential predictability of annular modes, tropical Atlantic and Indian Ocean variability and trends, and the monsoons (Chapter 4.2)
- Analyses of societal adjustment to climate variability and seasonal-to-interannual forecasts (Chapter 9.2)

Focus 1.3: Sharpen understanding of climate extremes through improved observations, analysis, and modeling, and determine whether any changes in their frequency or intensity lie outside the range of natural variability

- Improved observational databases, including paleoclimate and historical data records, and model simulations of past climate, to detect and analyze regional trends in extreme events (Chapter 4.4)
- Observational and statistical analyses to assess the relationships between extreme events and natural climate variations, such as ENSO, PDV, NAO/Northern Hemisphere Annular Mode and Southern Hemisphere Annular Mode, and ecosystem response (Chapters 4.4 and 8.2)

Focus 1.4: Increase confidence in the understanding of how and why climate has changed

- Development and extension of critical data sets (including paleoclimatic data) to improve analyses of climate variability and attribution of causes of climate change (Chapter 4.2)
- Integrated long-term global and regional data sets of critical water cycle variables such as evapotranspiration, soil moisture, groundwater, clouds, etc., from satellite and *in situ* observations for monitoring climate trends and early detection of climate change (Chapter 5.1)

Focus 1.5: Expand observations and data/information system capabilities

- Data requirements analysis and planning
  - Requirements analysis for a global integrated climatological, ecological, and land-use monitoring system, followed by design specifications for the monitoring system (completed with significant international collaboration) (Chapter 12)
  - Definition of the initial requirements for ecosystem observations to quantify feedbacks to climate and atmospheric chemistry, to enhance existing observing systems, and to guide development of new observing capabilities (Chapters 8.1 and 3.5)
  - Definition of the initial requirements for observing systems to monitor the health of ecosystems, including a new suite of indicators of coastal and aquatic ecosystem change (Chapter 8.2)
  - Identification and rescue of data that are at risk of being lost because of media deterioration, poor accessibility, or limited distribution. (Chapter 13)
- Data systems and products
  - Monitoring and data systems for water resources research and management (Chapter 5.4)
  - Development of high-resolution climate data products for climate-sensitive regions, based on monthly instrumental data and annual paleoclimatic data and climate forecasts (Chapter 4.1 and 4.5)
  - Implementation of climate quality data and metadata documentation, standards, and formatting policies that will make possible the combined use of targeted data products taken at different times, by different means, and for different purposes (Chapter 13)
  - Global high-resolution satellite remotely sensed data and land-cover databases and global maps of areas of rapid land-use and land-cover change and location and extent of fires (Chapter 6.1)

CCSP Goal 2: Improve quantification of the forces bringing about changes in the Earth's climate and related systems

Combustion of fossil fuels, changes in land cover and land use, and industrial activities produce greenhouse gases and aerosols and alter the composition of the atmosphere and important physical and biological properties of the Earth's surface. For example, the atmospheric concentration of carbon dioxide (CO_2) has increased approximately 30 percent since the mid-18th century (see Figure 2-2). The current level of atmospheric CO_2 has not been exceeded during at least the past 420,000 years (the span measurable in ice cores). Atmospheric concentrations of other greenhouse gases [methane (CH_4), tropospheric ozone (O_3), nitrous oxide (N_2O), and halocarbons (e.g., chlorofluorocarbons (CFCs), hydrochlorofluorocarbons (HCFCs), and hydrofluorocarbons (HFCs))] have also increased. For example, CH_4 concentrations have more than doubled and the concentrations of industrially produced molecules have increased against a zero natural background. In the case of chlorine, the amount of chlorine in the stratosphere has essentially quintupled due to this human release.

These changes have several important climatic effects. Greenhouse gases (which remain in the atmosphere for years to millennia) have a warming influence on climate, while aerosols (which usually remain in the atmosphere for weeks to months) have both warming and cooling effects that can be quantified only poorly at present (see Figure 2-3). Research conducted as part of CCSP will reduce uncertainty about the sources and sinks of GHGs and emissions of aerosols and their precursors, their climate effects, and the implications of controls on aerosol emissions for both climate and air quality at regional and local scales.

Changes in land cover and use affect climate directly—for example by altering the Earth's albedo, or reflectivity. They also affect ecosystems and biological processes that contribute to carbon sources and sinks, emissions of other greenhouse gases, and the potential for land use changes to alter regional hydrology. Several CCSP initiatives, including accelerated research on the carbon cycle under the Climate Change Research Initiative (CCRI), will develop needed information for policymaking and improved carbon management.

Emissions of some sets of industrial compounds also contribute to depletion of the stratospheric ozone layer and increased exposure to ultraviolet radiation (which contributes to skin cancers, eye diseases, and possibly other environmental problems). Many of these same ozone-depleting gases are also significant greenhouse gases. CCSP will monitor the recovery of the ozone layer and long-term changes in the industrial compounds and stratospheric conditions that bring about stratospheric ozone depletion.

Future human contributions to climate forcing and potential associated environmental changes will depend on rates and levels of population change, economic growth, development and diffusion of technologies, and other dynamics in human systems. These developments are unpredictable over the long time scales relevant for climate change research. However, "If…, then…" scenario experiments, if carefully constructed, can make it possible to explore the potential implications of different technological, economic, and institutional conditions for future emissions, climate, and living standards. Improving the approach to developing and using these scenarios, in cooperation with

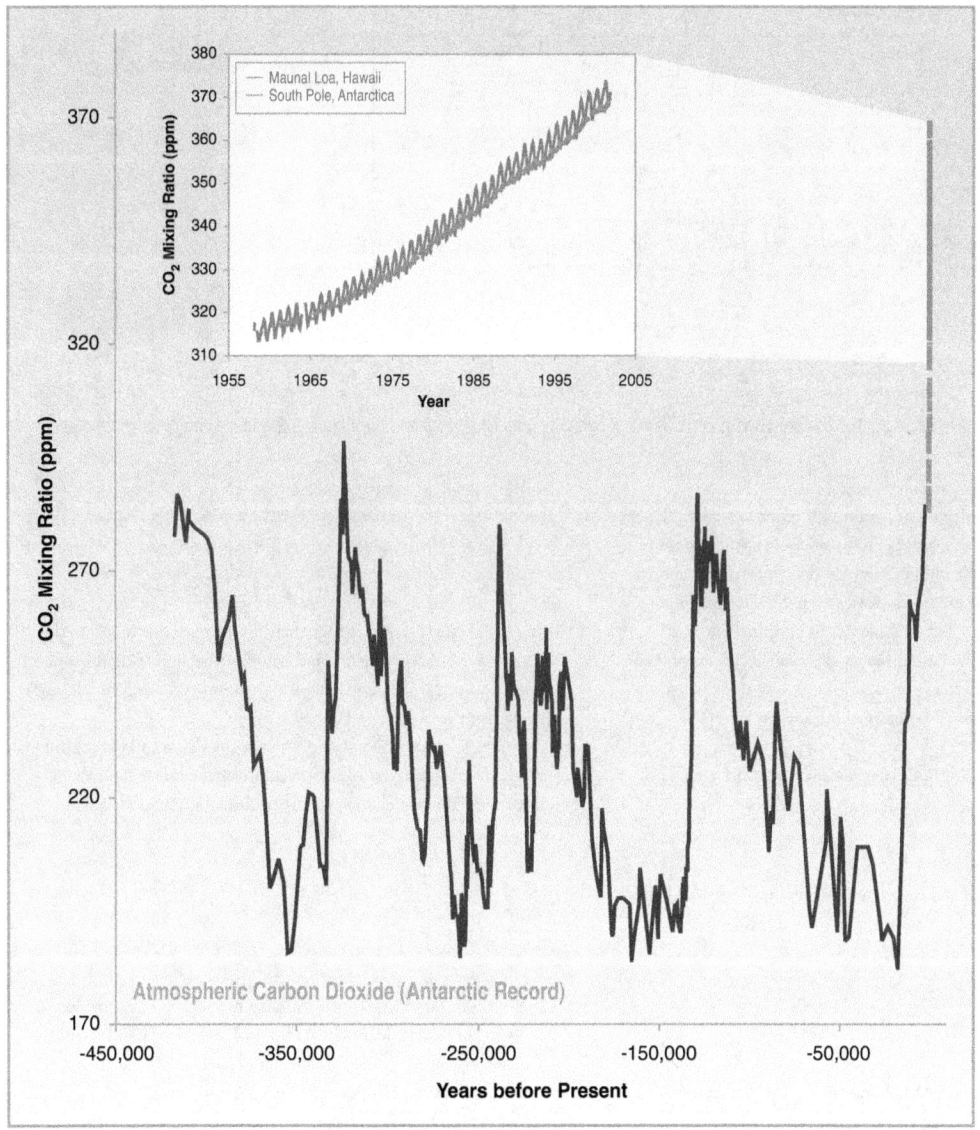

Figure 2-2: Direct measurements of atmospheric CO_2 at Mauna Loa, Hawaii and the South Pole, and CO_2 concentration as derived from the Vostok Antarctic ice core. Source: Adapted from IPCC (2001a), with Mauna Loa record updated by Dave Keeling and Tim Whorf, Scripps Institution of Oceanography.

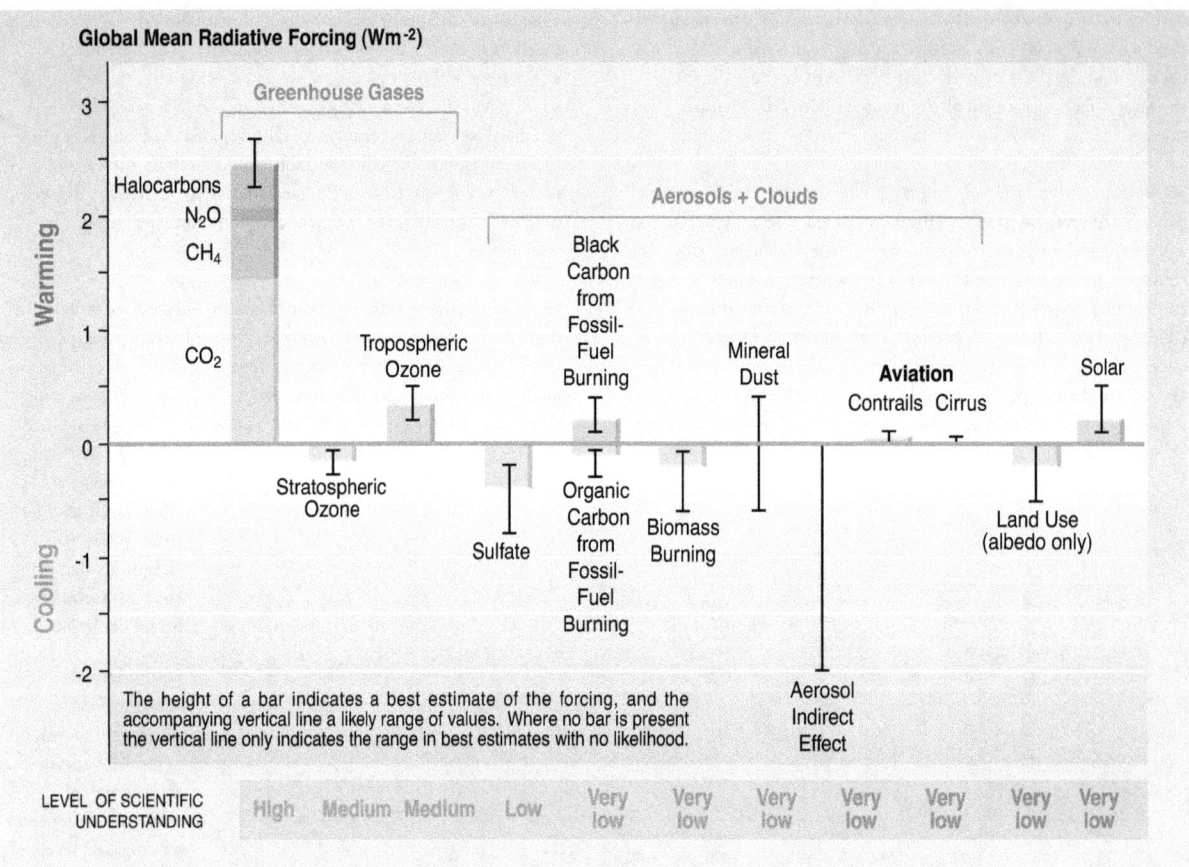

Figure 2-3: Global mean radiative forcing of the climate system and degree of certainty (see Chapter 3 and the Figure 3-2 Annex C entry for more detail). Source: Adapted from IPCC (2001a).

CCTP, is a priority for CCSP. The program will work with CCTP to explore the climate and related implications of different technology portfolios in order to provide information for decisionmaking and to enable the United States to continue its work within the United Nations Framework Convention on Climate Change. In conjunction with research on the carbon cycle and other biogeochemical cycles, improved scenario-based analyses hold the promise of increasing understanding of potential future forcing of climate, as called for by the NRC.

Examples of Key Research Activities

Focus 2.1: Reduce uncertainties about the sources and sinks of GHGs, emissions of aerosols and their precursors, and their climate effects

- Observationally assessed and improved uncertainty ranges of the radiative forcing of the chemically active greenhouse gases (Chapter 3.2)

Focus 2.2: Monitor the recovery of the ozone layer and improve the understanding of the interactions of climate change, ozone depletion, tropospheric pollution, and other atmospheric issues

- Updated trends of stratospheric ozone and ozone-depleting gases in the atmosphere (Chapter 3.4)
- Improved quantitative model evaluation of the sensitivity of the ozone layer to changes in atmospheric transport and composition related to climate change (Chapter 3.4)

Focus 2.3: Increase knowledge of the interactions among emissions, long-range atmospheric transport, and transformations of atmospheric pollutants, and their response to air quality management strategies

- *State of the Atmosphere* report that describes and interprets the status of the characteristics and trends associated with atmospheric composition, ozone layer depletion, temperature, rainfall, and ecosystem exposure (Chapter 3.5)
- A policy-relevant evaluation of the issues related to intercontinental transport, the impact of air pollutants on climate, and the impact of climate change on air pollutants (Chapter 3.5)

Focus 2.4: Develop information on the carbon cycle, land cover and use, and biological/ecological processes by helping to quantify net emissions of carbon dioxide, methane, and other greenhouse gases, thereby improving the evaluation of carbon sequestration strategies and alternative response options

- Improved coupled land-atmosphere models (jointly produced by the Carbon Cycle, Water Cycle, and Land-Use/Land-Cover Change research elements) and enhanced capability to assess the consequences of different land-use change scenarios (Chapters 5.2, 6.4, and 7.3)
- Report on the effects of land-use and land-cover changes on local carbon dynamics and the mitigation and management of greenhouse gases (Chapter 6.4 and 7.3)

	TOPICS FOR PRIORITY CCSP SYNTHESIS PRODUCTS	SIGNIFICANCE	COMPLETION
CCSP GOAL 2	Updating scenarios of greenhouse gas emissions and concentrations, in collaboration with the CCTP. Review of integrated scenario development and application.	Sound, comprehensive emissions scenarios are essential for comparative analysis of how climate may change in the future, as well as for analyses of mitigation and adaptation options.	within 2 years
	North American carbon budget and implications for the global carbon cycle.	The buildup of CO_2 and methane in the atmosphere and the fraction of carbon being taken up by North America's ecosystems and coastal oceans are key factors in estimating future climate change.	within 2 years
	Aerosol properties and their impacts on climate.	There is a high level of uncertainty about how climate may be affected by different types of aerosols, both warming and cooling, and thus how climate change might be affected by their control.	2-4 years
	Trends in emissions of ozone-depleting substances, ozone layer recovery, and implications for ultraviolet radiation exposure and climate change.	This information is key to ensuring that international agreements to phase out production of ozone-depleting substances are having the expected outcome (recovery of the protective ozone layer).	2-4 years

- Carbon cycle information to quantify the magnitude and variability of CO_2 sources, sinks, and fluxes to support analysis of carbon sequestration and management in terrestrial and oceanic systems (Chapter 7)
- Interactive global climate-carbon cycle models that explore the coupling and feedbacks between the physical and biogeochemical systems (Chapter 7.4)
- Improved projections of climate change forcings and quantification of dynamic feedbacks among the carbon cycle, human actions, and the climate system, with better estimates of errors and sources of uncertainty (Chapters 3.2 and 7.5)
- Scientific criteria and model tests of carbon management sustainability that take into account system interactions and feedbacks and analysis of options for science-based carbon management decisions and deployment by landowners (Chapters 7.6 and 3.2)

Focus 2.5: Improve capabilities to develop and apply emissions and related scenarios for conducting "If…, then…" analyses in cooperation with CCTP
- Quantify and project possible drivers of land-use change for a range of economic, environmental, and social values and develop regional, national, and global land-use and land-cover change projection models, incorporating advances in our understanding of drivers (Chapter 6.2 and 6.3)
- Linked ecosystem, resource management, and human dimensions models that enable scientific evaluation of a wide range of policy scenarios and assessment of effects on atmospheric CO_2 concentration and carbon sources and sinks (Chapters 7.3 and 8.1)
- Scenarios strengthened by an improved understanding of the interdependence among economic growth; population growth, composition, distribution, and dynamics (including migration); energy use in different sectors (e.g., electric power generation, transportation); advancements in technologies; and pollutant emissions (Chapter 9.1)

CCSP Goal 3: Reduce uncertainty in projections of how the Earth's climate and related systems may change in the future
While there is a great deal known about the mechanisms that affect the response of the climate system to changes in natural and human influences, many basic questions remain to be addressed. There is also a high level of uncertainty regarding precisely how much climate will change overall and in specific regions (see Figure 2-4). Current models project significantly different increases in global average surface temperature, from approximately 1°C to more than 5°C during the 21st century. This range of uncertainty incorporates both different estimates of climate sensitivity (the increase in temperature that results from a doubling of atmospheric concentrations of CO_2, for example) and a wide range in projections of future greenhouse gas emissions. A primary CCSP objective is to build on existing information and scientific capacity to sharpen qualitative and quantitative understanding through observations, data assimilation, and modeling activities.

CCSP-supported process research will address basic climate system properties and interactions, including improving characterization of the circulation and interaction of energy in the atmosphere and oceans. It will also seek to reduce uncertainties regarding a number of "feedbacks" or secondary changes caused by the initial influence that can either reinforce or dampen the initial effect of greenhouse gases and aerosols. These feedbacks include changes in the amount and distribution of water vapor, changes in extent of ice and the Earth's reflectivity, changes in cloud properties, and changes in biological and ecological systems that could significantly change emissions or absorption of greenhouse gases (see Figure 2-5).

A cause for concern about which there is considerable uncertainty is the potential for changes in extreme events, and rapid, discontinuous changes in climate. Such changes could have profound effects on the environment and human well-being because the time available for adaptation would be limited. The historical record of past climates revealed through the study of ancient ice cores and other paleoclimate data indicates that the climate system can change relatively rapidly in response to internal processes or rapidly changing external forcing. Increasing our understanding of the conditions that could give rise to events such as rapid changes in ocean circulation is a key aspect of CCSP-supported research.

Incorporation of this basic process-level knowledge into climate and ecosystems models holds promise for improving our ability to forecast and project climate phenomena. Climate models seek to quantitatively

a) Temperature Indicators

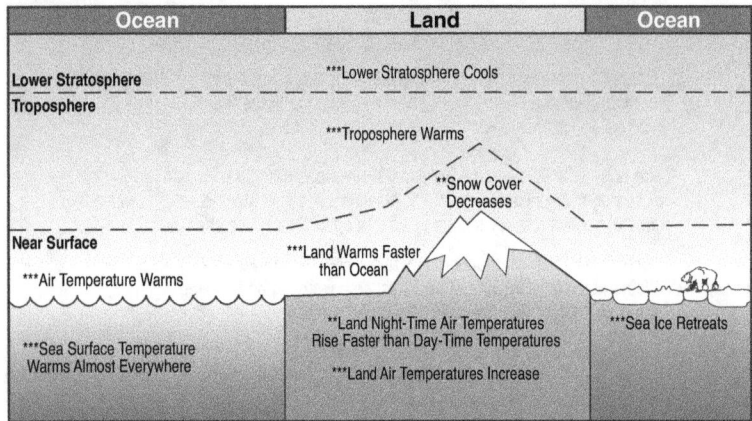

b) Hydrological and Storm-Related Indicators

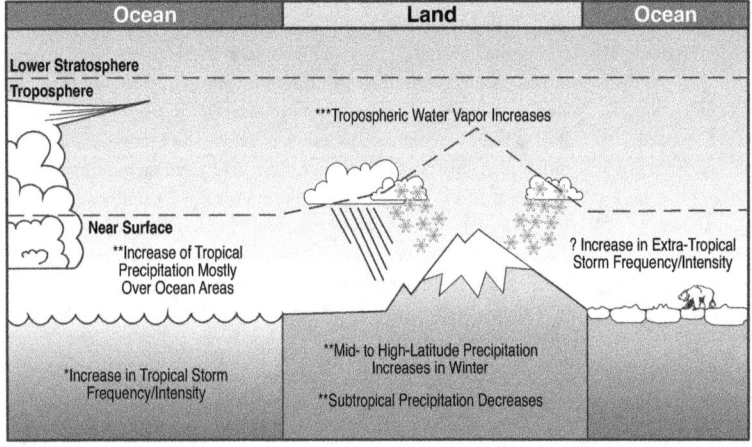

***	Virtually Certain
	(many models analyzed and all show it)
**	Very Likely
	(a number of models analyzed show it, or change is physically plausible and could readily be shown for other models)
*	Likely
	(some models analyzed show it, or change is physically plausible and could be shown for other models)
?	Medium Likelihood
	(a few models show it or results mixed)

Figure 2-4: Schematic of changes in temperature and hydrological indicators from projections of future climate changes from atmosphere-ocean general circulation models (AOGCMs). Source: Adapted from IPCC (2001a).

simulate the behavior of the climate system and its response to changes in forcing. They vary in complexity and uses, but in the most complex, various components of the climate system (atmosphere, oceans, biosphere, hydrosphere, and cryosphere) are coupled together to study the evolution of the entire Earth system. Research indicates that there are limits to the predictability of regional- and local-scale phenomena. At this point, modeled projections of the future regional impacts of global climate change are often contradictory and are not sufficiently reliable to be used as tools for planning. Alternative methods such as use of climate analog scenarios and weather generators may sometimes be employed for these purposes.

Examples of Key Research Activities

Focus 3.1: Improve characterization of the circulation of the atmosphere and oceans and their interactions through fluxes of energy and materials

- Improved high-resolution three-dimensional ocean circulation models and improved estimates of global air-sea-land fluxes of heat, moisture, and momentum needed to discern characteristics of ocean-atmosphere-land coupling and to assess the global energy balance (Chapters 4.2 and 5.1)
- Models of ocean uptake of carbon that integrate biogeochemistry, ocean circulation, and marine ecosystem responses (Chapters 7.2 and 8.2)

Focus 3.2: Improve understanding of key "feedbacks" including changes in the amount and distribution of water vapor, extent of ice and the Earth's reflectivity, cloud properties, and biological and ecological systems

- Results from process studies related to the indirect effects of aerosols on clouds available for future assessments of climate sensitivity to aerosols (Chapter 3.1 and 3.2)
- Quantification of the potential changes in the cryosphere, including the effects of permafrost melting on regional hydrology and the carbon budget, and the consequences for mountain snowpacks, sea ice, and glaciers (Chapter 5.2)
- Evaluation of the potential for dramatic changes in carbon storage and fluxes and quantification of important feedbacks from ecological systems to climate and atmospheric composition to improve the accuracy of climate projections (Chapters 7.5 and 8.1)

Focus 3.3: Increase understanding of the conditions that could give rise to events such as rapid changes in ocean circulation due to changes in temperature and salinity gradients

- State of understanding on the causes of abrupt changes, and probabilistic estimates of future risks of abrupt global and regional climate-induced changes, including the collapse of the thermohaline circulation, persistent ENSO conditions, abrupt sea-level rises, and the positive feedback associated with high-latitude ice (ice sheets and sea ice) and their impacts on albedo (Chapter 4.3)

Focus 3.4: Accelerate incorporation of improved knowledge of climate processes and feedbacks into climate models to reduce uncertainty in projections of climate sensitivity, changes in climate, and related conditions such as sea level

TOPICS FOR PRIORITY CCSP SYNTHESIS PRODUCTS	SIGNIFICANCE	COMPLETION
Climate models and their uses and limitations, including sensitivity, feedbacks, and uncertainty analysis.	Clarifying the uses and limitations of climate models at different spatial and temporal scales will contribute to appropriate application of these results.	within 2 years
Climate projections for research and assessment based on emissions scenarios developed through CCTP.	Production of these projections will help develop modeling capacity and will provide important inputs to comparative analysis of response options.	2-4 years
Climate extremes including documentation of current extremes. Prospects for improving projections.	Extreme events have important implications for natural resources, property, infrastructure, and public safety.	2-4 years
Risks of abrupt changes in global climate.	Abrupt changes have occurred in the past and thus it is important to evaluate what we know about the potential for abrupt change in the future.	2-4 years

(CCSP GOAL 3)

- Improved representation of processes (e.g., thermal expansion, ice sheets, water storage, coastal subsidence) in climate models that are required for simulating and projecting sea-level changes (Chapter 4.2)
- Estimates of the spatial and temporal limits of predictability of climate variability and change forced by human activities (Chapter 4.2)
- Incorporation of water cycle and carbon cycle processes, interactions, and feedbacks into an integrated Earth system modeling framework (Chapters 5.2 and 7.5)
- New observationally tested parameterizations for clouds and precipitation processes for use in climate models based on

cloud-resolving models developed in part through field process studies (Chapter 5.2)
- National and global models with a coupled climate-land use system (Chapter 6.4)
- Improved capability to include and accurately formulate terrestrial and marine ecosystem dynamics within local and regional climate models

Focus 3.5: Improve national capacity to develop and apply climate models

- Continue support of the two high-end climate modeling centers to respond to the need for scenario-driven

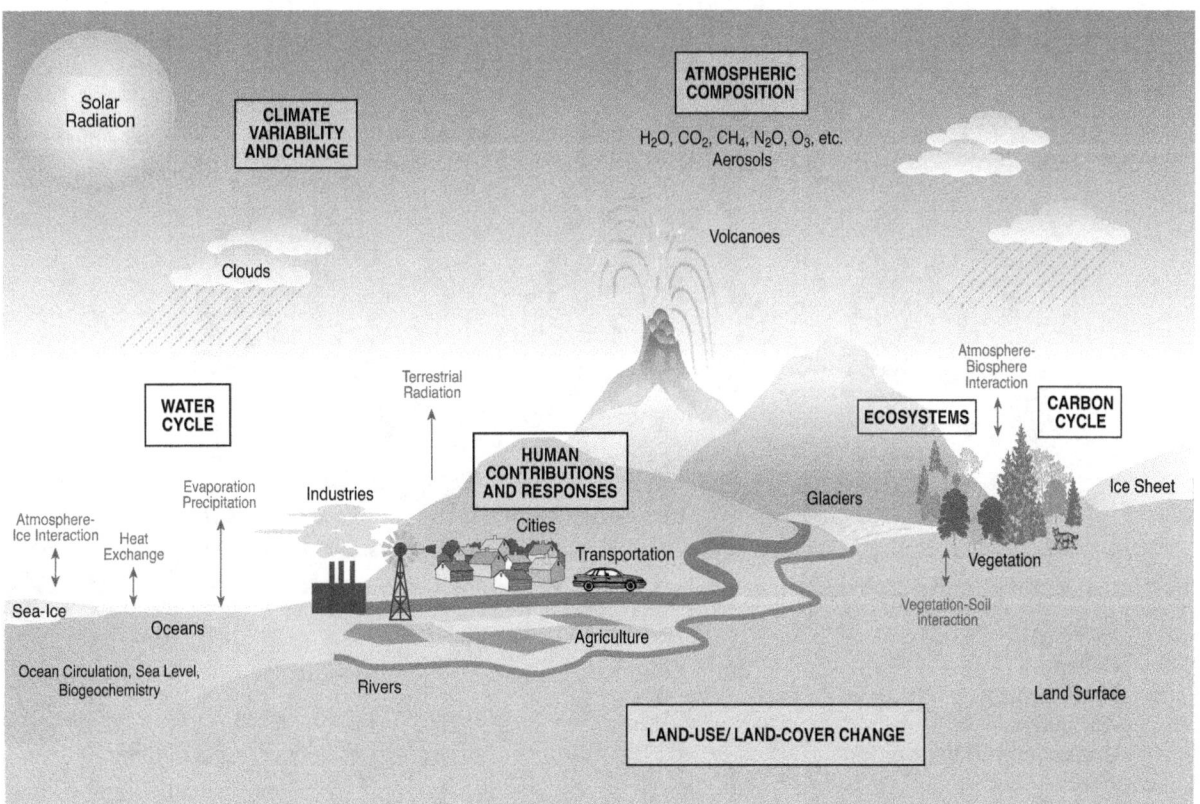

Figure 2-5: Major components needed to understand the climate system and climate change. Source: Adapted from IPCC (2001a).

climate modeling in support of assessments (Chapter 10)

* Support for a Common Modeling Infrastructure to optimize modeling resources and enable meaningful knowledge transfer among modelers (Chapter 10)

* Develop a program of focused model intercomparisons and conduct evaluation and analysis of the model sensitivities of major U.S. models as well as model validation with available observational data (Chapter 10)

CCSP Goal 4: Understand the sensitivity and adaptability of different natural and managed ecosystems and human systems to climate and related global changes

Seasonal-to-annual variability in climate has been connected to impacts on almost every aspect of human life: agricultural yields, water resources, energy demand and supply, transportation, price fluctuations, fishery yields, forest fires, human health and welfare, and many others. Long time scale natural climate cycles and potential future human-induced changes in climate could have additional effects, including altering the lengths of growing seasons, the sustainability of water resource management systems, the geographical ranges of plant and animal species, biodiversity, estuarine and ocean productivity, and the incidence of disturbance regimes that affect both natural and human-made environments. Potential benefits and risks have been identified for a number of systems and activities. Improving our ability to assess potential vulnerability and resilience to future

variations and changes in climate and environmental conditions could enable governments, businesses, and communities to reduce negative impacts and seize opportunities to benefit from changing conditions by adapting infrastructure, activities, and plans.

The potential effects of climate variability and change on ecosystems and human activities will not be determined solely by their sensitivity and adaptability, but also by multiple, cumulative interactions among physical, ecological, economic, and social conditions. For example, some crops and plants that might otherwise experience reductions in productivity as a result of changes in climate alone could actually experience increased growth and productivity as a result of increases in atmospheric CO_2 concentrations and nutrients (from deposition and runoff). Other species and ecosystems, which could adjust to climate change alone, might be endangered when land use and other factors interfere with adaptive mechanisms such as migration. Interacting factors must be identified and understood to develop accurate projections of effects. The CO_2 "fertilization effect" (increased plant growth due to higher atmospheric CO_2 concentrations), nitrogen deposition, disturbance (e.g., fire, pest infestations), land-cover fragmentation (see Figure 2-6), air pollution, and other factors affect the functioning (e.g., water use efficiency, biomass allocation) and composition of natural and managed ecosystems over long periods of time. Similarly, estuarine and coastal ecosystems face multiple-stressor problems associated with point source and non-point source pollution, increased sedimentation resulting from upstream land-use practices, invasive species and

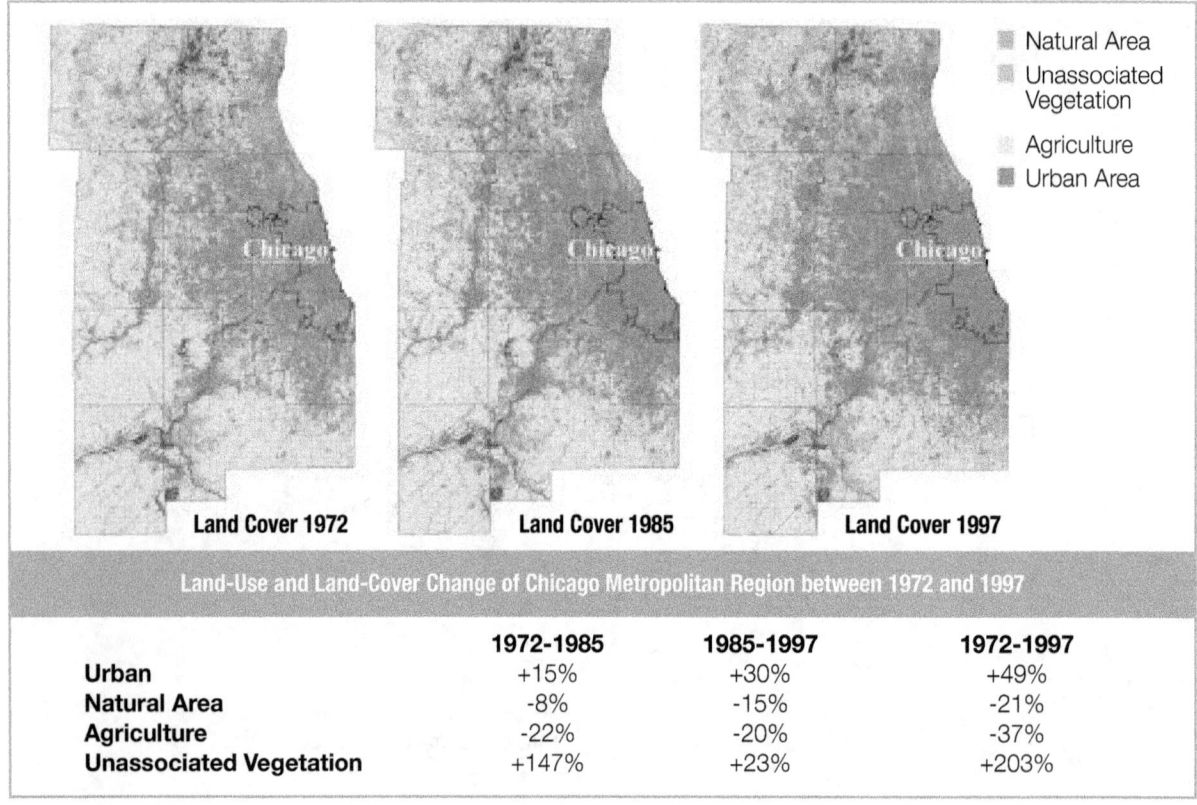

	Natural Area
	Unassociated Vegetation
	Agriculture
	Urban Area

Land Cover 1972   Land Cover 1985   Land Cover 1997

Land-Use and Land-Cover Change of Chicago Metropolitan Region between 1972 and 1997

	1972-1985	1985-1997	1972-1997
Urban	+15%	+30%	+49%
Natural Area	-8%	-15%	-21%
Agriculture	-22%	-20%	-37%
Unassociated Vegetation	+147%	+23%	+203%

Figure 2-6: Land-cover maps of the Chicago Metropolitan Region. These maps, created with Landsat imagery from 1972, 1985, and 1997, document changes in several categories of land cover and land use. Source: NASA and Y.Q. Wang, University of Rhode Island.

TOPICS FOR PRIORITY CCSP SYNTHESIS PRODUCTS	SIGNIFICANCE	COMPLETION
Coastal elevation and sensitivity to sea-level rise.	Evaluation of how well equipped society is to cope with potential sea-level rise can help reduce vulnerability.	within 2 years
State-of-knowledge of thresholds of change that could lead to discontinuities (sudden changes) in some ecosystems and climate-sensitive resources.	This approach seeks to determine how much climate change natural environments and resources can withstand before being adversely affected.	2-4 years
Relationship between observed ecosystem changes and climate change.	Earlier blossoming times, longer growing seasons, and other changes are being observed, and this report will explore what is known about why these events are happening.	2-4 years
Preliminary review of adaptation options for climate-sensitive ecosystems and resources.	Understanding of adaptation options can support improved resource management—whether change results from natural or human causes—and thus helps realize opportunities or reduce negative impacts.	2-4 years
Scenario-based analysis of the climatological, environmental, resource, technological, and economic implications of different atmospheric concentrations of greenhouse gases.	Knowing how well we can differentiate the impacts of different greenhouse gas concentrations is important in determining the range of appropriate response policies.	2-4 years
State-of-the-science of socioeconomic and environmental impacts of climate variability.	This product will help improve application of evolving ENSO forecasts by synthesizing information on impacts, both positive and negative, of variability.	2-4 years
Within the transportation sector, a summary of climate change and variability sensitivities, potential impacts, and response options.	Safety and efficiency of transportation infrastructure—much of which has a long lifetime—may be increased through planning that takes account of sensitivities to climate variability and change.	2-4 years

(Left margin: CCSP GOAL 4)

sea-level rise. Multiple changes affecting marine ecosystems include natural climate variations (e.g., ENSO cycles and PDV), fishing, and changes in ocean productivity brought on by uncertain causes. The interactions are even more complex when one examines the interactions of biophysical and socioeconomic systems. CCSP research will examine the implications of multiple interacting changes to improve projections of effects.

Evaluation of the potential impacts associated with different atmospheric concentrations of greenhouse gases and aerosols is an important input to weighing the costs and benefits associated with different climate policies. Further research is required to integrate still limited understanding of effects of different concentration levels, and the influence of human activities on concentrations, and to develop methods for aggregating and comparing impacts across different sectors and settings.

Examples of Key Research Activities

Focus 4.1: Improve knowledge of the sensitivity of ecosystems and economic sectors to global climate variability and change

- Results from field and modeling experiments to study the role of mountain environments on precipitation and runoff production (Chapter 5.3)
- Downscaling techniques, such as improved regional climate models, for improved evaluation of potential water resource impacts arising from climate variability and change (Chapter 5.3)
- Reports on past and projected trends in land cover or land use that are attributable to changes in climate (e.g., changes in forest type, changes in agriculture) (Chapter 6.4)
- Ecosystem observations, coupled physical-biological models, indicators, and reports to permit better assessment of the potential consequences of global and climatic changes on selected arctic, alpine, wetland, riverine, and estuarine and marine

ecosystems; selected forest and rangeland ecosystems; selected desert ecosystems; and the Great Lakes (Chapter 8.2)
- Development of data and predictive models determining the sensitivity of selected organisms and their assemblages to changes in ultraviolet-B (UV-B) radiation and other environmental variables relative to observations of UV-B radiation in terrestrial, aquatic, and wetland habitats (Chapter 8.2)
- Assessments of the potential economic impacts of climate change on the producers and consumers of food and fiber products (Chapters 8.2 and 9.2)

Focus 4.2: Identify and provide scientific inputs for evaluating adaptation options, in cooperation with mission-oriented agencies and other resource managers

- Data and models demonstrating how different adaptation strategies in croplands and commercial forests affect the sustainability of these ecosystems in changing environmental conditions and their ability to meet human demands (Chapter 8.3)
- Initiation of development of decision support tools relevant to regions where abrupt changes or threshold ecological responses may occur, especially high-altitude and high-latitude ecosystems and transitional zones between ecosystems (i.e., ecotones) such as forest-grassland, agriculture-native prairie, riparian and coastal zones, coastal-oceanic boundaries, and rural-urban interfaces (Chapter 8.3)
- Data sets and spatially explicit models for examining effects of management and policy decisions on a wide range of terrestrial and marine ecosystems to predict the efficacy and tradeoffs of management strategies at varying scales, and preliminary comparisons of the effectiveness of selected management practices (Chapter 8.3)
- Analysis of potential impacts of climate variability and change on transportation infrastructure and operations, and preliminary

development of decision support tools relevant to transportation decisionmakers (Chapter 9.2)

Focus 4.3: Improve understanding of how changes in ecosystems (including managed ecosystems such as croplands) and human infrastructure interact over long periods of time

- Evaluations of the effects of water cycle variability and change on trends in water quality conditions (Chapter 5.4)
- Data from field experiments quantifying aboveground and belowground effects of elevated CO_2 concentration in combination with other environmental changes on the structure and functioning of agricultural, forest, and aquatic ecosystems (Chapters 7.5 and 8.2)
- Examination of responses of ecosystems (including any changes in nutrient cycling) to combinations of elevated CO_2 concentration, warming, and altered hydrology, with data collection underway (Chapter 8.2)
- Analysis of the global occurrence, extent, and impact of major disturbances (e.g., fire, insects, drought, and flooding) on land use and land cover (Chapter 6.1)
- Identification of the regions in the United States where land-use and climate change may have the most significant implications for land management (Chapter 6.5)
- Spatially explicit ecosystem models at regional to global scales, to improve our understanding of contemporary and historical changes in ecosystem structure and functioning, and synthesis of known effects of increasing CO_2, warming, changes in precipitation, and other factors (e.g., increasing tropospheric ozone) on terrestrial ecosystems (Chapter 8.2)
- Estimates of the value of the ancillary benefits (e.g., enhanced wildlife habitat and improved water quality) that could result from implementing various mitigation and adaptation activities within the forestry sector (Chapters 8.3 and 9.2)
- Next phase of assessments of the health effects of combined exposures to climatic and other environmental factors (e.g., air pollution) (Chapter 9.4)

CCSP Goal 5: *Explore the uses and identify the limits of evolving knowledge to manage risks and opportunities related to climate variability and change*

The first four CCSP goals focus on the key issues of understanding natural climate conditions and variability, forces that influence climate, climate responses, and implications for human systems and natural environments. CCSP research will focus on this full range of issues in order to lay the basis for informed discussion and decisionmaking. The fifth CCSP goal focuses on the way that this scientific information is used for support of decision needs.

Over the last decade, USGCRP has enlisted federal and external experts to develop a variety of products and resources for decision support. These include methods and tools for integrated assessment (including a variety of quantitative integrated assessment models); state-of-science syntheses and assessments of climate variability and change; international assessments of stratospheric ozone depletion; applications of Earth science observations and data; experiments for application of ENSO forecasts in a variety of regions and economic sectors; and advisory committee assessments of the potential vulnerabilities and opportunities arising from climate change in different regions and sectors of the United States.

These products of the first decade have evoked much commentary, both positive and negative. Formal evaluations of the processes and products of some of these activities have been conducted by the NRC. CCSP's decision support activities will respond to the relevant recommendations of the NRC in reports such as *Global Environmental Change: Research Pathways for the Next Decade* (NRC, 1999a), *Climate Change Science: An Analysis of Some Key Questions* (NRC, 2001a), and *The Science of Regional and Global Change: Putting Knowledge to Work* (NRC, 2001e). The program will also encourage further evaluation and learning from these experiences in order to structure decision support processes and products that use existing knowledge to the best effect while communicating the limits of this knowledge.

CCSP activities in support of its fifth goal will improve the nation's and global community's understanding of the nature and extent of the challenges inherent in climate change and develop improved resources for evaluating options for adaptation and mitigation. Fulfilling this goal will require development of a variety of resources including observations, databases, data and model products, scenarios, a variety of visualization products, and improved approaches for interacting with users. Research in many of the program elements holds promise of yielding decision support information and products. Improved resources for decision support also require a concerted focus on the processes through which information is developed and applied to support decisionmaking in different areas.

Areas of decisionmaking to be supported by CCSP include adaptive management, planning, and policy. Each of these categories will have its own unique set of stakeholders and will require different decision support processes and tools that will be recognized in the implementation of CCSP decision support activities. CCSP will engage both decisionmakers and members of the research community in the development of processes well suited to both the state-of-science and decisionmaking needs. In addition to supporting decisionmaking, the decision support activities of CCSP will also help to identify key knowledge gaps and provide feedback to researchers that will guide the evolution of the CCSP agenda.

Examples of Key Research Activities

Focus 5.1: Support informed public discussion of issues of particular importance to U.S. decisions by conducting research and providing scientific synthesis and assessment reports

- Contribute to and participate in international assessments on climate change, ozone depletion, and other issues as they arise (multiple chapters)
- Produce a number of synthesis reports and assessments that integrate research results focused on identified science and decision issues in association with stakeholders and researchers to meet the requirements of the Global Change Research Act (multiple chapters)

Focus 5.2: Support adaptive management and planning for resources and physical infrastructure sensitive to climate variability and change; build new partnerships with public and private sector entities that can benefit both research and decisionmaking

- Increased partnerships with existing user support institutions, such as state climatologists, regional climate centers, agricultural

	TOPICS FOR PRIORITY CCSP SYNTHESIS PRODUCTS	SIGNIFICANCE	COMPLETION
CCSP GOAL 5	Uses and limitations of observations, data, forecasts, and other projections in decision support for selected sectors and regions.	There is a great need for regional climate information; further evaluation of the reliability of current information is crucial in developing new applications.	within 2 years
	Best-practice approaches to characterize, communicate, and incorporate scientific uncertainty in decisionmaking.	Improvements in how scientific uncertainty is evaluated and communicated can help reduce misunderstanding and misuse of this information.	within 2 years
	Decision support experiments and evaluations using seasonal to interannual forecasts and observational data.	Climate variability is an important factor in resource planning and management; improved application of forecasts and data can benefit society.	within 2 years

extension services, resource management agencies, and state and local governments to accelerate uses of climate knowledge (Chapter 11)
- Assess the adequacy of existing operational climate monitoring networks to provide regional decision support, and to identify major data gaps in addressing critical regional and policy issues, such as drought planning and response (Chapter 4.5)
- Conduct and analyze decision support experiments using observations, integrated data sets, forecasts of seasonal climate variability, and longer term model projections (multiple chapters)
- Develop and apply frameworks for assessing the uses and limits of current regional-scale information (multiple chapters)

Focus 5.3: Support policymaking by conducting comparative analyses and evaluations of the socioeconomic and environmental consequences of response options
- Improve stakeholder involvement in articulating and framing all aspects of policy support activities including question formulation, study design, and transparent review involving stakeholders (Chapter 11)
- Strengthen capacity and conduct "If…, then…" scenario analyses (Chapter 11 and through cooperation with CCTP)
- Develop and apply additional methods for integrated assessment and comparative analyses (Chapters 9 and 11)

"Critical Dependencies" and Approaches to Integration

The interdisciplinary research elements of CCSP partition the overall Earth system and the issue of climate change into discrete and manageable sets of research problems. While partitioning the problem is necessary for both research and program management purposes, it carries with it the potential to divert attention from critical questions that are beyond the scope of the individual research elements, instead emphasizing lower priority issues that reside completely within a single research area. Thus, a key challenge facing CCSP is engaging in integrated planning that "puts back together" the pieces of the Earth system and fosters problem-driven interdisciplinary research. This is especially true for "critical dependencies," topics for which progress in one research element is only possible if related research is first completed in other areas.

CCSP will need to foster integration across research elements and disciplines; among basic research and supporting activities such as observations, modeling, and data management; in the development of comprehensive climate models; and between participating departments and agencies if it is to achieve its objectives.

"Critical Dependencies" among the Research Elements

Scientifically, integration is essential for understanding how the Earth system functions and will evolve in response to future forcing. This is because of the inherent interconnectedness among components of the Earth system. In particular, the existence of feedback loops in the Earth system creates the need to work across disciplinary boundaries. For some issues, it will be necessary to coordinate and integrate across all of the CCSP program elements. One example that illustrates how the different research elements may help address a key uncertainty is given in Box 2-2, focused on the question of the evolution of the global distribution of atmospheric methane.

Because the components of the Earth system interact continuously, research in one program element often has a *critical dependency* with work in other elements. Critical dependencies can involve insights from process studies, data flows, model components, and other research and operational activities.

The most obvious set of critical dependencies is in the area of observational data. In many cases, observations of the underlying physical state of the Earth system (e.g., temperature, precipitation history) are required before questions about climate or global change can be addressed. Another example is information on land cover and land use, which also forms the basis for many studies of hydrology, biogeochemical cycling, and ecosystems research. Chapters 3-9 describe both the inputs that each research element needs from other program elements and the products that each expects to produce to support goals in other research areas. A challenge for CCSP will be to coordinate production of information required to satisfy these critical dependencies so that they meet requirements and are sequenced properly. CCSP will coordinate development of implementation plans for the research elements to meet this challenge. Box 2-3 lists illustrative examples of "critical dependencies" from the research elements.

BOX 2-2

INTEGRATING ACROSS RESEARCH ELEMENTS TO IMPROVE PROJECTIONS OF METHANE EMISSIONS AND CONCENTRATIONS

In order to improve understanding of current and projected atmospheric concentrations of methane, it is essential to integrate results obtained through the different CCSP science elements. This includes pulling together information on the initial concentration, the sources of additional emissions, the processes that transport it, and the removal processes. Each of the research elements will contribute the following information:

- **Atmospheric Composition**. Measurement of methane's distribution in the atmosphere is by definition the starting point for assessing the evolution of its future concentrations. Since the major removal process for methane from the atmosphere is its oxidation by hydroxyl (OH), assessment of the distribution of OH, including its spatial and temporal variation, is also required. The inverse modeling that is used to help infer source distributions is also part of this research element.
- **Climate Variability and Change**. The climate system circulates methane from its source regions and distributes it uniformly around the globe. Climate also determines temperature, which affects both methane emissions and the rate at which it is removed through oxidation by OH (one of the more temperature-dependent reactions that affects atmospheric composition). This

research element also provides estimates of future changes in surface temperatures at high latitudes. These estimates are important for determining the likelihood that methane tied up in the surface at high latitudes could be released.

- **Carbon Cycle**. Although methane is less abundant than carbon dioxide, its contribution to atmospheric radiative forcing is significant due to its much greater radiative impact per molecule. Carbon cycle studies estimate the overall release and uptake of methane, provide the context necessary for interpreting measurements and estimates of methane release and removal, and incorporate advanced understanding of process controls on methane uptake and release from the land and oceans into carbon cycling models.
- **Water Cycle**. Water plays a crucial role in helping to establish the conditions under which methane is emitted, and it also is a crucial precursor to OH, which removes methane from the atmosphere. Water availability also affects the magnitude and rate of emissions from agricultural sources (e.g., rice cultivation), landfills, and other land-based sources.
- **Land-Cover/Land-Use Change**. Some land uses are associated with methane emissions. For example,

methane is emitted from rice cultivation and natural wetlands, and is removed from the atmosphere by vegetation and soils. Improved understanding of how land is used in different locations, as well as how land uses might change over time, is crucial to estimating current and projecting future emissions.

- **Ecosystems**. The production of methane from biological processes can be traced to microbial processes, and the ecosystems research element provides the knowledge base for carrying out relevant studies of the biology that underlies methane production and its emission to the atmosphere.
- **Human Impacts**. Methane is emitted by many human activities, including agricultural practices such as raising livestock and growing rice, waste disposal practices in landfills and other sites, and the production and transportation of coal, natural gas, and oil. The amount of methane emitted by these activities depends on basic characteristics of the emission processes as well as the management practices that are employed. These have changed and will continue to change over time and hence must be considered in estimating current and projecting future emissions.

Coordination of Research and Supporting Activities

Just as meeting CCSP objectives will require integration across the research elements described in Chapters 3-9, so will it require coordination and integration of the different approaches employed by the scientific community in addressing these major challenges—basic research, surface-based networks, field-based process studies, global satellite observations, computational modeling, data management, assessment, and decision support. It is worth emphasizing that many of the surface-based networks and some of the global satellite observations are carried out as part of operational monitoring programs not previously considered as part of USGCRP, and which may not even now be identified in the CCSP

budget. Clearly, our nation's effort to provide comprehensive, integrated answers to the questions posed in this plan will require integrated use of all such data and approaches.

Basic research provides the intellectual framework of the entire global change research enterprise. From laboratory descriptions of critical physical, chemical, and biological processes, to determination of appropriate standards and calibration procedures, to approaches for simulation, analysis, visualization, etc., of data, research serves as the basis on which advances in observing and modeling depend. It also includes the necessary social science and related research that provides the data and models needed to include effects of human inputs and response into the CCSP program.

BOX 2-3

SELECTED EXAMPLES OF CRITICAL DEPENDENCIES TO BE SUPPORTED BY EACH RESEARCH ELEMENT

Chapter 3. Atmospheric Composition
- Radiative forcing input parameters to climate model simulations— *Climate Variability and Change*
- Characterization of other composition-climate processes (e.g., impact of aerosols on cloud formation and precipitation)—*Water Cycle*

Chapter 4. Climate Variability and Change
- Cloud/water vapor feedback processes in the context of the coupled climate system models—*Water Cycle*
- Environmental condition predictions/ estimates—*Land-Use/Land-Cover Change, Carbon Cycle, and Ecosystems*

Chapter 5. Water Cycle
- Long-range prediction capability for drought and flood risks (seasonal-to-interannual time scales)— *Climate Variability and Change*
- Quantification of the potential impact of changes in permafrost on regional hydrology and the carbon budget—*Carbon Cycle*

Chapter 6. Land-Use/ Land-Cover Change
- Analysis of the effects of historical and contemporary land use on carbon storage and release across environmental gradients— *Carbon Cycle*
- Quantification and projection of possible drivers of land-use change for a range of economic, environmental, and social values—*Ecosystems, Carbon Cycle, and Human Contributions and Responses*

Chapter 7. Carbon Cycle
- Quantitative estimates of carbon fluxes from managed and unmanaged ecosystems in North America and surrounding oceans, with regional specificity—*Water Cycle, Ecosystems, Land-Use/Land-Cover Change, and Climate Variability and Change*
- Evaluation of the environmental effects of mitigation options— *Ecosystems, Human Contributions and Responses, and Land-Use/Land-Cover Change*

Chapter 8. Ecosystems
- Information about ecological inputs to atmospheric composition through greenhouse gas exchanges between the atmosphere and ecosystems— *Atmospheric Composition*
- Information about energy exchanges between ecosystems and the atmosphere for inclusion in general circulation models— *Climate Variability and Change*
- Analysis of effects of global change on ecosystem goods and services— *Human Contributions and Responses*

Chapter 9. Human Contributions and Responses to Environmental Change
- Improved modeling frameworks that better link general circulation, ecological, and economic models of the agricultural and forestry sectors— *Water Cycle, Ecosystems, Land-Use/Land-Cover Change, and Climate Variability and Change*

Surface-based monitoring networks provide a distributed, accurate, and potentially high-frequency method for obtaining environmental data. They also allow the network managers to concentrate observing power in regions where it is most needed. As noted above, many of these are managed as part of operational monitoring programs and do not report as part of USGCRP. They also provide critical validation information for satellite-derived observations.

Field-based process studies allow the combined use of surface-based measurements and transportable platforms (ships, aircraft, balloons, etc.), and also provide opportunities to define intermediate spatial scales between the local scales best studied with surface-based networks and the larger spatial scales typically observed with satellite instruments. By virtue of being able to comprehensively sample a large number of environmental variables associated with a particular issue, they are excellent for providing data sets that can be used to quantitatively test process models that ultimately support the development of parameterizations that can be incorporated into larger scale models.

Global satellite observations allow for frequent observations over most if not all of the Earth's surface, and allow one to make measurements in regions for which *in situ* data are not available. Successful implementation of satellite observing programs requires validation with surface-based networks and with more focused *in situ* observations. Process representation in models will be derived from knowledge gained in the field-based process studies. It is important that research observations be integrated with those of operational monitoring programs not traditionally considered part of USGCRP.

Computational modeling provides a way in which information about the potential future of one or more elements of the Earth system can be expected to evolve, based on assumptions about naturally occurring and human-induced forcings. The models to be used in this way can be tested in a retrospective sense by comparing observed data on Earth system evolution (obtained from surface- and space-based observing systems as well as field-based process studies) and the known forcings; conversely, the models can also be used in an inverse set with the observations to infer present and historical forcings. Computational models can also be integrated with observational data in data assimilation systems to provide accurate and geophysically (and increasingly biogeochemically) consistent data sets of the distribution and fluctuations of key environmental variables. The assimilation process is also critical for initializing the global models and for their evaluation.

Data management is a critically important CCSP component for two reasons. First, it is necessary to provide active, long-term stewardship for the large volumes and multiplicity of types and sources of data. Second, it provides the means to distribute the data to a growing group of users, going far beyond the traditional scientific research community to the broader set of users who are making the policy and resource management decisions.

Scientific assessment provides the opportunity for researchers to go back and critically assess the status of their knowledge, sharpen the questions they are trying to answer, then provide a "feedback loop" to the research process so that the research of the future will better identify and address key uncertainties and knowledge gaps. Part of the assessment process may involve detailed study of the observational data sets, process knowledge, or modeling systems developed under CCSP to resolve discrepancies and competing approaches. End-to-end assessment is broader in that it can involve going back to the basic building blocks of a research area, assessing status and progress, and then looking at the implications of the results of this analysis for a variety of end states, including many relevant decision support applications. Assessment used to support decisions is an iterative analytic process that engages both researchers and interested stakeholders in the evaluation and interpretation of the interactions of natural and socioeconomic systems. These assessments typically consist of four elements:

problem formulation, analysis, characterization of consequences, and communication of results.

Decision support is the set of processes that includes interdisciplinary research, product development, communication, and operational services that provide timely and useful information to address challenges and questions confronting those who need to make policy and/or resource management decisions. In addition to the emphasis on the quality and robustness of scientific information for providing decision support, there is typically a premium on timeliness, resolution, comprehensiveness, and effective communication of levels of scientific confidence and certainty.

Models as Integrating Tools

Computational models are particularly important in integrating our knowledge of the Earth system, providing both a quantitative way of assessing the accuracy of the system and of projecting its future evolution (see Figure 2-7). It is important that the models used for projecting future conditions include the full range of processes linking the Earth system's components, and thus the feedback processes that are so important in determining the Earth's behavior. Such models become complex, requiring the integration efforts of large teams of scientists (both Earth system scientists and computational scientists), and require significant amounts of computing capability, together with human and data resources for the archiving and distributing of their results.

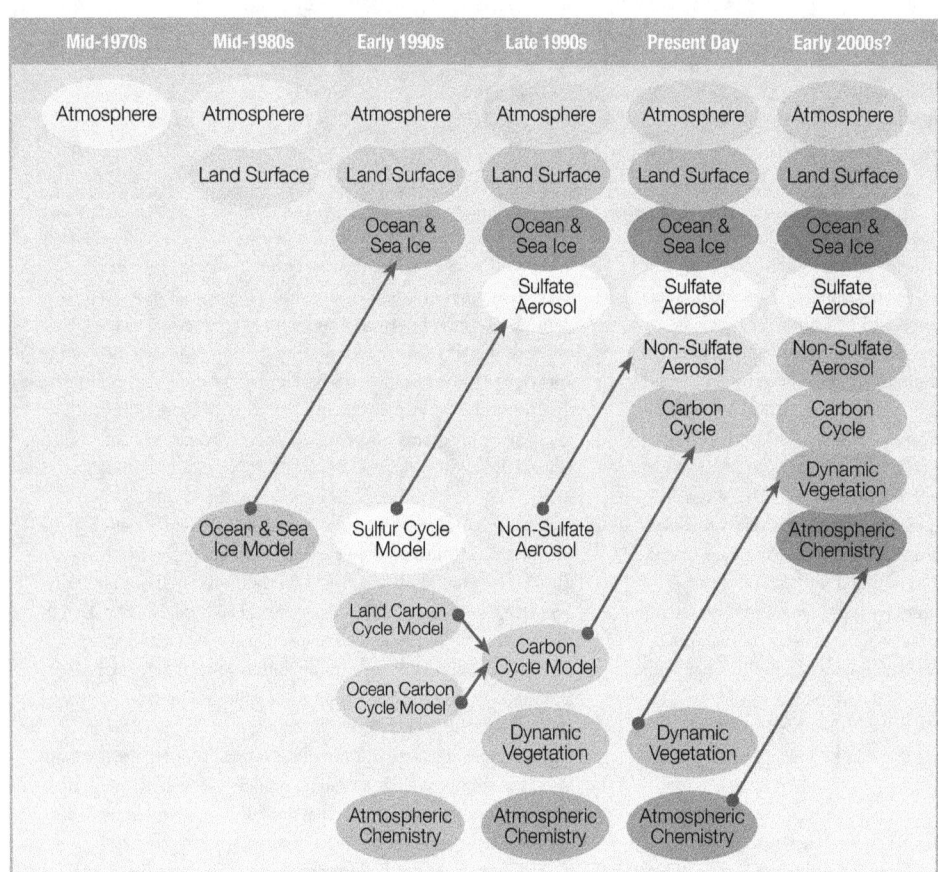

Figure 2-7: The development of climate models over the last 25 years showing how the different components are first developed separately and later coupled into comprehensive climate models. Source: IPCC (2001a).

TABLE 2-1
Summary of Synthesis and Assessment Products – Topics to be Covered

CCSP GOAL 1	Extend knowledge of the Earth's past and present climate and environment, including its natural variability, and improve understanding of the causes of observed changes
within 2 years	Temperature trends in the lower atmosphere—steps for understanding and reconciling differences.
within 2 years	Past climate variability and change in the Arctic and at high latitudes.
2-4 years	Re-analyses of historical climate data for key atmospheric features. Implications for attribution of causes of observed change.
CCSP GOAL 2	Improve quantification of the forces bringing about changes in the Earth's climate and related systems
within 2 years	Updating scenarios of greenhouse gas emissions and concentrations, in collaboration with CCTP. Review of integrated scenario development and application.
within 2 years	North American carbon budget and implications for the global carbon cycle.
2-4 years	Aerosol properties and their impacts on climate.
2-4 years	Trends in emissions of ozone-depleting substances, ozone layer recovery, and implications for ultraviolet radiation exposure and climate change.
CCSP GOAL 3	Reduce uncertainty in projections of how the Earth's climate and related systems may change in the future
within 2 years	Climate models and their uses and limitations, including sensitivity, feedbacks, and uncertainty analysis.
2-4 years	Climate projections for research and assessment based on emissions scenarios developed through CCTP.
2-4 years	Climate extremes including documentation of current extremes. Prospects for improving projections.
2-4 years	Risks of abrupt changes in global climate.
CCSP GOAL 4	Understand the sensitivity and adaptability of different natural and managed ecosystems and human systems to climate and related global changes
within 2 years	Coastal elevation and sensitivity to sea-level rise.
2-4 years	State-of-knowledge of thresholds of change that could lead to discontinuities (sudden changes) in some ecosystems and climate-sensitive resources.
2-4 years	Relationship between observed ecosystem changes and climate change.
2-4 years	Preliminary review of adaptation options for climate-sensitive ecosystems and resources.
2-4 years	Scenario-based analysis of the climatological, environmental, resource, technological, and economic implications of different atmospheric concentrations of greenhouse gases.
2-4 years	State-of-the-science of socioeconomic and environmental impacts of climate variability.
2-4 years	Within the transportation sector, a summary of climate change and variability sensitivities, potential impacts, and response options.
CCSP GOAL 5	Explore the uses and identify the limits of evolving knowledge to manage risks and opportunities related to climate variability and change
within 2 years	Uses and limitations of observations, data, forecasts, and other projections in decision support for selected sectors and regions.
within 2 years	Best-practice approaches for characterizing, communicating, and incorporating scientific uncertainty in decisionmaking.
within 2 years	Decision support experiments and evaluations using seasonal-to-interannual forecasts and observational data.

These models should have a number of important attributes:

- They should adequately represent the state of the Earth system and its prior evolution, in particular over the recent past, when there is a more comprehensive set of observations with which to compare the models and where better knowledge of the types and amounts of naturally occurring and human-induced forcings is available.
- They should be sufficiently comprehensive that they can be applied confidently over the period of interest. This is particularly crucial for longer simulations. For example, detailed representation of the partitioning of carbon among the atmosphere, the ocean, and terrestrial ecosystems will be crucial to representing

future distributions of atmospheric carbon dioxide and its concomitant contribution to radiative forcing and climate change. Human interactions must also be included in models operating over time scales that are influenced by human activities. The CCSP strategy for adding comprehensiveness to these models is outlined in some detail in Chapter 10 of this plan.

- They should be constructed to provide information at the desired spatial resolution.
- They should be able to be run in sufficient number and in sufficiently short time that their results can be provided to scientific, policy, and decisionmaking users.

Integration between and among Agencies

Within CCSP, multiple agencies have responsibilities in a number of the areas outlined above, so it is crucial that the agencies work together to optimize results, minimize duplication, and capitalize on their expertise, experience, and capabilities. The way in which this will be done is outlined in detail in Chapter 16 for federal agencies, while the relationship between U.S. and international activities is discussed in Chapter 15. The details of the CCSP structure, including the creation of interagency working groups, community-based science steering groups, and management-level interactions within the CCSP agencies and at higher levels are explained in Chapter 16.

In this section we note as examples some of the things that agencies working together through CCSP will strive to do to assure the necessary linkages between program elements:

- Make integrated use of infrastructure, especially for surface-based networks, airborne campaigns, and ship-based expeditions, to both reduce cost and maximize return by enhancing the availability of integrated data sets
- Ensure common data formats and interoperable data systems so that scientific, management, and policy-oriented data users can integrate desired data sets with minimal difficulty
- Collaborate on calibration systems and relevant intercomparisons to be sure that related observing capability can be consistently tied to measurement activities and to recognized laboratory standards
- Integrate process-oriented, surface- and airborne-based studies and network observations with calibration and validation activities for satellite programs
- Develop integrated modeling frameworks that will enhance opportunities for interchange of components and intercomparison of model results (both with other models and with observations)
- Issue joint solicitations where the optimal way to engage the research community in addressing issues of scientific importance.

Synthesis and Assessment Products

CCSP will provide a variety of synthesis and assessment products on an ongoing basis as discussed. These products will support both policymaking and adaptive management. The overall approach to decision support is described in Chapter 11 of this plan. This section integrates the earlier discussion in this chapter and summarizes the synthesis and assessment products that CCSP will generate.

The 1990 Global Change Research Act provides the overall framework for the conduct and management of the interagency research program

on climate and global change, and Section 106 of the act defines requirements for scientific assessments.[2] To comply with the terms of Section 106, CCSP will produce assessments that focus on a variety of science and policy issues important for public discussion and decisionmaking. The assessments will be composed of syntheses, reports, and integrated analyses that CCSP will complete over the next 4 years. The subjects to be addressed are listed in Table 2-1. This approach takes account of the need for assessments on the full range of issues spanning all CCSP objectives and will provide a "snapshot" of knowledge of the environmental and socioeconomic aspects of climate variability and change. The products will support specific groups or decision contexts across the full range of issues addressed by CCSP, and where appropriate, CCTP.

National policymaking focuses on many issues, but in this case involves integrating information from forcings and climate response to impacts and the costs and benefits of different response strategies. Synthesis of scientific data and results with economic information, demographic trends, technical specifications, etc., is essential for producing useful information on responding to the challenges posed by climate variability and change. How much information needs to be integrated depends on the specific issue or question that is being addressed and the type of decision that is being supported.

One highly important issue to be addressed at a national level is the challenge of developing new technologies to reduce the projected growth in greenhouse gas emissions. This requires integrating information from the natural sciences, engineering, and the economic and social sciences. CCSP will work closely with CCTP in incorporating information on portfolios of different technologies in evaluations of the implications of different response options through "If…, then…" scenario analyses, integrated assessment, and other approaches to comparative evaluation. This will facilitate understanding needs for development of new technologies.

There are uncertainties associated with all of this information, so when knowledge is integrated, information about the degree of certainty/uncertainty associated with it must also be carried forward. CCSP's decision support activities will address this essential requirement by communicating uncertainties in a manner appropriate to the decision context. CCSP's work program also includes research to improve methods for taking account of uncertainty in research and analysis, assessing the implications of uncertainty for decisionmaking, and communicating uncertainty to different audiences.

The comprehensive structure of CCSP's goals will help to ensure that the program sponsors research and provides support for decisionmaking on the full range of issues associated with variability and change in climate and related systems.

---

[2]"On a periodic basis (not less frequently than every 4 years) the Council, through the Committee, shall prepare and submit to the President and the Congress an assessment which:
1) Integrates, evaluates, and interprets the findings of the Program and discusses the scientific uncertainties associated with such findings
2) Analyzes the effects of global change on the natural environment, agriculture, energy production and use, land and water resources, transportation, human health and welfare, human social systems, and biological diversity
3) Analyzes current trends in global change, both human-induced and natural, and projects major trends for the subsequent 25 to 100 years."

CHAPTER 2 AUTHORS

Lead Authors
Richard H. Moss, CCSPO
Margarita Conkright Gregg, CCSPO
Jack Kaye, NASA
James R. Mahoney, DOC

CHAPTER 3 Atmospheric Composition

CHAPTER CONTENTS

Question 3.1: What are the climate-relevant chemical, microphysical, and optical properties, and spatial and temporal distributions, of human-caused and naturally occurring aerosols?

Question 3.2: What are the atmospheric sources and sinks of the greenhouse gases other than CO_2 and the implications for the Earth's energy balance?

Question 3.3: What are the effects of regional pollution on the global atmosphere and the effects of global climate and chemical change on regional air quality and atmospheric chemical inputs to ecosystems?

Question 3.4: What are the characteristics of the recovery of the stratospheric ozone layer in response to declining abundances of ozone-depleting gases and increasing abundances of greenhouse gases?

Question 3.5: What are the couplings and feedback mechanisms among climate change, air pollution, and ozone layer depletion, and their relationship to the health of humans and ecosystems?

National and International Partnerships

The composition of the atmosphere—its gases and particles—plays a critical role in connecting human welfare with global and regional changes because the atmosphere links all of the principal components of the Earth system. The atmosphere interacts with the oceans, land, terrestrial and marine plants and animals, and the frozen regions (see Figure 3-1). Because of these linkages, the atmosphere is a conduit of change. Emissions from natural sources and human activities enter the atmosphere at the surface and are transported to other geographical locations and often higher altitudes. Some emissions undergo chemical transformation or removal while in the atmosphere or interact with cloud formation and precipitation. Some natural events and human activities that change atmospheric composition also change the Earth's radiative (energy) balance. Subsequent responses to changes in atmospheric composition by the stratospheric ozone layer, the climate system, and regional chemical composition (air quality) create multiple environmental effects that can influence human health and natural systems.

Atmospheric composition changes are indicators of many potential environmental issues. Observations of trends in atmospheric composition are among the earliest harbingers of global changes. For example, the decline of the concentrations of ozone-depleting substances, such as the chlorofluorocarbons (CFCs), has been the first measure of the effectiveness of international agreements to end production and use of these compounds.

A principal feature of the atmosphere is that it acts as a long-term "reservoir" for certain trace gases that can cause global changes. The long removal times of some gases, such as carbon dioxide (CO_2, >100 years) and perfluorocarbons (PFCs, >1,000 years), imply that any associated global changes could persist over decades, centuries, and millennia—affecting all countries and populations.

An effective program of scientific inquiry relating to changes in atmospheric composition must include two major foci. The first is a focus on Earth system interactions: How do changes in atmospheric composition alter and respond to the energy balance of the climate system? What are the interactions between the climate system and

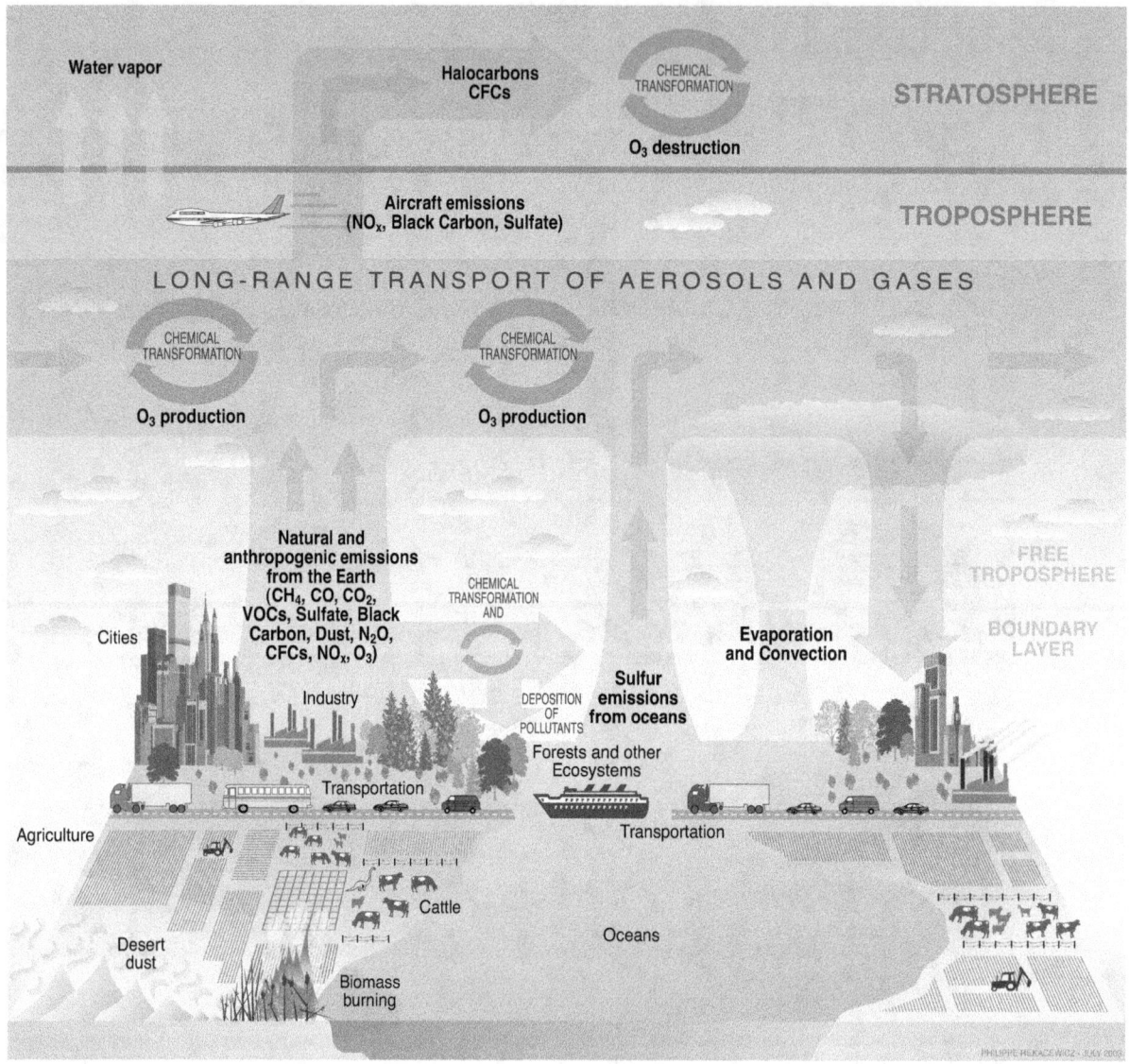

Figure 3-1: Schematic of chemical and transport processes related to atmospheric composition. These processes link the atmosphere with other components of the Earth system, including the oceans, land, and terrestrial and marine plants and animals.

stratospheric ozone? What are the effects of regional pollution on the global atmosphere and the effects of global climate and chemical change on regional air quality? The second is a focus on Earth system and human system linkages: How is the composition of the global atmosphere, as it relates to climate, ozone depletion, ultraviolet radiation, and pollutant exposure, altered by human activities and natural phenomena? How quantitative is the knowledge of the major sources of emissions to the atmosphere? What are the atmospheric composition changes that could affect human health and natural ecosystems?

The *overall research approach* for understanding the role of atmospheric composition is an integrated application of long-term systematic observations, laboratory and field studies, and modeling, with periodic assessments of understanding and significance to decisionmaking.

Most of the activities related to atmospheric composition research are part of *national and international* partnerships, some of which are noted at the end of this chapter. Such partnerships are necessitated by the breadth and complexity of current issues and because the atmosphere links all nations. The overall research approach is based on the substantial body of knowledge and understanding available from the work of many international scientists. The status of understanding is reported as part of cooperative international assessment activities (e.g., IPCC, 2001a,b,d; WMO, 1999, 2003).

In looking ahead at what the specific policy-relevant information needs associated with atmospheric composition will be, five broad challenges are apparent, with goals and examples of key research objectives outlined below.

Question 3.1: What are the climate-relevant chemical, microphysical, and optical properties, and spatial and temporal distributions, of human-caused and naturally occurring aerosols?

State of Knowledge

Research has demonstrated that atmospheric particles (aerosols) can cause a net cooling or warming tendency within the climate system, depending upon their physical and chemical characteristics. Sulfate-based aerosols, for example, tend to cool, whereas black carbon (soot) tends to warm the system (see Figure 3-2). In addition to these direct effects, aerosols can also have indirect effects on radiative forcing (e.g., changes in cloud properties). When climate models include the effects of sulfate aerosol, the simulation of global mean surface temperatures is improved. One of the largest uncertainties about the net impact of aerosols on climate is the diverse warming and cooling influences of the very complex mixture of aerosol types and their spatial distributions. Further, the poorly understood impact of aerosols on the formation of both water droplets and ice

crystals in clouds also results in large uncertainties in the ability to project climate changes (see Figure 3-2). More detail is needed globally to describe the scattering and absorbing optical properties of aerosols from regional sources and how these aerosols impact other regions of the globe.

Illustrative Research Questions

The relationship of aerosols to climate change is complex because of the diverse formation and transformation processes involving aerosols (see Figure 3-3). This complexity underlies many of the important research questions related to aerosols.

- What are the global sources (e.g., oceanic, land, atmospheric) of particle emissions (e.g., black carbon/soot, dust, and organic compounds), and their spatial and temporal variability?
- What are the regional and global sources of emissions of aerosol precursor gases [e.g., sulfur dioxide (SO_2), dimethyl sulfide (CH_3SCH_3), ammonia (NH_3), and volatile organic carbon (VOC)]?
- What are the global distributions and optical characteristics of the different aerosol components, and how do they directly and

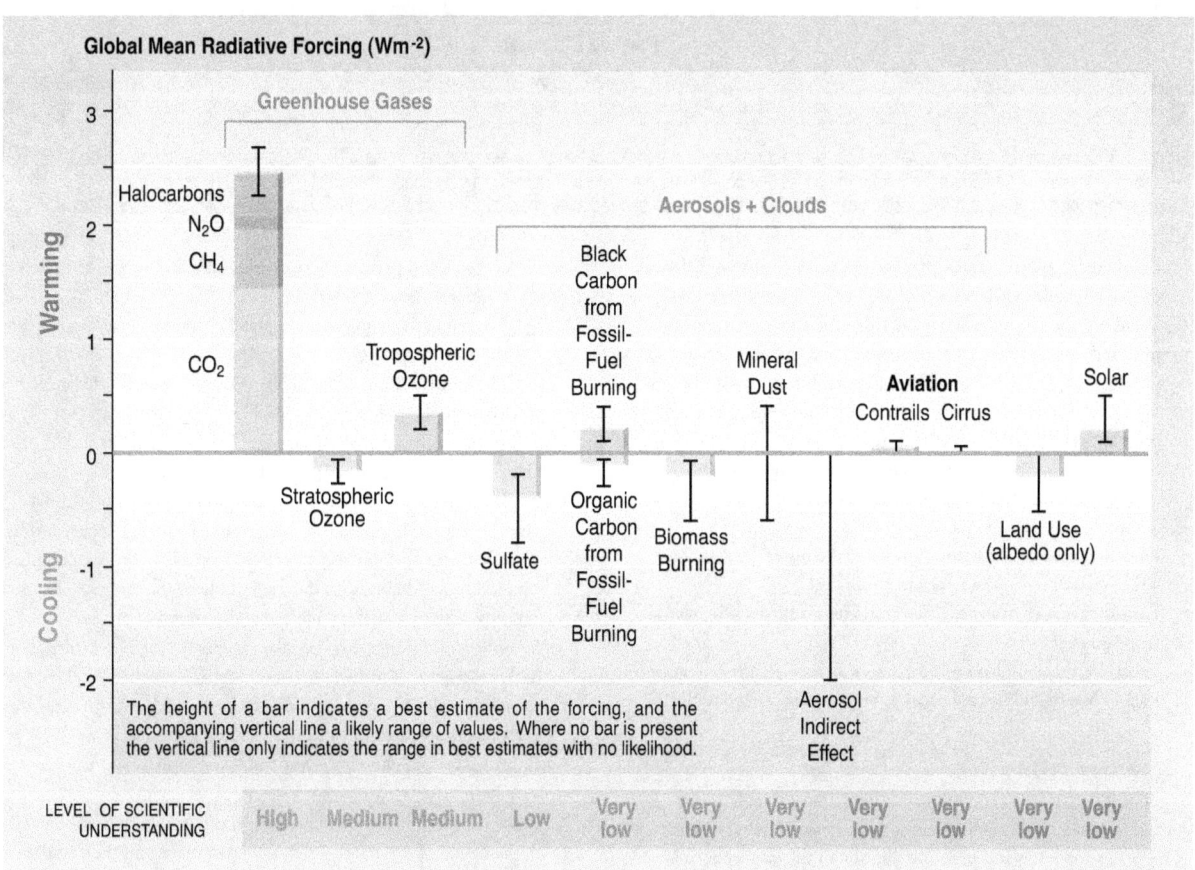

Figure 3-2: Schematic comparing several factors that influence Earth's climate on the basis of their contribution to radiative forcing between 1750 and 2000. Two principal categories of radiative forcing factors are the greenhouse gases and the combination of aerosols and clouds. The rectangular bars represent a best estimate of the contributions of these forcings, some of which yield warming and some cooling, while the vertical line about the rectangular bars indicates the range of estimates. A vertical line without a rectangular bar denotes a forcing for which no best estimate can be given owing to large uncertainties. Scientific understanding of aerosol effects is very low, as shown on the horizontal axis. Source: IPCC (2001d). For more information, see Annex C.

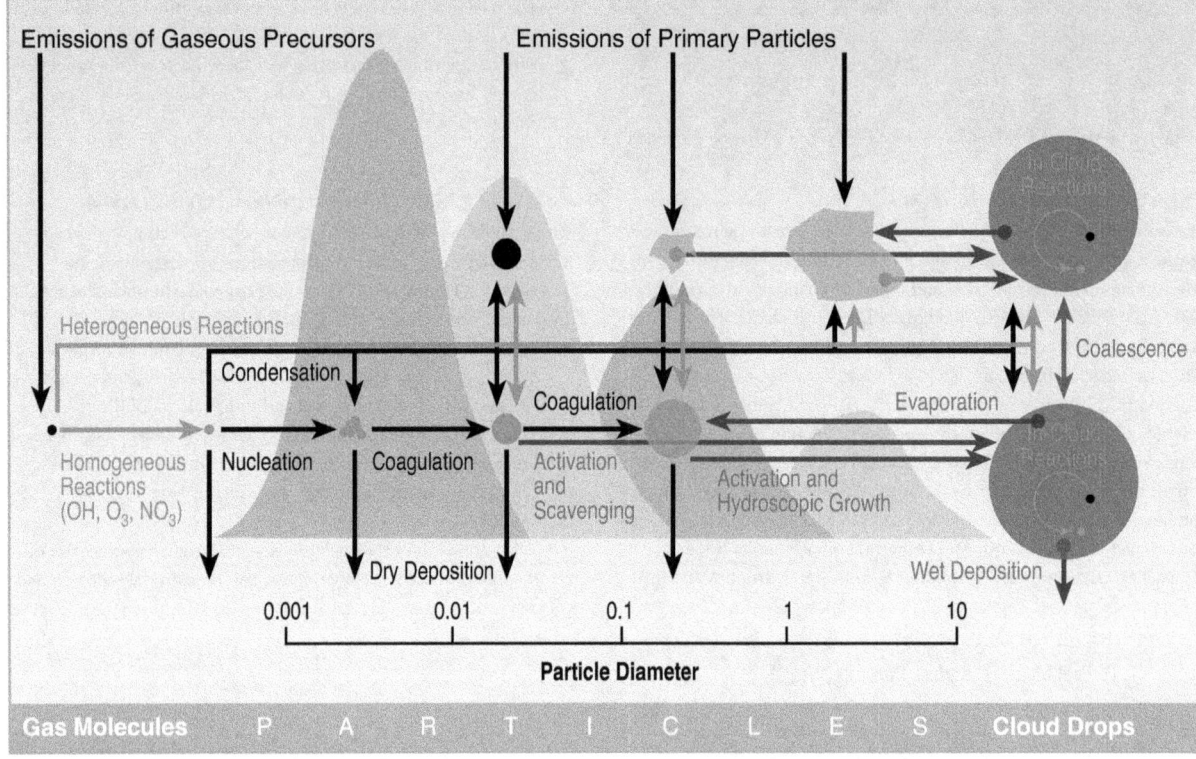

Figure 3-3: Schematic of the processes that cause the formation and transformation of aerosol particles in the atmosphere. Aerosols may be emitted directly into the atmosphere or be formed there from the emissions of gaseous precursors. Particles grow by condensation of gases and by coagulation with other particles, and their number and composition can influence the formation and radiative characteristics of clouds. For more information, see Annex C.

indirectly (e.g., cloudiness changes) affect the radiative balance of Earth's atmosphere?

- What are the processes that control the spatial and temporal distributions and variability of aerosols and that modify aerosol physical, chemical and optical properties during transport (see Figures 3-1 and 3-3), and how well do models simulate these processes and resulting spatial distributions?
- What are the effects of changes in aerosol abundance and composition on cloud formation and persistence, precipitation amounts, and cloud radiative properties (see Chapter 5)?
- How do aerosols and aerosol precursors emitted by aviation affect cloudiness in the upper troposphere?
- How do aerosols affect the chemical composition of the global troposphere?
- What are the abundances and sources of aerosols that affect human health and natural ecosystems (see Chapters 8 and 9)?

Research Needs

Significant research is required to complete our understanding of aerosols and their role in climate change processes. The representation of aerosol properties and their distribution in the atmosphere is highly complex. The needs outlined below describe important future steps to improve our understanding:

- Expand the use of space-based, airborne, and ground-based instruments and laboratory studies to provide better data for aerosols containing sulfate, black carbon/soot, dust, or organic compounds in the following areas:
  - Distributions (primarily from space-based instruments) of

aerosols and precursor gases over land and oceans and their temporal variabilities

  - The physical, chemical, and optical properties of anthropogenic and natural aerosols
  - The contributions of aerosols to Earth's radiative balance.
- Enhance field and laboratory studies of the processes that influence aerosol distributions and characteristics, including those involved in indirect (e.g., cloud) effects.
- Improve comprehensive climate model simulations to estimate aerosol-induced temperature changes and associated uncertainties, and the direct and indirect effects of aerosols with an emphasis on establishing bounds on the magnitude of the indirect effects.
- Intensify efforts to determine the composition of organic aerosols and develop simpler instruments for the characterization and measurement of carbon-containing aerosols.
- Establish realistic aerosol and precursor source-strength estimates for specific aerosol compositions for the industrial era.
- Improve aerosol chemistry and transport models and carry out simulations for aerosol source-strength scenarios.
- Make comprehensive comparisons of the geographic and height dependence of simulated aerosol distributions and their radiative characteristics with results from field and satellite data (e.g., with multi-wavelength polarimetric measurements), with an emphasis on regions that can best test the reliability of current model simulations (e.g., using the extensive North American emissions database).
- Assess aerosol abundances and variability in the paleoaerosol record (e.g., ice, bog, and lake core data).

These research needs are strongly linked to those of the Climate Variability and Change (Chapter 4), Water Cycle (Chapter 5), Land-Use/Land-Cover Change (Chapter 6), and Carbon Cycle (Chapter 7) elements.

Milestones, Products, and Payoffs

- Improved description of the global distributions of aerosols and their properties [2-4 years].
- Empirically tested evaluation of the capabilities of current models to link emissions to (i) global aerosol distributions and (ii) the chemical and radiative properties (and their uncertainties) of aerosols [2-4 years]. Specific products and payoffs include:
  - Better estimates of the radiative forcing of climate change for different aerosol types and the uncertainties associated with those estimates. Aspects of these capabilities will be addressed in collaboration with the Climate Variability and Change research element (Chapter 4). These results will contribute to the preparation of Climate Change Science Program (CCSP) climate projections as decision support resources.
  - Better defined potential options for changing radiative forcing within a short time, in contrast to the longer response times associated with CO_2 and other greenhouse gases. The relatively short atmospheric residence times of aerosols form the basis of these options.

- An estimate of the indirect climate effects of aerosols (e.g., on clouds) that is improved beyond the estimate in the last Intergovernmental Panel on Climate Change report (IPCC, 2001a) [2-4 years].
- Characterization of the impact of human activities and natural sources on global and regional aerosol distributions [beyond 4 years].
- Better understanding and description of the physical and chemical processes (and their uncertainties) that form, transform, and remove aerosols during long-range atmospheric transport [beyond 4 years].

Question 3.2: What are the atmospheric sources and sinks of the greenhouse gases other than CO_2 and the implications for the Earth's energy balance?

State of Knowledge

The increasing concentrations of chemically active greenhouse gases, such as methane (CH_4), tropospheric ozone (O_3), nitrous oxide (N_2O), and halocarbons [e.g., CFCs, PFCs, hydrochlorofluorocarbons (HCFCs), hydrofluorocarbons (HFCs), and sulfur hexafluoride (SF6)], represent primary contributions to the radiative forcing of the climate system (see Figure 3-2). Anthropogenic emissions of CO_2 are addressed in Chapter 7. Water vapor, considered the most dominant greenhouse gas (albeit as a feedback), is discussed as part of the hydrological cycle (see Chapter 5).

The natural and anthropogenic emissions sources leading to the observed growth rates of CH_4 (the second-most influential anthropogenic greenhouse gas) and N_2O abundances are qualitatively understood but poorly quantified (e.g., the amount of CH_4 emitted by rice agriculture). Trends in tropospheric ozone (the third-most influential anthropogenic greenhouse gas) are not well determined because they are driven by a mix of natural and anthropogenic emissions, including CH_4 and regional pollutants such as the reactive nitrogen oxides (NO_x). The atmospheric concentrations and sources of the CFCs and HCFCs have been well studied because of their role in stratospheric ozone depletion. Some important greenhouse gases are removed from the troposphere and stratosphere in reaction with the hydroxyl radical (OH) (e.g., CH_4, HFCs, HCFCs) or by photolysis (e.g., N_2O). Reactions that remove a greenhouse gas control its lifetime in the atmosphere and, hence, its contribution to radiative forcing of climate.

BOX 3-1

ATMOSPHERIC COMPOSITION

FY04 CCRI Priority - Aerosols
The Climate Change Research Initiative (CCRI) will leverage existing U.S. Global Change Research Program (USGCRP) research to address major gaps in understanding climate change. Uncertainties related to the effects of aerosols on climate are large, with both warming and cooling effects possible depending on the nature and distribution of the aerosol.

The CCRI will advance the understanding of the distribution of all major types of aerosols and their variability through time, the different contributions of aerosols from human activities, and the processes by which the different contributions are linked to global distributions of aerosols. The CCRI will support research to improve understanding of the processes by which trace gases and aerosols are transformed and transported in the atmosphere. Studies of how atmospheric chemistry, composition, and climate are linked will be emphasized, including those processes that control the abundance

of constituents that affect the Earth's radiation budget, such as tropospheric methane, ozone, and aerosols.

The global distributions of a limited number of atmospheric parameters (including climatically relevant parameters such as ozone and aerosols) and their variabilities will be obtained from satellite observations over long periods of time along with more comprehensive suites of observations over briefer time periods. Satellite data recently obtained and to become available for the first time for methane, tropospheric ozone, and tropospheric aerosols will be analyzed and interpreted in the context of global models and assimilation systems.

The studies will provide an observational- and model-based evaluation of the radiative forcings associated with aerosol direct and indirect effects. These forcing results will contribute to the CCSP climate projections for research and assessment.

Illustrative Research Questions

Driven by the need to have a better predictive understanding of the relationship between the emission sources of these radiatively active gases and their global distributions and radiative forcing, several questions continue to face the research community. More quantitative information is needed to help answer these questions, which include:

- What is the inventory of anthropogenic (i.e., energy, industry, agriculture, and waste) and natural (e.g., ecosystems) emissions sources of CH_4 and N_2O on global and regional scales (see Chapters 7, 8, and 9)?
- What are the causes and related uncertainties of the observed large variations in the growth rates of CH_4 and N_2O abundances?
- What are the global anthropogenic and natural (both biogenic and lightning-related) sources of reactive nitrogen oxides, which are precursors of troposphere ozone?
- How can paleorecords of greenhouse gases and climate variability be used to understand the potential for future climate change (see Chapter 4)?
- What are the trends in mid-tropospheric ozone, particularly in the Northern Hemisphere, and how well can the trends be attributed to known chemical and transport processes?
- How is the oxidative capacity of the atmosphere changing (e.g., OH abundances) and how do these changes affect the radiative forcing impact of greenhouse gas emissions?
- How do changes in the abundance of a principal greenhouse gas (e.g., CH_4) cause a feedback that alters its lifetime or that of another greenhouse gas (see Chapter 7)?
- How do the increasing abundances of greenhouse gases influence the distribution of atmospheric water vapor and its feedback in the radiative balance of the climate system (see Chapter 5)?

Research Needs

Field and laboratory studies, satellite observations, and diagnostic transport/chemical modeling are needed to fully address these questions. Examples of these activities are:

- Satellite, field, laboratory, and modeling studies to develop, evaluate, and improve inventories of global emissions and the potential for emission reductions for CH_4, carbon monoxide (CO), N_2O, and NO_x from anthropogenic and natural sources. This need will be addressed in collaboration with the carbon cycle research element (Chapter 7).
- Global monitoring sites to continue recording the growth rate of CH_4 and its variability. This need will be addressed in collaboration with the Carbon Cycle research element (Chapter 7).
- Field and modeling studies to reduce the uncertainty in the air-sea exchange rate of key gases (e.g., N_2O, CH_3SCH_3, short-lived halocarbons) over important regions of the world's oceans.
- Satellite observations to provide estimates of global distributions of tropospheric ozone and some of its precursors (e.g., NO_x).
- Model studies to simulate past trends in tropospheric ozone to improve the understanding of its contribution to radiative forcing over the past ~50 years.
- Satellite and field studies to characterize how regional- and continental-scale changes in ozone precursor emissions alter global tropospheric ozone distributions, thereby providing tests of and improvement in the representation of ozone-related processes in models.

- Laboratory studies to expand the quantitative descriptions of tropospheric chemical processes, thereby facilitating the continued development of reliable climate models.
- Field and model studies to quantify how changes in NO_x, CH_4, CO, water vapor, and ozone could alter the abundance of the hydroxyl radical, which controls the lifetime of many principal greenhouse gases.

Milestones, Products, and Payoffs

- Observationally assessed and improved uncertainty ranges of the radiative forcing of the chemically active greenhouse gases [2-4 years]. Aspects of this product will be addressed in collaboration with the Climate Variability and Change research element (Chapter 4) and those specifically related to CH_4 and CO_2 with the Carbon Cycle research element (Chapter 7). These improved ranges will be used in formulating future scenarios of radiative forcing, which will be part of CCSP climate projections. As a result, there will be a broader suite of choices (i.e., in addition to CO_2) for decisionmakers to influence anthropogenic radiative forcing, particularly in coming decades.

Question 3.3: What are the effects of regional pollution on the global atmosphere and the effects of global climate and chemical change on regional air quality and atmospheric chemical inputs to ecosystems?

State of Knowledge

Emissions from rapidly industrializing regions of the world have the potential to impact air quality and ecosystem health in regions far from the sources. Paleochemical data from ice cores and snow document past perturbations and demonstrate that even remote areas, such as Greenland, are influenced by worldwide emissions. The anthropogenic contribution to the nitrogen cycle from fossil-fuel combustion and fertilizer production now rivals in magnitude the natural input from nitrogen-fixing organisms and lightning. This additional nitrogen input to the biosphere illustrates how human activities could have important consequences for ecosystem structure and function. The importance of the effect of regional pollution on global tropospheric chemistry has been recognized for some time. Now, the importance of understanding the reverse effect—that of global-scale transport of pollutants or global change on regional air quality—is increasing. A well-recognized example is the enhancement of background global ozone concentrations by anthropogenic emissions.

Illustrative Research Questions

This emerging picture is shaping several questions of importance to society. Some examples of these are as follows:

- What are the impacts of climate change and long-range transport of regional air pollution on water resources, human health, food-producing areas, and ecosystems (see Chapters 6, 8, and 9)?
- How do El Niño-Southern Oscillation (ENSO)-related drought and fires affect regional and global aerosol haze and air quality?
- How do interactions between the biogeochemical cycles of the macronutrients (e.g., carbon and nitrogen) affect greenhouse gas

abundances in the atmosphere and the radiative forcing of the climate system (see Chapters 6 and 8)?

- How do regional changes in atmospheric composition due to biomass burning affect the abundances of greenhouse gases and global nutrient cycles (see Chapters 6 and 8)?
- How do the primary and secondary pollutants from the world's megacities and large-scale, non-urban emissions (e.g., agriculture, ecosystems, etc.) contribute to global atmospheric composition?
- What are, and what contribute to, North American "background" levels of air quality—that is, what levels of pollution are beyond national control?
- What controls the long-range transport, accumulation, and eventual destruction of persistent organic pollutants or the long-range transport, transformation, and deposition of mercury?

Research Needs

These questions are being addressed by measurements of key tropospheric constituents from satellites, airborne platforms, and surface sites. Model analyses and simulations are used to provide a regional and global context for the measurement data set and to address future scenarios. Near-term goals include the following:

- Quantify North American inflow and outflow of reactive and long-lived gases and aerosols using observations with increasing spatial and temporal resolution and project future changes.
- Understand the balance between long-range transport and transformation of pollutants.
- Build and evaluate models that couple the biogeochemical cycles of elements with specific emphasis on carbon and nitrogen compounds.
- Continue baseline observations of atmospheric composition over North America and globally.
- Carry out a detailed global survey of vertically resolved distributions of tropospheric ozone and its key precursor species.
- Carry out studies with atmospheric chemistry models coupled to general circulation models to improve the understanding of the feedbacks between regional air pollution and global climate change.

Milestones, Products, and Payoffs

- A simulation of the changes in the impacts of global tropospheric ozone on radiative forcing over the past decade brought about by clean air regulations [2-4 years]. Aspects of this product will be addressed in collaboration with the Climate Variability and Change research element (Chapter 4).
- Estimates of atmospheric composition and related processes to be used in assessments of the vulnerability of ecosystems to urban growth and long-range chemical transport [beyond 4 years]. Aspects of this product will be addressed in collaboration with the Ecosystems research element (Chapter 8).
- An evaluation of how North American emissions contribute to and influence global atmospheric composition [beyond 4 years].
- A 21st century chemical baseline for the Pacific region against which future changes can be assessed [2-4 years].

Question 3.4: What are the characteristics of the recovery of the stratospheric ozone layer in response to declining abundances of ozone-depleting gases and increasing abundances of greenhouse gases?

State of Knowledge

The primary cause of stratospheric ozone depletion observed over the last 2 decades is an increase in the concentrations of industrially produced ozone-depleting chemicals (e.g., CFCs). The depletion has been significant, ranging from a few percent per decade at mid-latitudes to greater than 50 percent seasonal losses at high latitudes. Notable is the annually recurring Antarctic ozone hole, as well as smaller, but still significant, winter/spring ozone losses recently observed in the Arctic. Reductions in atmospheric ozone levels lead to increased exposure to ultraviolet radiation at the surface, which has harmful consequences for plant and animal life, and human health. In response to these findings, the nations of the world ratified the *Montreal Protocol on Substances That Deplete the Ozone Layer* and agreed to phase out the production of most ozone-depleting chemicals.

Ground-based and satellite measurements show that concentrations of many of these compounds are now beginning to decrease in the atmosphere (WMO, 1999, 2003). As the atmospheric burden of ozone-depleting chemicals falls in response to international efforts, stratospheric ozone concentrations should begin to recover in coming decades (see Figure 3-4). However, because of the ongoing changes in atmospheric composition and climate parameters, which began before the onset of stratospheric ozone depletion, the exact course and timing of ozone recovery in the coming decades is not fully known.

Illustrative Research Questions

- How will changes in the atmospheric abundances of greenhouse gases, such as CO_2, N_2O, and CH_4, and the resulting changes in the radiation and temperature balance (e.g., stratospheric cooling), alter ozone-related processes in the stratosphere (see Chapter 7)?
- How will changes in physical climate parameters (e.g., stratospheric winds and temperatures) affect the production and loss of stratospheric ozone and how might these changes in ozone cause a feedback that would alter climate parameters (see Chapter 4)?
- What are the ozone-depleting and radiative forcing properties of the new chemicals chosen to be substitutes for the now-banned ozone-depleting substances?
- How might climate change affect the abundances of very short-lived halocarbon gases of natural origin (e.g., from the oceans) and their contribution to stratospheric ozone depletion?

Research Needs

Improving our understanding of the complex interaction between the ozone layer and the climate system requires further investigations of the processes that interconnect ozone, water vapor, reactive trace constituents (notably chlorine and bromine compounds and reactive nitrogen oxides), aerosols, and temperature. Research needs include the following:

- Extend interagency and international satellite observations of ozone trends, with an emphasis on detecting and attributing ozone recovery.

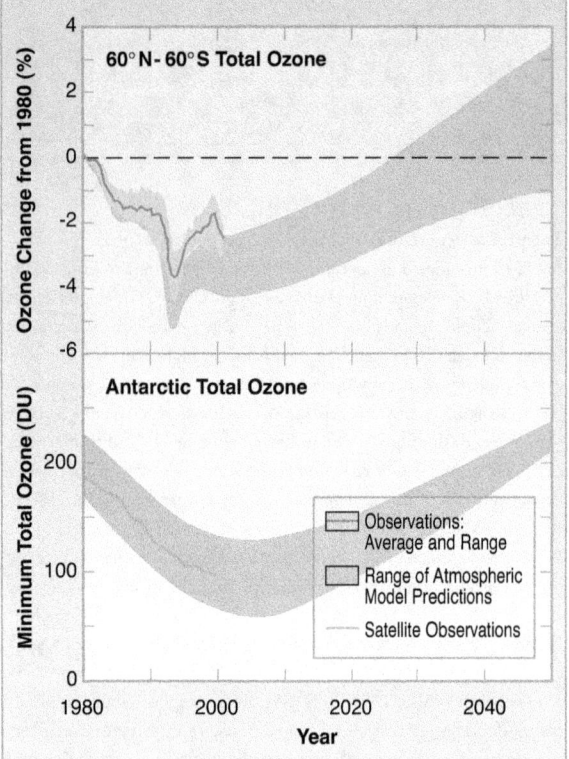

Figure 3-4: Schematic illustrating the estimated recovery of stratospheric ozone in the coming decades. Observations show the decline of global total ozone (top panel) and minimum values of total ozone over Antarctica (lower panel) beginning in 1980. As the abundances of ozone-depleting gases decline in the atmosphere, ozone values are expected to increase and recover towards 1980 and earlier values. The estimated recovery of ozone is shown based on predictions by several atmospheric models which use different assumptions about the composition and meteorology of the future stratosphere. The research needs as outlined in Section 3.4 address reducing the range of uncertainty in ozone recovery estimates. Source: Adapted from WMO (2003). For more information, see Annex C.

- Continue global monitoring of the changes in the abundances of ozone-depleting substances and their substitutes and assessing compliance with the Montreal Protocol.
- Evaluate the impacts on stratospheric ozone and climate forcing of proposed substitute gases with laboratory chemistry and atmospheric models to provide early information to industry prior to large investments in production.
- Maintain and expand measurements of the ozone vertical profile throughout the troposphere and stratosphere in order to obtain a more precise evaluation of the radiative forcing associated with ozone and a more precise detection of stratospheric ozone recovery.
- Use focused aircraft, balloon, and ground-based campaigns, satellite data sets, and chemical transport modeling activities to address:
  - Cross-tropopause transport processes to better understand the ozone-depleting role of the very short-lived (days to months) halocarbon gases that have either been observed in the atmosphere or are proposed for applications that may lead to atmospheric release

- The response of polar stratospheric cloud formation and ozone-loss chemistry to trends in water vapor and temperature in the stratosphere
  - Stratospheric transport processes to better understand ozone layer responses to climate change.
- Continue monitoring the trends in ultraviolet radiation, particularly in regions of high radiation exposure and high biological sensitivity.

Milestones, Products, and Payoffs
- Updated trends in ozone and surface ultraviolet radiation, analysis of compliance with the provisions of the Montreal Protocol, and model forecasts of ozone recovery [2-4 years].
- Improved quantitative model evaluation of the sensitivity of the ozone layer to changes in atmospheric transport and composition related to climate change [2-4 years].
- Contribute new findings to the 2006 update of the international scientific assessment of stratospheric ozone depletion [2-4 years].
  - This sixth in the series of "operational" products of the ozone science community is key to addressing accountability with respect to ozone depletion—namely, is the outcome expected from international actions being observed?

> **Question 3.5**: What are the couplings and feedback mechanisms among climate change, air pollution, and ozone layer depletion, and their relationship to the health of humans and ecosystems?

State of Knowledge
This question is intended to underscore the explicit need to better understand the relationships that exist between research issues that historically have been treated separately. Understanding the potential consequences of changes in atmospheric composition cuts across a range of important societal issues and scientific disciplines. Atmospheric composition links climate change, air quality, and ozone depletion with human and ecosystem health. The links involve physical processes and feedback mechanisms—some of which are newly recognized and most of which require further study.

Understanding these connections is important for understanding climate change on regional and global scales. For example, research has demonstrated that stratospheric ozone depletion not only causes increased exposure to ultraviolet radiation at the surface, but also exerts a cooling influence on the global climate. Conversely, climate-related changes may cool the lower stratosphere and increase ozone layer depletion in polar regions. Tropospheric ozone, of concern primarily as a component of smog, is not only a local health risk, but also exerts a warming influence on the global climate. Emissions of SO_2 from fossil-fuel combustion not only lead to the formation of acid rain, but also contribute to sulfate-aerosol haze, which exerts a cooling influence on the global climate system.

It is clear that these issues, which initially may have been treated separately by scientists and policymakers, now require a more integrated approach that addresses multiple sources, stresses, and impacts (NASA, 2002).

Illustrative Research Questions

- How do changes in the anthropogenic emissions of NO_x, CO, and VOCs affect the abundance of CH_4 and ozone on regional and global scales (see Chapters 7 and 9)?
- How is air quality affected by changes in climate and weather patterns—for example, changes in ozone resulting from a change in cloudiness, temperature, precipitation, etc. (see Chapter 4)?
- How do the regional and global radiative forcings of aerosols respond to changes in aerosol precursor gases (e.g., sulfur gases, ammonia, and VOCs)?
- What are the common stresses that climate change, ozone layer depletion, and regional air quality exert on humans and ecosystems and do these stresses interact synergistically (see Chapters 8 and 9)?

Research Needs

Research needs, in addition to those given in Sections 3.1 to 3.4, linking atmospheric composition issues to the health of humans and ecosystems are:

- Build and evaluate diagnostic/prognostic models of the coupled climate, chemistry, transport, and ecological systems at the local, regional, and global scales. This need will be addressed in part by products of the Ecosystems research element (Chapter 8).
- Identify and quantify how ecosystems are affected by human activities that change the chemical composition of the atmosphere.
- Build and evaluate models that efficiently represent the behavior of biogeochemical systems and link these models with decisionmaking frameworks.
- Carry out multiple-issue state-of-understanding scientific assessments, in partnership with a spectrum of stakeholders and with the aim of characterizing integrated "If…, then…" options.

Milestones, Products, and Payoffs

- Strengthened processes within the national and international scientific communities providing for integrated evaluations of the impacts on human health and ecosystems caused by the intercontinental transport of pollutants, the impact of air pollutants on climate, and the impact of climate change on air pollutants [2-4 years]. The evaluations will be useful in developing integrated control strategies to benefit both regional air quality and global climate change, and the local attainment of air quality standards [2-4 years]. Aspects of this product will be addressed in collaboration with the Climate Variability and Change (Chapter 4), Carbon Cycle (Chapter 7), and Human Contributions and Responses (Chapter 9) research elements.
- In 2006, the U.S. atmospheric research community will produce the first *State of the Atmosphere* report that describes and interprets the status of the characteristics and trends associated with atmospheric composition, ozone layer depletion, temperature, rainfall, and ecosystem exposure (see Chapters 12 and 13) [2-4 years].
- Diagnostic/prognostic models of the coupled climate, chemistry/transport, and ecological systems [beyond 4 years].

National and International Partnerships

The Atmospheric Composition research focus is linked via cooperation, co-planning, and joint execution to several national and international planning and coordinating activities. A few examples follow:

- **Interagency Programs**: Joint planning [e.g., with the National Aerosol-Climate Interactions Program (NACIP)] is a principal strategy for achieving CCSP objectives.
- **Committee on Environment and Natural Resources - Air Quality Research Subcommittee (AQRS)**: Joint research is conducted on the global/continental scales within the CCSP and on the regional/local scales within the AQRS (e.g., global influences on the "natural background" of air pollutants and linkages with the stakeholders via the AQRS).
- **International Global Atmospheric Chemistry (IGAC)**: IGAC, a core project of the International Geosphere-Biosphere Programme, coordinates several international projects focused on the chemistry of the global troposphere and its impact on the Earth's radiative balance (e.g., the new Intercontinental Transport and Chemical Transformation project, involving Asian, North American, and European researchers).
- **Surface Ocean-Lower Atmosphere Study (SOLAS)**: SOLAS, a new core project of the International Geosphere-Biosphere Programme, will coordinate several international projects focused on the exchange of climate-relevant gases and particles between the troposphere and oceans.
- **World Climate Research Programme/Stratospheric Processes and their Role in Climate (WCRP/SPARC)**: SPARC coordinates international cooperation in research on climate-related aspects of stratospheric science, including efforts aimed at understanding long-term trends in the composition of the stratosphere and upper troposphere.
- **International Assessments**: Atmospheric composition and its role in the radiative forcing of climate and ozone depletion is a key component of international assessments on these topics—performed under the auspices of the Intergovernmental Panel on Climate Change and the United Nations Environment Programme/World Meteorological Organization, respectively. Each involves the participation of a large international body of scientists that represents atmospheric and cross-cutting disciplines.

CHAPTER 3 AUTHORS

Lead Authors
Daniel L. Albritton, NOAA
Philip L. DeCola, NASA
Donald E. Anderson, NASA
David W. Fahey, NOAA
James F. Gleason, NASA
Terry J. Keating, USEPA
Dina W. Kruger, USEPA
Michael J. Kurylo, NASA and NIST
Joel M. Levy, NOAA
Peter Lunn, DOE
Jarvis Moyers, NSF
Anne-Marie Schmoltner, NSF
Henry F. Tyrrell, USDA
Darrell A. Winner, USEPA

Contributors
Ronald J. Ferek, ONR
Mary Gant, HHS

4 Climate Variability and Change

CHAPTER

CHAPTER CONTENTS

Question 4.1: To what extent can uncertainties in model projections due to climate system feedbacks be reduced?

Question 4.2: How can predictions of climate variability and projections of climate change be improved, and what are the limits of their predictability?

Question 4.3: What is the likelihood of abrupt changes in the climate system such as the collapse of the ocean thermohaline circulation, inception of a decades-long mega-drought, or rapid melting of the major ice sheets?

Question 4.4: How are extreme events, such as droughts, floods, wildfires, heat waves, and hurricanes, related to climate variability and change?

Question 4.5: How can information on climate variability and change be most efficiently developed, integrated with non-climatic knowledge, and communicated in order to best serve societal needs?

National and International Partnerships

Climate variability and change profoundly influence social and natural environments throughout the world, with consequent impacts on natural resources and industry that can be large and far-reaching. For example, seasonal-to-interannual climate fluctuations strongly affect the success of agriculture, the abundance of water resources, and the demand for energy, while long-term climate change may alter agricultural productivity, land and marine ecosystems, and the resources that these ecosystems supply. Recent advances in climate science are beginning to provide information for decisionmakers and resource managers to better anticipate and plan for potential impacts of climate variability and change. Further advances will serve the nation by providing improved knowledge to enable more scientifically informed decisions across a broad array of climate-sensitive sectors.

Climate research has indicated that, globally, it is very likely that the 1990s were the warmest decade in the instrumental record, which extends back to the 1860s (see Figure 4-1); large climate changes can occur within decades or less, yet

last for centuries or longer; and the increase in Northern Hemisphere surface temperatures during the 20th century likely exceeds the natural variability of the past 1,000 years (IPCC, 2001a,d). Placing instrumental records in the context of longer term variability through paleoclimate analyses has played a key role in these findings. Moreover, observational evidence together with model simulations incorporating a comprehensive suite of natural and anthropogenic forcings indicate that "…the changes observed over the last several decades are likely mostly due to human activities, but we cannot rule out that some significant part of these changes is also a reflection of natural variability" (see Figure 4-2) (NRC, 2001a). All climate models used in the most recent Intergovernmental Panel on Climate Change (IPCC) assessment project that global mean temperatures will continue to increase in the 21st century and will be accompanied by other important environmental changes, such as sea-level rise, although the magnitudes of the projected changes vary significantly depending on the specific models and emissions scenarios (IPCC, 2001a,d).

Climate research has also significantly advanced our knowledge of the temporal and spatial patterns of climate variability. Substantial improvements in our ability to monitor the upper tropical Pacific

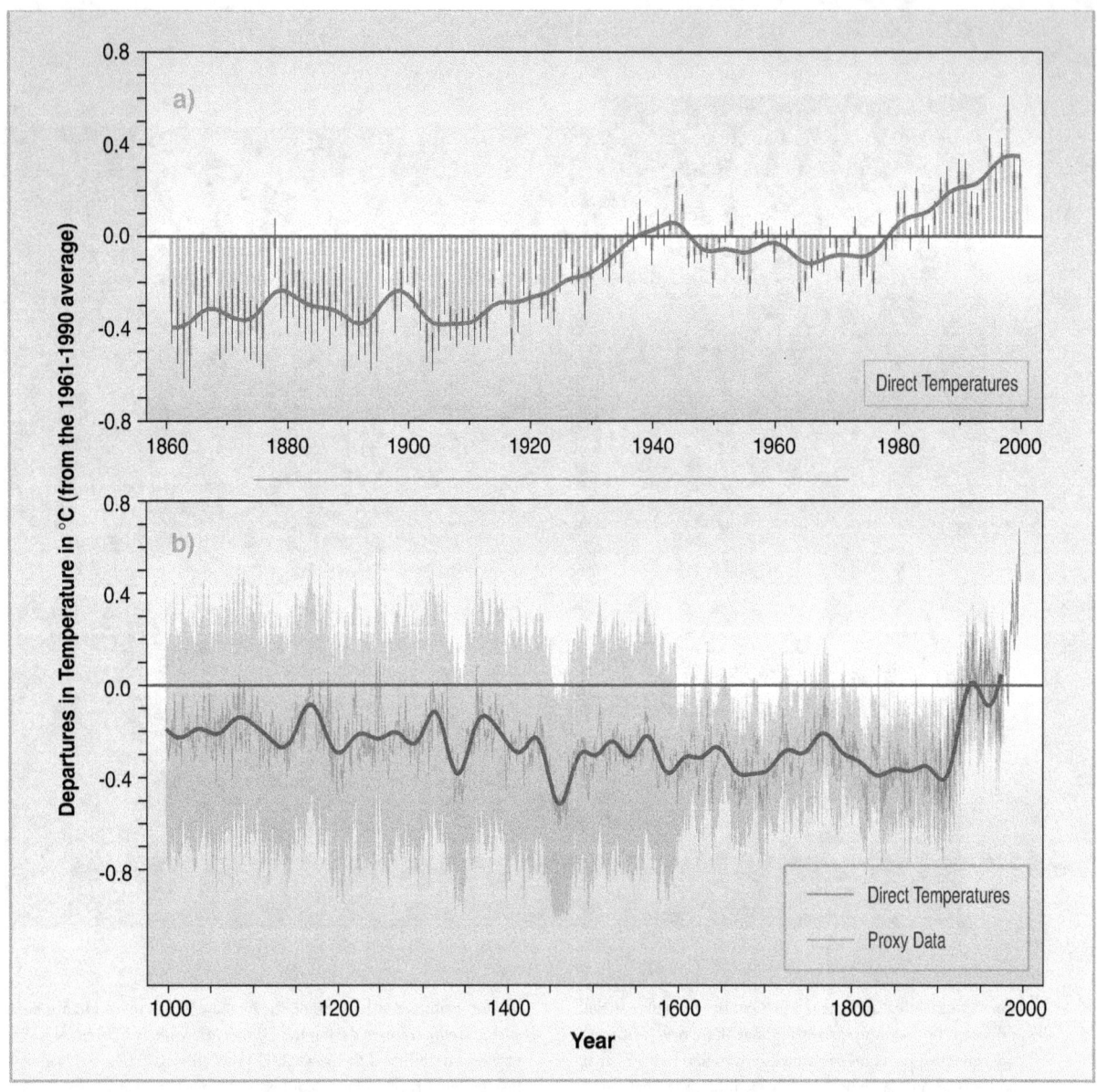

Figure 4-1: *Top Panel*: Changes in the Earth's surface temperature over the period of direct temperature measurements (1860–2000). The departures from global mean surface temperature are shown each year by the red bars (with *very likely* ranges as thin black lines) and approximately decade-by-decade by the continuous red line. *Bottom Panel*: Proxy data (year-by-year blue line with *very likely* ranges as gray band, 50-year-average purple line) merged with the direct temperature measurements (red line) for the Northern Hemisphere. The proxy data consist of tree rings, corals, ice cores, and historical records that have been calibrated against thermometer data. Source: IPCC (2001d). For more information, see Annex C.

Ocean now provide the world with an "early warning" system that shows the development and evolution of El Niño-Southern Oscillation (ENSO) events as they occur. This improved observational system, together with an increased understanding of the mechanisms that produce ENSO, has led to useful climate forecasts at lead times of up to several months. This developing capability has given the world an unprecedented opportunity to prepare for and reduce vulnerabilities to the impacts of ENSO, and thereby provided direct social and economic benefits as returns on climate science investments.

Research supported by the U.S. Global Change Research Program (USGCRP) has played a leading role in these scientific advances,

which have provided new climate information to help society better anticipate and prepare for potential effects of climate variability and change. While progress in this area has been impressive, there still remain many unresolved questions about key aspects of the climate system, including some that have enormous societal and environmental implications. For example, we are just beginning to understand how climate variability and change influence the local and regional occurrence and severity of extreme events such as hurricanes, floods, droughts, and wildfires. In many parts of the world, including the United States, such events are tied to ENSO variability, which has undergone significant changes in the past, perhaps in response to relatively subtle changes in forcing. A better

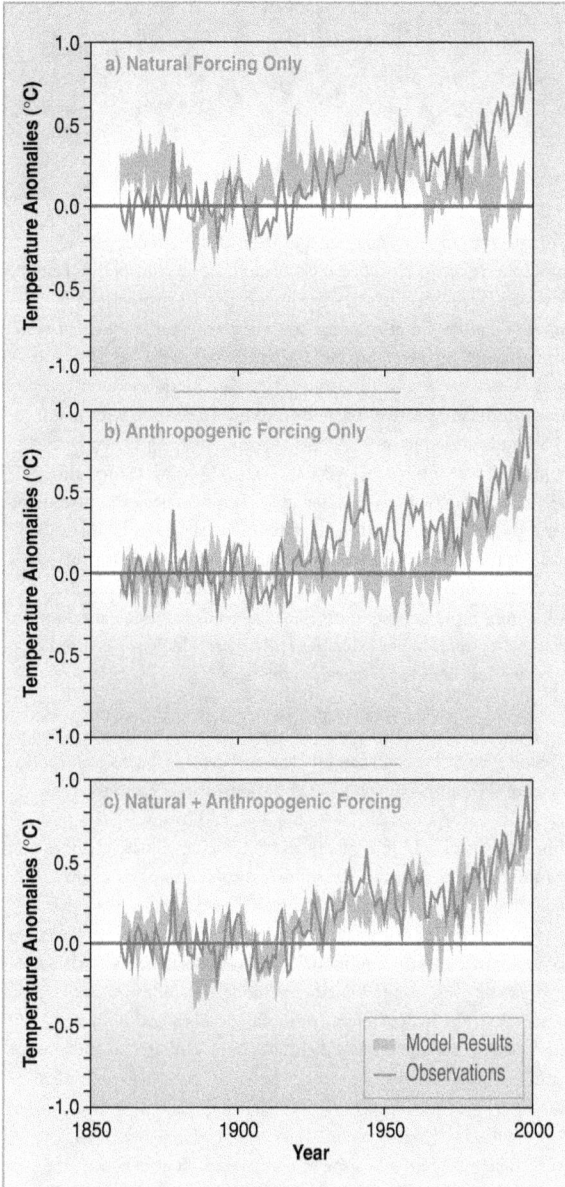

Figure 4-2: Climate model simulations of the Earth's temperature variations compared with observed changes for (a) natural forcing due to solar variations and volcanic activity; (b) anthropogenic forcing from greenhouse gases and an estimate of sulfate aerosols; and (c) both natural and anthropogenic forcing included. The model results show that the forcings included are sufficient to explain the observed changes, but do not exclude the possibility that other forcings may also have contributed. Source: IPCC (2001d). For more information, see Annex C.

understanding of ENSO behavior under different climate states is therefore needed.

We have also identified several major recurrent natural patterns of climate variability other than ENSO, but do not yet know to what extent they are predictable. Our predictive capabilities at local and regional scales show promise in some regions and for some

phenomena, but in many instances are still quite poor. We have yet to obtain confident estimates of the likelihood of abrupt climate transitions, although such events have occurred in the past (NRC, 2002). Perhaps most fundamentally, we do not yet have a clear understanding of how these natural climate variations may be modified in the future by human-induced changes in the climate, particularly at regional and local scales, and how emerging information about such changes can be used most effectively to evaluate the vulnerability and sustainability of human and natural systems (see, e.g., Figure 4-3).

The transformation of knowledge gained from climate research into information that is useful for societal decisions presents many challenges, as well as significant new opportunities. The process of understanding climate impacts and using climate information requires a detailed understanding of the interactions of climate, natural systems, and human institutions. Thus, to obtain maximum benefits from advances in knowledge of climate variability and change, it will be essential to forge new relationships between the climate research community, social scientists, and the rapidly expanding base of public and private sector users of climate information. For continued progress over the next decade, research on climate variability and change will focus on answering two overarching questions:

- How are climate variables that are important to human and natural systems affected by changes in the Earth system resulting from natural processes and human activities?
- How can emerging scientific findings on climate variability and change be further developed and communicated in order to better serve societal needs?

Providing decision-relevant answers to these questions will require a research infrastructure that includes: a sustained, long-term observing system of a quality necessary for climate research and assessments (Chapter 12); a highly focused and adequately funded modeling activity to analyze and integrate climate observations and support climate predictions and projections (Chapter 10); and a research-based infrastructure to develop partnerships among climate scientists, other natural scientists (e.g., biologists), social scientists, and public/private-sector decisionmakers to accelerate the production and applications of climate knowledge (Chapter 11). In addition,

Figure 4-3: Wind-blown dust buried farms and equipment, killed livestock, and caused human death and misery during the Dust Bowl drought of the 1930s. Source: *Monthly Weather Review*, June 1936 (courtesy NOAA).

coordinated research management will be required to ensure a broad-based and collaborative research program spanning academic institutions, government and private laboratories, and public and private sector expertise, to sustain research into climate variability and change and to provide advanced graduate and post-doctoral training for the next generation of scientists (Chapter 16).

In developing a research strategy, it is vital to recognize that the problems of climate variability and change are intrinsically connected: for example, regional impacts of climate change will depend directly on the variability of the global climate system. Moreover, future climate variability (e.g., frequency of ENSO events) will depend in part on changes in the mean climate. Therefore, problems of climate variability and change cannot be cleanly separated, and the success of understanding each will require improved understanding of both. The overall scientific strategy described in this chapter includes:

- Carefully designed, implemented, and managed observing system elements that will directly improve our knowledge of the climate system and will drive improvements in climate models
- Systematic, ongoing programs of climate data collection, integration, and analysis
- Process studies to elucidate critical processes that govern the climate system, but which in many cases are poorly understood and modeled
- Building a research infrastructure, such as the Earth System Modeling Framework, that supports collaborations among climate scientists and climate modeling centers
- Improving capabilities to assess climate information needs and to provide needed information to decisionmakers at local, regional, and national levels.

Advances will require improvements in paleoclimatic data as well as modern observational data systems, because in general the latter have been present for too short a time to extract robust features of climate variability on decadal or longer time scales. For example, in the Arctic, few climate stations have records extending back beyond 50 years, but paleoenvironmental analyses indicate that both the magnitude and spatial extent of 20th century Arctic warming may be unprecedented over the past 400 years. Paleoclimatic analyses also reveal the occurrence of decades-long mega-droughts at lower latitudes, including large portions of the United States (NRC, 2002).

The Climate Variability and Change research element will play a central integrating role in the Climate Change Science Program (CCSP). As indicated by the numerous linkages, the Climate Variability and Change research element will provide the array of advanced climate prediction and projection products that the other CCSP elements will utilize. This can be achieved only through the continued development of core climate system models that integrate the observational, analytical, and specialized modeling capabilities planned within the other CCSP elements, in order to provide improved information necessary to respond to the scientific and decisionmaking needs of the overall program. The overarching questions in the areas of climate variability and change can be addressed most effectively by focusing attention on five key science questions and their associated research objectives, as described below.

Question 4.1: To what extent can uncertainties in model projections due to climate system feedbacks be reduced?

State of Knowledge

Climate system feedbacks, such as from clouds, water vapor, atmospheric convection, ocean circulation, ice albedo, and vegetation, produce large uncertainties in climate change projections by modulating the direct response to radiative perturbations that result from changing greenhouse gas concentrations, solar variability, or land-cover changes. State-of-the-art climate models exhibit a large range in the cumulative strengths of these feedbacks, with major U.S. models used in recent IPCC assessments lying at nearly the opposite ends of this range (IPCC, 2001a). A key issue for climate science is the extent to which the range in model projections resulting from the differences in climate system feedbacks can be quantified and reduced. Important feedbacks include relatively fast processes on time scales of minutes to months (e.g., clouds and turbulent ocean mixing). Such rapid processes also affect models used for seasonal-to-interannual climate predictions, which can be used as effective test beds for research in this area.

All major U.S. climate models fail to accurately simulate certain climate system processes and their associated feedbacks in response to natural or anthropogenic perturbations. The oceans store and transport energy, carbon, nutrients, salt, and freshwater on multiple time scales and help to regulate and determine climate changes on a continuum of time scales. Yet some critical ocean phenomena, including ocean mixing and large-scale circulation features that determine the rate of storage and transport, remain as key challenges to understand, assess, and model. Other critical processes that are inadequately represented in climate models include atmospheric convection, the hydrological cycle, and cloud radiative forcing processses. Although observed changes in incoming solar radiation, a natural climate forcing, are small relative to changes in net radiative forcing by greenhouse gases (IPCC, 2001a,b), there is some evidence that feedbacks within the climate system may magnify otherwise weak solar variability. In spite of many research efforts over the past several decades and longer, the physical processes responsible for such feedbacks remain uncertain.

The cumulative effect of these processes influences the magnitude, rate, and spatial distributions of the climate response to natural or anthropogenic forcing. Modeling deficiencies are related both to limits in understanding the physics of the climate system and insufficient fine-scale treatment of the key processes. They contribute to uncertainties in projections of climate change, and thereby hinder the development of adequate response strategies and formulation of environmental and energy policies. High-priority research will focus on several sub-questions:

- *4.1.1*—What are the key climate system feedbacks that determine the transient and equilibrium responses for a specified radiative forcing?
- *4.1.2*—How and to what extent can uncertainties in these feedbacks be quantified?
- *4.1.3*—How sensitive are climate change projections to various

strategies for limiting changes in anthropogenic forcing, such as by enhancing biogeochemical sequestration or changing land use and cover?

- *4.1.4*—How can satellite, instrumental, and paleoclimatic observations of the Earth's past variations in climate be used to quantify and reduce uncertainties in feedbacks and provide bounds for the major elements of climate change projections for the next century?
- *4.1.5*—To what extent are climate changes as observed in instrumental and paleoclimate records related to volcanic and solar variability, and what mechanisms are involved in producing climate responses to these natural forcings?
- *4.1.6*—How may information about climate sensitivity and feedbacks be used to develop effective strategies for the design and deployment of observational systems?

Research Needs

U.S. research into climate forcing, feedbacks, and sensitivity is conducted at a few major modeling centers, federal and private laboratories, and universities. The intellectual quality of the research is outstanding, with many new and innovative ideas for model development and applications. However, the infrastructure and the observational data are currently inadequate to implement and evaluate these ideas cooperatively among the various modeling groups. Steps are being taken to address these deficiencies, but more must be done.

In order to optimize modeling resources and enable meaningful collaborations among modelers, it is necessary to develop and maintain a common and flexible infrastructure at the major modeling centers. By adopting common coding standards and system software, researchers will be able to test ideas at any of the major modeling centers, and the centers themselves will be able to easily exchange parameterizations as well as entire modules so that all groups benefit. The CCSP-supported Earth System Modeling Framework is an important start in this direction.

Additional infrastructure needs include: continuing enhancements of computational resources to keep pace with increasing model complexity (e.g., chemistry, biology); higher resolution, multi-century climate model simulations run from many different initial states (i.e., as ensembles) to help understand climate variability and change of the 18th, 19th, and 20th centuries and quantify probabilities of future climate events; additional software engineers for developing and managing model codes and building common and flexible infrastructure; and modeling center outreach scientists to aid and enable collaborations with external researchers. Benefits will include more efficient and rapid transfer of research results into applications, thereby achieving savings in human resource and dollar costs (Chapter 10).

Climate research also requires sustained, high-quality environmental observations. Long-term climate observing systems (e.g., ARGO floats and ocean profilers, aerosol-radiation-cloud observatories), satellite data, retrospective data (instrumental and paleoclimatic), field observations, and increased fidelity of current operational data streams and improved reanalyses of historical data will all be needed to produce data sets designed for climate change detection studies,

trend analyses, process research, and model development and testing (see also Question 4.2 and Chapter 12). Moreover, incorporation of observational data into modeling through improved data assimilation methods and more advanced models will address the reliability and uncertainties of these frameworks as well as facilitate the design of observing networks.

Further modeling research is required to improve simulations of seasonal-to-interannual variability in global models used for climate projections and to apply these models to improve seasonal-to-interannual climate predictions. Because many of the most important effects of global change will be felt at regional to local scales, improved capabilities of the global models to simulate and predict seasonal-to-interannual variability at these scales will be important both for validating the credibility of the models and building confidence among decisionmakers regarding the use of these models in global change projections.

A new research mode for accelerating improvements in climate models will be tested and evaluated with Climate Process and Modeling Teams (CPTs, see Box 4-1). CPTs are intended to complement rather than replace single investigators or other collaborative research on climate sensitivity and feedbacks. Pilot CPTs will begin in FY2003-2004.

Milestones, Products, and Payoffs
Products

- Refined estimates of the role of climate feedback processes in affecting climate sensitivity and improvements in their representation in climate models, leading to a narrowing of the range of climate model projections (Questions 4.1.1 and 4.1.2) [2-4 years and beyond].
  - For cloud and water vapor feedbacks, the Water Cycle research element will provide theoretically based cloud-resolving models, mesoscale models of cloud processes, cloud/precipitation process research, cloud energy budgets, and satellite data sets (e.g., CloudSat). The Climate Variability and Change research element will be responsible for cloud/water vapor feedback processes in the context of the coupled climate system models (e.g., the use of cloud-resolving models to test cloud parameterizations in climate models).
- Improved estimates of the climate response to different emissions [e.g., carbon dioxide (CO_2), aerosols, black soot] and land-use scenarios (Question 4.1.3) [2-4 years and beyond].
  - The capabilities of current climate models to link emissions to global atmospheric distributions of concentrations of pollution, including the chemical and heating and cooling properties of embedded atmospheric aerosols, will be addressed in cooperation with the Atmospheric Composition research element, as will assessments of the ability of the models to simulate observed radiative forcing of chemically active greenhouse gases, including improved uncertainty ranges. Aspects of this research will also be conducted cooperatively with the Land-Use/Land-Cover Change research element.
- New and improved climate data products, including: assimilated data from satellite retrievals and other remotely sensed and *in*

BOX 4-1

CLIMATE VARIABILITY AND CHANGE

FY04 CCRI Priority - Cloud and Water Vapor Feedbacks and Ocean Circulation and Mixing Processes

The Climate Change Science Program will address targeted climate processes known to be responsible for large uncertainties in climate predictions and projections. A new paradigm for conducting the research—Climate Process and Modeling Teams—will be used and evaluated.

Important processes that are inadequately represented in climate models include atmospheric convection, the hydrological cycle, and clouds and their net radiative forcing. Water vapor is the most important of the greenhouse gases, and clouds affect both vertical heating profiles and geographic heating patterns. Results from climate models suggest that there will be an overall increase in water vapor

as the climate warms. However, scientists know neither how the amounts and distributions of water vapor and clouds will change as the total water vapor in the atmosphere changes, nor how the associated changes in radiative forcing and precipitation will affect climate. Improved representation of the distribution of and processes involving water vapor in climate models is therefore critical to improving climate change projections.

Ocean mixing plays a pivotal role in climate variability and change, and is a primary source of uncertainty in ocean climate models. The highly energetic eddies of the ocean circulation are not well resolved and cannot be sustained for the multiple thousands of years of simulations required to assess coupled climate sensitivity. This leaves the problem of parameterization of eddy fluxes as a

key issue for improving coupled model simulations.

Accelerating improvements in climate models requires coordinated observational, process, and modeling programs by teams of scientists—that is, CPTs, an approach first proposed by U.S. CLIVAR (a complete description of CPTs can be found on its website, <http://www.usclivar.org>). CPTs will rapidly identify, characterize, and ultimately reduce uncertainties in climate model projections as well as determine observational requirements for critical processes. For problems that are generic to all climate models (e.g., cloud processes and ocean mixing), the CPTs will consist of teams of climate process researchers, observing system specialists, and modelers working in partnership with designated modeling centers.

situ data for model development and testing; consistent and regularly updated reanalysis data sets suitable for climate studies; centuries-long retrospective and projected climate system model data sets; high-resolution data sets for regional studies [e.g., Atmospheric Radiation Measurement (ARM) site data to initialize and evaluate cloud-resolving models]; and assimilated aerosol, radiation, and cloud microphysical data for areas with high air pollution, such as urban centers throughout the world (Question 4.1.4) [2-4 years and beyond].
- Some of these data will be collected, quality-controlled, and integrated in cooperation with the Atmospheric Composition and Water Cycle research elements, and will support these and the Carbon Cycle research element.
- Increased understanding and confidence in attribution of the causes of recent and historical changes in the climate (Questions 4.1.4 and 4.1.5) [2-4 years and beyond].
  - The Atmospheric Composition research element will provide emission and atmospheric concentration data and the Land-Use/Land-Cover Change research element will provide historical land-use change time series and land-use scenarios. Selected Climate Variability and Change data and analyses will be provided for the Atmospheric Composition research element's assessment of the impacts of tropospheric ozone on radiative forcing brought about by clean air regulations enacted during the last decade.
- Targeted paleoclimatic time series as needed, for example, to establish key time series of observations and natural forcing mechanisms as benchmarks of climate variability and change (Question 4.1.5) [2-4 years and beyond].

- Improved effectiveness of global and regional observing systems, including deployment of new systems and re-deployment of existing systems, based on guidance provided by modeled climate sensitivities and feedbacks (Question 4.1.6) [2-4 years and beyond].
- Policy-relevant information on climate sensitivities and the uncertainties in climate model projections due to climate system feedbacks, in support of the IPCC and other national and international assessments (all 4.1 questions) [2-4 years and beyond].

Payoffs
- More efficient and rapid transfer of research results into applications [2-4 years and beyond].
- Increased confidence in estimates of the global and regional manifestations of future changes in climate [beyond 4 years].

Question 4.2: How can predictions of climate variability and projections of climate change be improved, and what are the limits of their predictability?

State of Knowledge

One of the major advances in climate science over the past decade has been the recognition that much of climate variability is associated with a relatively small number of recurrent spatial patterns, or climate modes. These include, in addition to ENSO, the North Atlantic Oscillation (NAO), the northern and southern hemisphere annular modes (NAM, SAM), Pacific Decadal Variability (PDV),

Tropical Atlantic Variability (TAV), the Tropical Intra-Seasonal Oscillation (TISO), and monsoon systems. At present, there is limited understanding of the physical mechanisms that produce and maintain natural climate modes, the extent to which these modes interact, and how they may be modified in the future by human-induced climate changes. These limitations in knowledge introduce major uncertainties in climate predictions, climate change projections, and estimates of the limits of climate predictability, especially for regional climate (see also Question 4.1). They directly hinder our capabilities to address many of the "If…, then…" questions posed by decisionmakers.

Simulations of past climate conditions for which forcing estimates have been obtained provide an effective and practical means for assessing the scientific credibility of climate models. They enable detailed investigations of whether climate models realistically reproduce past climate states and responses in key environmental variables, such as sea level. They may also be used to evaluate how well the models simulate the various naturally recurring modes of climate variability. Process research that includes enhanced and extended observations—such as over the Pacific, Atlantic, and Indian Oceans basins and adjacent land and ice regions—provides a critical means for evaluating physical mechanisms and feedbacks, validating models, and assessing the corresponding effects on regional climate.

The extent to which skillful regionally specific climate predictions and climate change projections can be provided is an issue of fundamental practical importance. Various approaches have been proposed, including high-resolution global models, nested global-regional models, probabilistic information derived from ensembles with either individual or multiple climate models, and statistical downscaling. Much additional work is required to determine optimal methods and the feasibility of downscaling climate information to regional-to-local levels. Here, research on short-term climate variability (e.g., due to ENSO) can provide valuable insights. Developing capabilities to reproduce regional manifestations of interannual climate variability in climate models will also be crucial for establishing credibility with scientists and decisionmakers regarding longer term climate change scenarios (see, e.g., Figure 4-4).

High priority research will seek to answer the following questions:
- *4.2.1*—How can advances in observations, understanding, and modeling of ocean-atmosphere-land interactions be used to further improve climate predictions on seasonal to decadal time scales?
- *4.2.2*—What are the time scales for changes in climate variability following major changes in the land surface, oceans, or sea ice, and how does this "memory" contribute to climate predictability on multi-year to decadal time scales?
- *4.2.3*—What are the projected contributions from different components of the climate system to future sea-level changes,

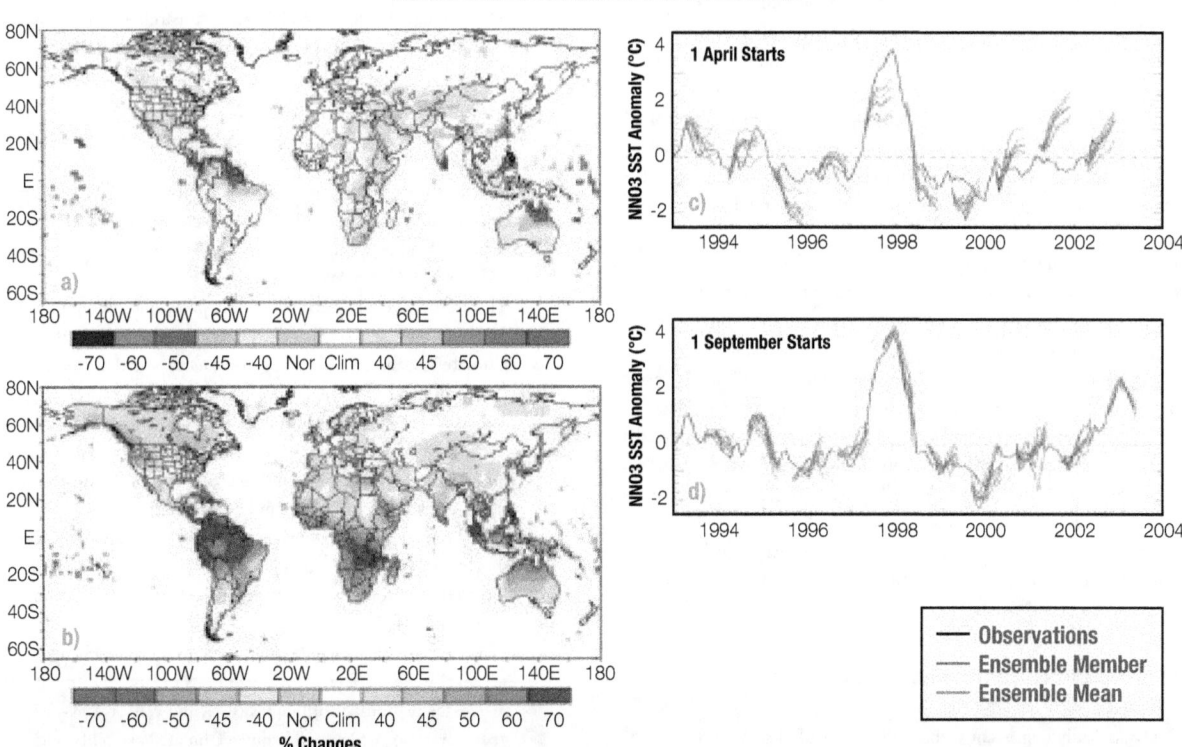

Figure 4-4: *Left Panel*: Multi-model derived probability forecasts of the most likely category for (a) precipitation and (b) temperature for January-March 2003 (models run December 2002). Positive values indicate forecast probabilities in the above normal category (upper one-third of climatological temperature or precipitation distributions, respectively) while negative values indicate probabilities in the below normal category (lower one-third of the distributions). *Right Panel*: Ensemble predictions with a coupled model system of eastern Pacific sea surface temperature (SST) anomalies 9 months in advance, with predictions starting in April (top) and September (bottom) of each year. The spread of the six-member ensembles, which is particularly evident in the 1 April starts, indicates the relatively large uncertainties in predicting SST anomalies so far in advance. Sources: (a) International Research Institute for Climate Prediction, and (b) NASA Seasonal-to-Interannual Prediction Project.

what are the uncertainties in the projections, and how can they be reduced?

- *4.2.4*—What is the potential for improved representation of natural modes of climate variability, such as ENSO, PDV, NAO, TAV, TISO, monsoons, and the annular modes, and how might this knowledge be used to extend and improve climate predictions?

- *4.2.5*—How might human-induced changes that affect the climate system, such as changes in atmospheric composition, aerosols, ground cover and land use, and natural forcing from solar variability and volcanic activity, alter climate forcing and hence climate variability and predictability on global and regional scales?

- *4.2.6*—How do current and projected climate changes compare with past changes and variations in climate in terms of patterns, magnitudes, and regional manifestations? For example, is the magnitude and time scale of the observed 20th century warming of the Arctic unprecedented in the last 1,000 to 10,000 years?

Research Needs

Essential research needs include the development and support for long-term, sustained climate modeling and observing capabilities (see also Question 4.1 and Chapters 10 and 12). These include remote-sensing data sets, global and regional reanalyses, and retrospective data including new high-resolution paleoclimate data sets. Field observations and process studies are necessary for improving understanding and modeling of the physical mechanisms responsible for climate feedbacks, evaluating the extent to which climate models successfully replicate these mechanisms, and determining observational requirements for critical processes. Additional research is required to develop improved methodologies to determine from global model projections changes in regional climate and seasonal-to-interannual variability. Vital constraints that must be considered include the water cycle (Chapter 5) and global energy balance.

Focused research efforts, such as the CPTs described earlier in this chapter, can play an important role in accelerating improvements in global climate models. Sea-level observations, geodetic reference frame measurements, ice sheet and glacier volume estimates, as well as advances in modeling are required to further refine sea-level change projections. Other research needs include producing data sets from ensembles of extended model simulations, and an updated, consistent reanalysis suitable for climate diagnostic analysis, attribution, and detection, including, if feasible, all of the 20th century. Moreover, access to model products, predictions, and tailored value-added products/information must be provided to the decisionmaking community to foster progress in utilizing prediction capabilities (see Question 4.5 and Chapter 11).

Milestones, Products, and Payoffs
Products

- Dynamically consistent global time series of observations (e.g., regularly updated and extended global climate reanalyses; 50-year long, 1° ocean data assimilation products) (Question 4.2.1.) [2-4 years].
  - Some aspects of this work will be conducted in collaboration with the Water Cycle research element. Selected data products will provide the Carbon Cycle research element information required for carbon cycle model development. Climate

model simulations of the 18th, 19th, and 20th centuries, and projections for the next couple of centuries that contain gridded values of all climate (ocean, atmosphere, sea ice, etc.) variables, also will be available for use by several other science elements.

- Extended model-based data sets to assess predictability and develop new approaches to improving seasonal-to-interannual climate predictions (Question 4.2.1) [2-4 years and beyond].

- Improved predictions of El Niño-La Niña, particularly the onset and decay phases (Question 4.2.1) [2-4 years and beyond].

- Improved probability forecasts of regional manifestations of seasonal climate anomalies resulting from ENSO (Questions 4.2.1 and 4.2.4) [less than 2 years and beyond].

- A paleoclimatic database designed to evaluate the ability of state-of-the-art climate models to simulate observed decadal- to century-scale climate change, responses to large changes in climate forcing, and abrupt climate change (Questions 4.2.2, 4.2.5, and 4.2.6) [2-4 years].
  - The Carbon Cycle research element will provide information on feedbacks to the climate from large changes in carbon storage and fluxes.

- Development and extension of critical data sets to improve analyses of climate variability and attribution of causes of climate change (Questions 4.2.2 and 4.2.6) [2-4 years and beyond].

- Improved high-resolution, three-dimensional ocean circulation models (Questions 4.2.1–4.2.4) [2-4 years and beyond].
  - The Climate Variability and Change research element will lead in development of suitable ocean models for climate purposes, but will rely on other elements for model subcomponents (e.g., Carbon Cycle for biogeochemistry, Ecosystems for biological, Water Cycle for important forcings such as continental freshwater runoff).

- Improved estimates of global air-sea-land fluxes of heat, moisture, and momentum needed to discern characteristics of ocean-atmosphere-land coupling and to assess the global energy balance (Questions 4.2.1–4.2.4) [2-4 years].
  - The Water Cycle research element (providing precipitation products and land-surface fluxes) will work jointly with Climate Variability and Change on this product. Climate Variability and Change, working with the Water Cycle research element, will coordinate selected field measurements with those of the Carbon Cycle research element to derive more complete three-dimensional time series for flux and transport studies of carbon and ocean nutrients.

- Improved representation of processes (e.g., thermal expansion, ice sheets, water storage, coastal subsidence) in climate models that are required for simulating and projecting sea-level changes (Question 4.2.3) [2-4 years and beyond].
  - In addition to the above factors, local to regional changes in sea level will be affected by movements in land due to natural processes or human influences (e.g., the removal of groundwater), which will require Climate Variability and Change to coordinate in this area with the Water Cycle and Land-Use/Land-Cover Change research elements.

- Improved understanding and parameterizations of key mechanisms for seasonal-to-decadal variability, based on process studies together with modeling research and analyses (Question 4.2.4) [2-4 years and beyond].

- Improvements in the representation of major modes of climate variability in climate change projection models and predictions of regional patterns of different modes of climate variability (Question 4.2.4) [beyond 4 years].

- An assessment of potential predictability beyond ENSO (e.g., associated with PDV, NAO, annular modes, tropical Atlantic and Indian Ocean variability and trends, and the monsoons) (Question 4.2.4) [beyond 4 years].

- Estimates of the spatial and temporal limits of predictability of climate variability and change forced by human activities (Question 4.2.5) [beyond 4 years].

- Policy-relevant information on natural climate variability and the potential predictability of climate variability and change in support of national and international science assessments [2-4 years and beyond].

Payoffs

- An improved ability to separate the contributions of natural versus human-induced climate forcing to climate variations and change, resulting in more credible answers to "If…, then…" policy-related questions [beyond 4 years].

- Increased understanding of changes in natural variability and potential impacts on predictability that may result from anthropogenic forcing [beyond 4 years].

- Research to address Questions 4.1 and 4.2 will provide more reliable and useful climate prediction products and other essential support to U.S. and international decisionmakers and resource managers, and will assist climate assessment efforts by increasing understanding of critical processes required to evaluate and improve major climate models (see Chapter 10) [2-4 years and beyond].

> **Question 4.3:** What is the likelihood of abrupt changes in the climate system such as the collapse of the ocean thermohaline circulation, inception of a decades-long mega-drought, or rapid melting of the major ice sheets?

State of Knowledge

Analyses of the paleoclimate record—the record of the Earth's environmental history derived from sources such as ice cores, tree rings, and lake and ocean sediments—provide compelling evidence for past abrupt climate changes. In some locations, changes of up to 16°C in temperature and a factor of two in precipitation have occurred within decades to years, yet lasted for centuries and longer (NRC, 2002). Paleoclimate data indicate that these changes have been manifested by significant shifts in the baseline climate and in the character and patterns of variations about average conditions. The rapidity of such changes poses major challenges to the vulnerability and adaptability of societies and ecosystems.

Previous paleoclimate research has provided significant advances in our understanding of the general structure and geographic extent of past abrupt climate changes. Much past research has focused on colder climate conditions, and a challenge for the future will be to understand the potential for abrupt change in the context of an overall warming climate. Abrupt climate changes may be associated

with the crossing of a climatic threshold, the onset of nonlinear responses, or feedbacks in the climate system. To date, however, the causes of past abrupt changes are not fully explained or understood. In addition, present climate models fail to adequately capture the magnitude, rapidity, and geographical extent of past abrupt changes. Consequently, at this time climate models cannot be used with confidence to estimate the potential for future abrupt changes (NRC, 2002). Improved knowledge of the causes for abrupt changes, and the ability to project their future probabilities, will provide policymakers with an improved scientific basis to evaluate risks of future abrupt changes and, as needed, to develop strategies to reduce vulnerabilities. Major questions include:

- *4.3.1*—How common are abrupt changes, based on paleoclimate records?

- *4.3.2*—What are the observational requirements needed to resolve the spatial and temporal patterns of past abrupt changes, and to answer questions about differences in timing of the changes in different parts of the globe?

- *4.3.3*—What are the primary natural mechanisms for producing abrupt climate changes?

- *4.3.4*—When might future abrupt changes be expected to occur, and what would be the expected global and regional manifestations of such changes?

Figure 4-5: The retreat of the South Cascade Glacier in Washington's Cascade Mountains shown in photographs from (a) 1928 and (b) 2000. Since 1957, the glacier has retreated about 700 meters, losing about one-fifth of its length and one-third of its volume. Source: USGS.

- *4.3.5*—What is the nature and extent of abrupt climate change in the Holocene? Are these stochastic events or the result of periodic forcing?
- *4.3.6*—What is the potential for high-impact climate changes, such as much drier and warmer summers over the mid-continents of North America and Eurasia, accelerated Arctic warming, and more intense coastal storm surges and coastal erosion due to rising sea levels?
- *4.3.7*—What would be the environmental consequences of extreme warming in the Arctic, and what would be the expected feedbacks on global climate?

Research Needs

Improved paleoclimatic data sets, expanded observing and monitoring systems and rigorous paleoclimate modeling studies will be required to identify the causes and mechanisms of past abrupt changes. Efforts should be focused on key regions or phenomena that may be especially vulnerable or contribute most strongly to abrupt climate change, such as the tropics, the Arctic and Antarctic regions, and the ocean thermohaline circulation. Significant research into how to numerically model the full three-dimensional circulation of the ocean will be required in order to accurately project impacts and time scales for abrupt changes, which range from interannual ENSO variability to centennial-millennial fluctuations in the ocean circulation. Key research needs include better understanding of the relationships between abrupt change and:

- Oceanic circulation, especially related to deepwater formation
- Sea-ice transport and processes, particularly where they interact with deepwater formation
- Land-ice behavior, including conditions beneath ice sheets
- Modes of atmospheric variability and how they are altered by changes in mean climate conditions
- The hydrological cycle, including storage, runoff, and permafrost changes. The Water Cycle research element will provide studies and modeling of cryosphere hydrological processes to complement the efforts of Climate Variability and Change to address cryosphere-atmosphere-ocean coupling and feedbacks.

Milestones, Products, and Payoffs

Products

- Databases of drought and mega-drought occurrences in North America (Questions 4.3.1 and 4.3.6) [2-4 years].
- Online database of annual-to-decadal resolution paleoclimatic time series

and maps of Arctic climate variability over the past 2,000 years (Questions 4.3.2, 4.3.6, and 4.3.7) [2-4 years].
- Improved understanding of thresholds and nonlinearities in the climate system, especially for coupled atmosphere-ocean, ocean thermocline and deepwater, hydrology, land surface, biogeochemical cycle, and ice processes (Questions 4.3.3 and 4.3.5) [beyond 4 years]. Advances require cooperation with the Water Cycle and Carbon Cycle research elements.
- Improvements in the accuracy, management, and synthesis of paleoclimatic data that can be combined with instrumental observations and climate models to address the nature and likelihood of future abrupt climate changes (Questions 4.3.2 and 4.3.4) [beyond 4 years].
- Policy-relevant information of the state of understanding on the causes of abrupt changes, and probabilistic estimates of future risks of abrupt global and regional climate-induced changes, including the collapse of the thermohaline circulation, persistent ENSO conditions, and abrupt sea-level rises, in support of national and international assessments (Questions 4.3.4 and 4.3.6) [2-4 years and beyond].

Payoff
- Increased use and effectiveness of paleoclimate data and analyses of abrupt change to better inform environmental decisions and adaptation strategies [4 years and beyond].

BOX 4-2

CLIMATE VARIABILITY AND CHANGE

**FY04 CCRI Priority -
Polar Feedbacks**
The Climate Change Research Initiative (CCRI) will leverage existing USGCRP research to address major gaps in understanding climate change. Polar systems may be especially sensitive to climate change and might provide early indications of climate change as well as interact with climate variability and change through several important feedback processes.

The CCRI will support research to improve understanding of processes that determine the behavior of slowly varying elements of the physical climate system, especially the oceanic and cryospheric portions. Particular foci include the processes by which ice-covered regions of the high-latitude Earth behave, the processes by which the distribution of sea ice varies, and the way in which knowledge of ocean circulation can be enhanced through use of global observations of ocean state and forcing parameters. The development and testing of new

capabilities for measuring climatic properties, such as ocean surface salinity, mixed layer depth, and ice sheet thickness will also be carried out.

The CCRI will support the obtaining of systematic data sets for a limited number of Earth system parameters such as ice thickness, extent, and concentration in the case of sea ice, and mass balance and surface temperatures in the case of land ice and snow cover. It will shortly enable the initiation of regular observations of ice-sheet thickness. Data assimilation systems using satellite data that provide for accurate, geophysically consistent data sets will also be carried out through this program. The polar feedbacks research will also contribute to decision support through cryospheric observations and associated models that enable the initialization and verification of climate models, and the reduction in uncertainty of model output. The models will also provide real-time information for use by the U.S. Navy and commercial maritime interests in high-latitude regions.

Question 4.4: How are extreme events, such as droughts, floods, wildfires, heat waves, and hurricanes, related to climate variability and change?

State of Knowledge

One of the highest priorities for decisionmakers is to determine how climate variations, whether natural or human-induced, alter the frequencies, intensities, and locations of extreme events (NRC, 1999a). There is now compelling evidence that some natural climate variations, such as ENSO, PDV, and the NAO/NAM, can significantly alter the behavior of extreme events, including floods, droughts, hurricanes, and cold waves (IPCC, 2001a,b). Studies of long-term trends in extreme events show that in many regions where average rainfall has been increasing, these trends are evident in extreme precipitation events (there continues to be debate on how to define an extreme precipitation event). For other high-impact phenomena, such as tropical storms/hurricanes, no compelling evidence yet exists for significant trends in frequency of occurrence (IPCC, 2001a,b).

A question central to both short-term climate predictions and longer term climate change is how climate variability and change will alter the probability distributions of various quantities, such as of temperature and precipitation, as well as related temporal characteristics (e.g., persistence), and hence the likelihood of extreme events (see Figure 4-6). A key challenge is to develop improved methods for modeling or downscaling climate information to the scales required for extreme event analysis (IPCC, 2001a). Further, understanding of the processes by which climate variability and change modulate extreme event behavior is incomplete. Major research questions include:

- *4.4.1*—What is the range of natural variability in extreme events, by phenomena and region?
- *4.4.2*—How do frequencies and intensities of extreme events vary across time scales?
- *4.4.3*—What are observed and modeled trends in extreme events and how do they compare?

Figure 4-6: Heavy rains and high surf from storms associated with the 1998 El Niño event produced severe erosion along the California coast, leading to major property losses. Source: Paul Neiman, Environmental Technology Laboratory, NOAA.

- *4.4.4*—How are the characteristics of extreme events changed by natural climate variations, for example, by ENSO, PDV, NAO/NAM and SAM?
- *4.4.5*—To what extent are changes in the statistics of extreme events predictable?
- *4.4.6*—How are behaviors of extreme events likely to change over this century, and what are the mechanisms that would be expected to produce these changes?
- *4.4.7*—How can the emerging findings on climate-extreme event links be best developed and communicated to evaluate societal and environmental vulnerability and opportunities?

Research Needs

Progress in this area will require two key steps. First, it will be necessary to advance scientific understanding and quantitative estimates of how natural climate variations such as ENSO, NAO/NAM, SAM, or PDV alter the probabilities of extreme events (e.g., floods, droughts, hurricanes, or storm surges). Second, it will be essential to improve understanding of how human-induced climate change may alter natural variations of the atmosphere, ocean, land surface, and cryosphere, and hence the behavior of extreme events in different regions.

Key data requirements include the development of improved climate-quality data and reference data sets and higher resolution model reanalyses to support analyses of extreme event variability and trends. High-resolution observations together with focused process studies will be essential for scientific evaluation of regional model simulations, especially in regions with significant topographic variations, such as mountainous and coastal regions. Higher resolution paleoclimatic data will also be necessary to improve descriptions and understanding of how natural climate variations have in the past altered drought, mega-drought, flood, and tropical storm variability (see Question 4.3). Improving hydrological extreme event risk estimates will require improved hydrological data sets and advances in coupled climate-land surface-hydrology models (see Chapters 5, 6, 11, and 12).

Empirical and diagnostic research will be required to ascertain relationships between natural climate modes, boundary forcing mechanisms (e.g., SST variations, land surface and cryospheric changes), and extreme events; to clarify the physical bases for these relationships; and to evaluate the veracity of model simulations and projections. Model sensitivity experiments will significantly advance understanding of how natural climate modes, boundary variations, and human-induced climate trends alter the probabilities of extreme events. Further development of regional climate modeling and improved downscaling techniques will be necessary to provide information at the scales needed by resource managers and decisionmakers.

Continuing development of ensemble-based approaches, and the capabilities to produce large ensembles from climate models, will be essential in order to improve probability estimates of extreme events for either short-term climate predictions or longer term climate projections. Because extreme events can have societal and environmental impacts, it will be essential to identify key climate information needed to better anticipate and plan for such events (see Question 4.5 and Chapters 9 and 11).

Milestones, Products, and Payoffs

Products

- Improved observational databases, including paleoclimate and historical data records, and model simulations of past climate to detect and analyze regional trends in extreme events and to assess whether changes in the frequencies of extreme events lie within or outside the range of natural variability (Questions 4.4.1-4.4.3) [2-4 years and beyond].
- Observational and statistical analyses to assess the relationships between extreme events and natural climate variations, such as ENSO, PDV, NAO/NAM, and SAM (Question 4.4.4) [2-4 years and beyond].
- Improved diagnostic capabilities to better interpret the causes of high-impact climate events, such as droughts or unusually cold or warm seasons (Question 4.4.4) [2-4 years and beyond].
    - Aspects of this research require data, information, and analyses of regional hydrological processes related to floods and droughts, to be provided by the Water Cycle ressearch element.
- Assessments of potential predictability and forecasts of probabilities of extreme events associated with natural climate variations (Questions 4.4.4 and 4.4.5) [2-4 years and beyond].
    - For floods and droughts, the Water Cycle research element will focus on the role of hydrological feedbacks and the predictability of extreme events on weather time scales. The Climate Variability and Change research element will focus on variations and changes that generate conditions favorable for extreme events, assess the predictability of these events, and develop products useful for applications (e.g., extreme event outlooks) on seasonal and longer time scales.
- Documented impacts of climate extremes on regions and sectors, and evaluations of the implications should climate change in the future (Question 4.4.6) [2-4 years and beyond].
- Policy-relevant information on past variability and trends in extreme events, and probabilistic estimates of possible future changes in frequencies, intensities, and geographical distributions of extreme events in support of national and international assessments (Question 4.4.1 and 4.4.6) [2-4 years and beyond].

Payoffs

- Improved anticipation of and response to extreme climate events (e.g., to reduce regional impacts of ENSO or more rapidly respond to emerging droughts) [2-4 years and beyond].
- Increased understanding of and capabilities to project the regional manifestations of extreme climate events, to provide a sounder scientific basis for policymakers to develop strategies to minimize potential vulnerabilities [beyond 4 years].

Question 4.5: How can information on climate variability and change be most efficiently developed, integrated with non-climatic knowledge, and communicated in order to best serve societal needs?

State of Knowledge

Research in this area focuses on making climate knowledge more useful and responsive to the needs of decisionmakers, policymakers, and the public. Climate information, when integrated together with knowledge of non-climatic factors, can reduce costs and risks related to climate variability and change while increasing management and decisionmaking opportunities across a broad range of sectors, from local and regional to global scales (NRC, 1999a; IPCC, 2001b).

For example, pilot efforts in sustained regional integrated science research, such as the NOAA-supported Regional Integrated Science and Assessments (RISA) projects, NASA Regional Earth Science Application Centers (RESACs), and NOAA-supported International Research Institute for Climate Prediction (IRI) for areas outside of the United States, have provided opportunities to apply climate information in decision processes in climate-sensitive sectors, including agriculture, water, energy (e.g., hydropower), and forest (wildfire) management. USEPA also sponsors regional science and assessment projects. Finally, specialized entities outside the research domain, such as state climatologists, regional climate centers, and agricultural cooperative networks, have served as partners and liaisons by identifying and communicating climate information needs and requirements between the climate research and service communities and a broad array of users.

With continuing population growth and increasing demands on environmental resources, the need to more effectively identify, develop, and provide climate information useful for society will become ever more vital. Even in the absence of human-induced climate changes, further research in this area provides new opportunities for resource managers and policymakers to develop strategies to reduce vulnerabilities to natural climate variability. Major questions include:

- *4.5.1*—What new climate information would provide the greatest potential for benefits, for different regions and sectors?
- *4.5.2*—How can climate information be best developed for use in adaptive management strategies?
- *4.5.3*—Can new climate indicators be developed to better assess climate vulnerability and resilience in climate-sensitive sectors such as agriculture, water, marine fisheries and other environmental resources, transportation, and the built environment, as well as other potential societal impacts (positive and negative), including on human health?
- *4.5.4*—What are potential entry points and barriers to uses of climate information?
- *4.5.5*—How can access to and communication of climate data and forecasts be improved in order to better serve the needs of the public, scientific community, decisionmakers, and policymakers?

Research Needs

In recent reports the National Research Council (NRC) identifies the "region" as a key scale for decisionmaking, and stresses the critical need to improve regional scientific capabilities and user interactions to better inform such decisions (NRC, 1999a, 2001e). As these reports emphasize, the impacts of climate variability and change will continue to be felt most directly at regional to local scales—for example, within natural boundaries associated with coastlines, mountains, or watersheds, within the context of demographics, ecosystems, and land use, and within the context of its economic and technological wherewithal. A central goal of research over the next decade will be to improve capabilities to identify, develop, and deliver climate information at regional to local scales in order to better meet societal needs.

A key challenge in this area is to continue developing the observational, diagnostic, and modeling expertise required to determine the impacts of climate variability and change at global and regional scales. The required basic science research must be complemented by a strong applied research component to ensure identification of key regional issues and impacts of multiple stresses on resource management, determine responsiveness to user needs, and develop objective means for measuring success. To be developed most effectively, these research efforts should be conducted as sustained, two-way partnerships that directly involve decisionmakers and other regional stakeholders. This will help to ensure that results of climate research are made most useful for applications. Further, user needs can provide important guidance for developing future research directions. Regional "test beds" or "enterprises" can serve as important foci for developing such partnerships, evaluating potential uses of climate information at regional scales, and performing analyses of regional climate impacts, vulnerabilities, adaptation, and mitigation options related to climate variability and change. They can also be used as demonstration projects for providing end-to-end delivery of climate information and evaluations of its uses, and for establishing an improved national decision support capability.

Because of the difficulties in evaluating the effectiveness of decisions with long lead times, important initial steps toward building confidence in the use of climate information can be made through focused research on shorter term decisions, such as those that occur on monthly and seasonal-to-interannual time scales (e.g., agriculture, water management, energy distribution, wildfire management). Improved information for supporting climate-sensitive decisions on these time frames will be critical for building credibility on the uses of climate data and projections to better inform difficult, long-term decisions. A focus on shorter term climate variability also provides opportunities to try various decision options.(e.g., through adaptive management strategies). Evaluating the effectiveness of strategies on shorter time scales will be useful for developing longer term policy options and decisions.

As climate knowledge improves, evaluation can be extended to multi-year and decadal time scales, which will provide an important bridge to policy and decision options related to longer term changes. In this regard, the decadal time scale offers a valuable bridge for research on climate-sensitive adaptive strategies across time scales. It links the management of the impacts of individual extreme events and interannual variations to longer term variations and can provide tangible observational regional analogs for climatic change.

To ensure that these efforts are efficient and cost-effective, it will be crucial to involve existing regional experts in climate information, applications, and user needs, such as state climatologists, regional climate centers, university extension agents, local weather service offices, and members of the private sector. These regional efforts must ultimately be coordinated effectively in order to provide information that will serve the needs of policymakers and decisionmakers at the national level (see Chapter 11).

Milestones, Products, and Payoffs
Milestones and Products
- Establishment of research teams involving climate and social

scientists and stakeholders in climate-sensitive regions to create focused, user-responsive partnerships (Questions 4.5.1–4.5.5) [less than 2 years and beyond].
- Increased partnerships with existing stakeholder support institutions, such as state climatologists, regional climate centers, agricultural extension services, resource management agencies, and state governments to accelerate uses of climate information (Questions 4.5.1–4.5.5) [less than 2 years and beyond].
- Assessments of the adequacy of existing operational climate monitoring networks to provide regional decision support, and to identify major data gaps in addressing critical regional and policy issues, such as drought planning and response (Questions 4.5.1 and 4.5.2) [2-4 years].
- Development of high-resolution climate products for climate-sensitive regions, based on monthly instrumental data, annual paleoclimatic data, and climate forecasts (Questions 4.5.1 and 4.5.3) [2-4 years and beyond].
- Documented regional impacts of climate variability, and development of reports on the potential implications of projected climate changes (Questions 4.5.1 and 4.5.5) [beyond 4 years].
- Development of a framework for assessing the effectiveness of current regional-scale climate science and services (Question 4.5.5) [2-4 years].
- Development of first-generation "test bed" integrated climate science and assessment decision support systems for subsets of user groups (e.g., farmers, ranchers, water managers, forest managers, fisheries managers, coastal zone managers, urban planners, and public health officials) in regions where user demand is already demonstrated (Question 4.5.5) [2-4 years and beyond].
- Policy-relevant information on uses and needs for climate information, and potential impacts of future climate variability and change at regional to local levels, in support of national and international assessments (Question 4.5.1-4.5.5) [2-4 years and beyond].

Payoffs
- Expanded decision support resources and the capacity to effectively apply climate knowledge [2-4 years and beyond].
- Increased public and decisionmaker use of research-based information on climate variations, forecasts, and impacts [2-4 years and beyond].
- Knowledge to develop and sustain effective climate services for all parts of the nation and to support national decisionmaking capabilities [2-4 years and beyond].

National and International Partnerships

Internationally coordinated research programs such as the World Climate Research Programme (WCRP) and its projects Climate Variability and Predictability (CLIVAR), Stratospheric Processes and their Role in Climate (SPARC), Climate and Cryosphere (CliC), the Global Energy and Water Cycle Experiment (GEWEX), as well as the International Geosphere-Biosphere Programme (IGBP) PAGES paleoscience project are critical for developing global infrastructure and research activities designed to ensure that global aspects of climate variability and change are addressed in a coordinated manner.

In particular, CLIVAR—the broadest of the WCRP programs (WCRP, 1995)—has a suite of vigorous activities that address numerous facets of the climate problem. For example, its Working Group on Seasonal-to-Interannual Prediction leads worldwide development and assessment of prediction approaches and forecast systems while the CLIVAR-WCRP Working Group on Coupled Modeling is fostering advancements in coupled modeling. In the Atlantic region, CLIVAR is actively coordinating and encouraging international (e.g., U.S., European, and South American) observational, analysis, and modeling activities that will advance understanding and predictions of the puzzling climate changes that impact this region. These research activities are identifying how the regional climate changes are manifested through features such as TAV and NAO/NAM. Furthermore, CLIVAR investigations are elucidating critical ocean-atmosphere-land-cryosphere (and with WCRP-SPARC, stratosphere-troposphere) coupled processes as well as critical inherent features such as the Atlantic Ocean thermohaline circulation that must be correctly modeled to project future climate changes. CLIVAR and WCRP are fostering numerous activities in the Americas that are addressing global issues (e.g., process studies focusing on the evolution and dynamics of the monsoons) as well as regional issues (e.g., extreme events, paleoenvironmental variability) and their implications for the global climate system.

Within the United States there are a number of partners that will coordinate implementation of the Climate Variability and Change strategic vision. U.S. CLIVAR has in place a nucleus of scientific and programmatic elements, but will need to strengthen ties with additional WCRP groups (e.g., U.S. CliC) as well as with focused model and assimilation system development (e.g., at the National Center for Atmospheric Research, Geophysical Fluid Dynamics Laboratory, and the Goddard Space Flight Center). The deep-ocean observation program—a joint U.S. CLIVAR-Carbon Cycle Study Program effort—is fostering a complementary international component that is providing an example of the benefits of program coordination. Additionally, NOAA's International Research Institute

of Climate Prediction is leading international development of climate predictions and their applications. Finally, NOAA's National Centers for Environmental Prediction (NCEP) will also contribute in many key areas including climate monitoring and diagnostics, forecasting, reanalyses, and high-resolution weather modeling. Stronger linkages with NCEP and other members of the weather modeling community will be most helpful in advancing capabilities to address issues of high societal relevance, such as downscaling information to regional and local scales and improving predictions and projections of extreme events. Finally, involving the cadre of regional climate centers and state climatologists will help ensure that regional and user expertise is represented in the development of effective frameworks for developing useful information.

CHAPTER 4 AUTHORS

Lead Authors
Randall Dole, NOAA
Jay Fein, NSF
David Bader, DOE
Ming Ji, NOAA
Anjuli Bamzai, NOAA
Tsengdar Lee, NASA
David Legler, CLIVAR
Milan Pavich, USGS
David Verardo, NSF

Contributors
Stephen Meacham, NSF
Anthony Socci, USEPA
James Todd, NOAA
David Rind, NASA
Ants Leetmaa, NOAA

CHAPTER

The water cycle is essential to life on Earth. As a result of complex interactions (see Figure 5-1), the water cycle acts as an integrator within the Earth/climate system, controlling climate variability and maintaining a suitable climate for life. The water cycle manifests itself through many processes and phenomena, such as clouds and precipitation; ocean-atmosphere, cryosphere-atmosphere, and land-atmosphere interactions; mountain snow packs; groundwater; and extreme events such as droughts and floods. Inadequate understanding of and limited ability to model and predict water cycle processes and their associated feedbacks account for many of the uncertainties associated with our understanding of long-term changes in the climate system and their potential impacts, as described by the Intergovernmental Panel on Climate Change (IPCC). For example, clouds, precipitation, and water vapor produce feedbacks that alter surface and atmospheric heating and cooling rates, and the redistribution of the associated heat

sources and sinks leads to adjustments in atmospheric circulation, evaporation, and precipitation patterns.

Because water cycle processes occur, are observed, and are studied at a wide variety of scales (watershed, basin, continental, global), understanding of the water cycle is extremely challenging. Characterizing the interactions between the land and the atmosphere will require capabilities, such as improved observations and regional climate models, to scale down global climate model fields and to scale up the effects of land surface heterogeneity. The interactions between oceans and the atmosphere manifest themselves as the slower modes of climate variability, as described in Chapter 4.

Clean water is an essential resource for human life, health, economic growth, and the vitality of ecosystems. From social and economic perspectives, the needs for water supplies adequate for human uses—such as drinking water, industry, irrigated agriculture,

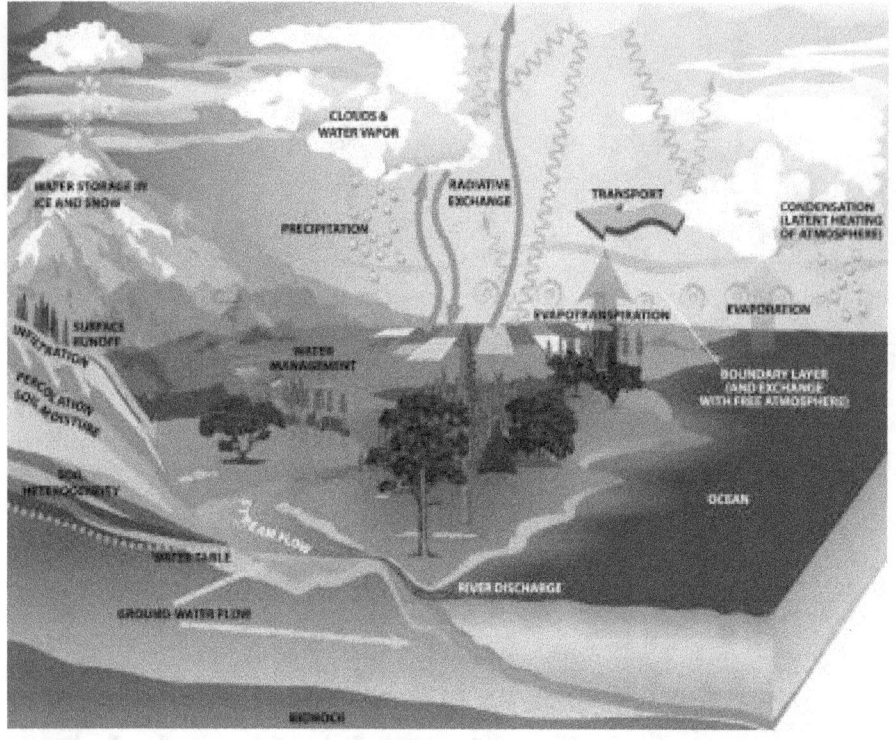

Figure 5-1: Conceptualization of the global water cycle and its interactions with all other components of the Earth-climate system. The water cycle involves water in all three of its phases [solid (snow, ice), liquid, and gaseous], operates on a continuum of time and space scales, and exchanges large amounts of energy as water undergoes phase changes and is moved dynamically from one part of the Earth system to another. These interactions with radiation and atmospheric circulation dynamics link the water and energy cycles of the Earth system. Source: Paul Houser and Adam Schlosser, NASA GSFC. For more information, see Annex C.

hydropower, waste disposal, and the protection of human and ecosystem health—are critical. Water supplies are subject to a range of stresses, such as population growth, pollution, and industrial and urban development. These stresses are exacerbated by climate variations and changes that alter the hydrologic cycle in ways that are currently not predicted with sufficient accuracy for decisionmakers. A number of these concerns and related questions and strategies are documented in a recent report on research needs and opportunities, *A Plan for a New Science Initiative on the Global Water Cycle* (USGCRP, 2001), which formed the basis for initial interagency planning related to the global water cycle.

Advances in observing techniques, combined with increased computing power and improved numerical models, now offer new opportunities for significant scientific progress. Furthermore, field studies and modeling initiatives like the Global Energy and Water Cycle Experiment (GEWEX) Continental-Scale International Project, observation systems such as the Tropical Rainfall Measuring Mission (TRMM, see Figure 5-2), and regional test beds such as the Cloud Atmospheric Radiation Testbed (CART)/Atmospheric Radiation Measurement (ARM) site have provided data and insights that accelerated improvements in model physics. Recently, for example, credible predictions of seasonal variations in the water cycle have been produced for the western United States and Florida. This activity has served as a basis for dialogue between the research community and decisionmakers on the latter's information needs and on opportunities for improving the adaptability of infrastructure and management practices to long-term changes and extremes. Along with the growing ability to provide advance notice of extreme hydrologic events, this forecast capability provides new options for social and economic development and resource and ecosystem management.

In addition, recently launched satellites such as Terra, Aqua, GRACE, and IceSAT, among others, will substantially increase the detailed data needed to better understand and model global and regional water cycle processes. The water cycle variables needed from satellite and *in situ* systems and field campaigns are included in the comprehensive list shown in Appendix 12.1. In addition, there are some central water cycle variables that will be featured in water cycle prediction efforts including clouds, precipitation, soil moisture, runoff, evaporation, and infiltration rate.

At the same time, considerable additional effort will be required to extract accurate regional and local climate predictions from global models. Furthermore, effective operational application of many of these new prediction and measurement capabilities is hampered by the lack of adequate networks for observing critical water cycle variables such as soil moisture, and the absence of effective coordination of terrestrial water observing activities.

To address the urgent need for better information on the water cycle, the Climate Change Science Program (CCSP) is planning its Water Cycle research program around two overarching questions:

- How does water cycle variability and change caused by internal processes, climate feedbacks, and human activities influence the distribution of water within the Earth system, and to what extent is this variability and change predictable?
- What are the potential consequences of global water cycle variability and change for society and the environment, and how can knowledge of this variability and change improve decisions dependent on the water cycle?

The following five questions address different aspects of these overarching questions. The first overarching question is dealt with in questions 5.1 to 5.3. Questions 5.4 and 5.5 relate to the second question. Further clarification of the science emphasis planned for each of the five areas is provided by the illustrative science questions. Linkages between the Water Cycle element and other CCSP elements are noted in parentheses after each illustrative question.

> **Question 5.1**: What are the mechanisms and processes responsible for the maintenance and variability of the water cycle; are the characteristics of the cycle changing and, if so, to what extent are human activities responsible for those changes?

State of Knowledge

The global water cycle encompasses the distribution and movement of water in its three phases throughout the Earth system and includes precipitation, surface and subsurface runoff, oceans, cloud cover, atmospheric water vapor, soil moisture, groundwater, and so on. The *Third Assessment Report* of the Intergovernmental Panel on

Climate Change (IPCC, 2001a) cites evidence of possibly significant changes in critical water cycle and related climate variables during the 20th century. These include a $0.6 \pm 0.2°C$ increase in global mean surface temperature, a 7-12% increase in continental precipitation over much of the Northern Hemisphere, massive retreats of most mountain glaciers, and later autumn freeze-up and earlier spring break-up dates for ice cover on many Northern Hemisphere lakes. Less certain, but potentially important possible changes, include a 2% increase in total cloud cover over many mid- to high-latitude land areas, increases in total area affected globally by combined extreme events including droughts and floods, and a 20% increase in the amount of water vapor in the lower stratosphere. Other studies suggest other conclusions, indicating that the question of significant changes may still be open for some water cycle variables.

Because there is a substantial range of natural variability in the climate system due to internal processes alone, it is difficult to distinguish natural excursions from the "norm" from changes that might be the result of forcing due to human activities, such as land-use change and aerosols (see Figure 5-3). The distribution and nature of atmospheric aerosols have an effect on both cloud radiative properties and the generation of precipitation. Moreover, the impact of increased upper tropospheric/lower stratospheric water vapor on the radiative balance and cloud structure is potentially quite large. These effects are not yet well enough understood to accurately project their impact on water and energy cycles or to predict the effects of climate change on regional water resources. Although significant advances have been made in modeling moderately sized watersheds, current global climate models cannot properly simulate many aspects of the global water cycle, such as precipitation amounts, frequency, and diurnal cycle, as well as cloud distribution and its influence on climate. Without appropriate models to conduct tests, it is difficult to attribute observed trends to human-induced climate changes or natural variability.

Illustrative Research Questions

- How have the characteristics of the water cycle changed in recent years, and to what extent are the changes attributable to natural variability and human-induced causes (Chapter 4)?
- What are the key mechanisms and processes responsible for maintaining the global water cycle and its variability over space and time scales relevant for climate (Chapter 4)?
- How are regional groundwater recharge, soil moisture, and runoff affected by changing global precipitation and vegetation patterns, and cryospheric processes (Chapter 8)?
- How have changes in land use and water management and agricultural practices affected trends in regional and global water cycles (Chapter 6)?
- How do aerosols, their chemical composition, and distribution affect cloud formation and precipitation processes, patterns, and trends (Chapter 3)?
- With what accuracy can local and global water and energy budgets be closed?
- What is the relative importance of local and remote factors in extreme hydrologic events such as droughts and floods (Chapter 4)?
- What are the characteristics of upper tropospheric/lower stratospheric water vapor and clouds and how are they affected by deep convection (Chapter 4)?

Figure 5-2: The TRMM Space-Based Observatory—a joint project of NASA and Japan's National Space Development Agency (NASDA)—monitors global tropical precipitation, sea surface temperature, hurricane structure, and other key aspects of the global water cycle. These particular renderings are visualizations of the vertical structure of a typical hurricane created using data from TRMM's Precipitation Radar (Hurricane Bonnie, August/September 1998). Source: X. Shiraz and Y. Morales, NASA Scientific Visualization Studio.

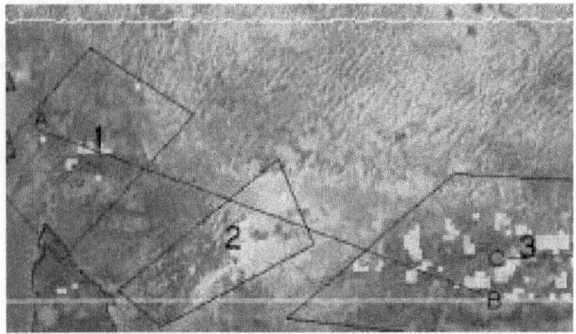

Figure 5-3: Observational evidence showing the effect of aerosols on cloud formation and precipitation processes over south Australia. The yellow patches within Area 2 have reduced droplet sizes, indicating the presence of pollution. White patches outside Area 2 indicate rainfall. Although measurements indicated ample water in the polluted clouds, the smaller droplet size may have prevented precipitation. Source: NASA GSFC (research results from Daniel Rosenfield, The Hebrew University of Jerusalem). For more information, see Annex C.

Research Needs

Although techniques for measuring water cycle variables have improved, the number of observations is limited and, in some cases, new sensors are needed. New satellite and *in situ* observing capabilities will be critical for detecting patterns and quantifying fluxes, especially in terrestrial variables such as soil moisture, and atmosphere-ocean fluxes. Existing *in situ* networks need to be maintained and enhanced, particularly those monitoring precipitation, river discharge, and snow pack. Data sets should be developed using historical and new observations to ensure consistency in the record and for sufficiently long time periods to assess climate variability. Network enhancements and open data access are needed to address water-related issues especially in areas that are currently underrepresented. Also needed are new data assimilation techniques to produce consistent data products for research and process studies from inhomogeneous and/or disparate observations. Appropriate paleoclimate data sets must be assembled to provide a long-term perspective on water cycle variability. New models are needed that can simulate critical water processes at resolutions that allow comparison with long-term data sets. Finally, a wide range of process studies must be conducted to provide understanding of the mechanisms that maintain the water cycle system.

Milestones, Products, and Payoffs

- In collaboration with the Observing and Monitoring research element, development of an integrated global observing strategy for water cycle variables, employing Observational System Simulation Experiments (OSSE) as appropriate [less than 2 years].
- Documentation of trends in key water cycle variables through data analysis and comparisons with model simulations to assess the mechanisms responsible for these trends [beyond 4 years].
- Planned satellite measurements and focused field studies to better characterize water vapor in the climate-critical area of the tropical tropopause (the boundary between the troposphere and the stratosphere) [2-4 years].
- Regional and global precipitation products that merge measurements from different satellite and other remote-sensing

data streams to support joint Water Cycle/Climate Variability and Change studies of ocean- and land-atmosphere coupling and the global energy balance [2-4 years].
- Integrated long-term global and regional data sets of critical water cycle variables such as evapotranspiration, soil moisture, groundwater, clouds, etc., from satellite and *in situ* observations for monitoring climate trends and early detection of potential climate change [2-4 years]. The Ecosystems, Carbon Cycle, and other CCSP research elements will use these data sets as inputs for their analyses and model development studies.
- Results from process studies related to the indirect effects of aerosols on clouds will be available for future assessments of climate sensitivity to aerosols [2-4 years].
- Improved regional water cycle process parameterizations based on process studies conducted over regional test beds to improve the reliability of climate change projections [beyond 4 years].
- Development of analyses for a *State of the Water Cycle* evaluation [beyond 4 years].

Question 5.2: How do feedback processes control the interactions between the global water cycle and other parts of the climate system (e.g., carbon cycle, energy), and how are these feedbacks changing over time?

State of Knowledge

Feedback processes operating between the global water cycle and other components of the Earth/climate system represent the response to external forcing, such as increases in atmospheric carbon dioxide (CO_2). For example, results from climate models suggest there will be an increase in water vapor as the climate warms. Water vapor is the dominant greenhouse gas in the atmosphere; therefore, an increase would result in a strong positive feedback on temperature. Clouds strongly influence the energy budget because of their impact on the radiative balance, but the net cloud-radiation feedback is uncertain. Quantifying the water vapor-cloud-radiation feedback is key to understanding climate sensitivity and the factors governing climate change.

Because the physical processes responsible for the vertical transport of water vapor, cloud formation, cloud-radiation interactions, and precipitation occur at scales that currently are not resolved by climate models, they are parameterized. Although progress has been made in developing and applying high-resolution cloud-resolving models, to date the benefits of these developments for parameterizing three-dimensional cloud distributions in climate models have not been fully realized.

Climate model results also indicate that temperature increases will be amplified in the Arctic due to feedbacks involving permafrost, snow, and ice cover. Should these amplified increases occur, melting continental snow and ice may result in changes in northern river runoff and ocean salinity, while thawing permafrost may lead to increased releases of methane (a greenhouse gas) to the atmosphere. Given the same greenhouse gas increases, individual climate models produce different rates of warming and drastically different patterns of circulation, precipitation, and soil moisture depending on how feedback processes are represented in the models. Basic understanding of feedback processes must be improved and incorporated into models.

Illustrative Research Questions

*   What is the sign and magnitude of the current water vapor-cloud-radiation-climate feedback effect (Chapter 4)?
*   How do changes in water vapor and water vapor gradients, from the stratosphere to the surface, affect radiation fluxes, surface radiation budgets, cloud formation and distribution, and precipitation patterns, globally and regionally (Chapter 4)?
*   How do freshwater fluxes to and from the ocean that affect the global ocean circulation and climate (precipitation, river discharge, sea-ice melt, evaporation) vary, and how may they be changing (Chapter 4)?
*   How do changes in global and regional water cycles interact with evapotranspiration, vegetation and the carbon cycle and vice versa (Chapters 7 and 8)?
*   What are the interactions between land surface changes and regional water cycles (Chapter 6)?
*   How might an intensification of the hydrological cycle, warming in the Arctic, and melting permafrost affect the production of methane and nitrous oxide (Chapters 4 and 7)?

Research Needs

Model development can be accelerated by acquiring data from interdisciplinary field studies over regional test beds, such as those shown in Figure 5-4, to provide a better understanding of scaling effects and the best way to include them in parameterizations. New parameterizations of water cycle/climate feedbacks (e.g., cloud-aerosol and land-atmosphere) and sub-grid-scale processes (e.g., clouds, precipitation, evaporation) will have to be developed and validated, and the sensitivity of global climate models to these new parameterizations will have to be evaluated. Research on water and clouds will have to be closely linked to investigations of aerosols. The development and implementation of instrument systems over selected, globally distributed, test beds is essential. The data products must be comprehensive and include groundwater, and they should have sufficient resolution to assess optimal sampling strategies for future observational campaigns and field programs over larger regions. Where appropriate, data and experimental field sites will be shared with the Ecosystems and Carbon Cycle research elements.

Milestones, Products, and Payoffs

*   New observationally tested parameterizations for clouds and precipitation processes for use in climate models based on cloud-resolving models developed in part through field process studies [2-4 years]. This will support Climate Variability and Change research element work on climate feedbacks.
*   Incorporation of water cycle processes, interactions, and feedbacks into an integrated Earth system modeling framework [2-4 years].

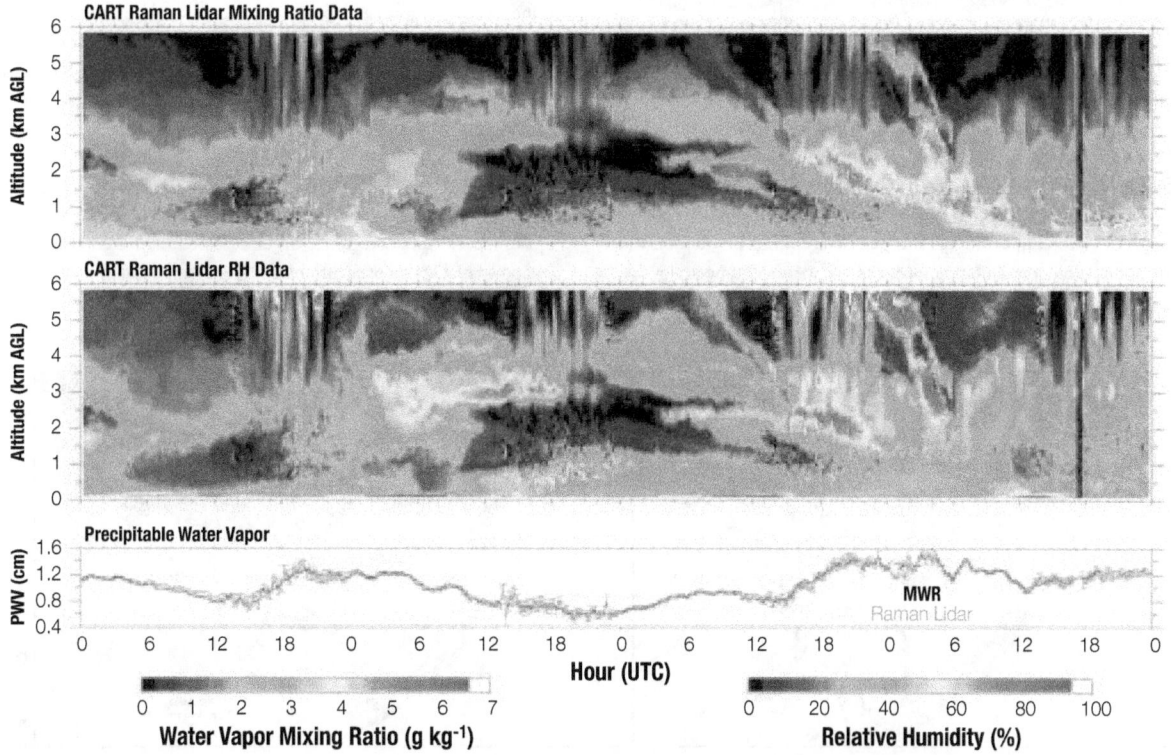

Figure 5-4: Water vapor measurements at the Atmospheric Radiation Measurement Program (ARM) Southern Great Plains (SGP) site from 29 November to 2 December 2002. *Top Panel*: Measured water vapor mixing ratio from the Raman lidar. These charts of the fundamental measured quantity from the lidar give unique information about the vertical and horizontal scales of turbulent fluxes that transport moisture. *Middle Panel*: Relative humidity calculated from the mixing ratio and associated temperature data at the SGP. *Bottom Panel*: Comparison between the integrated water vapor from the Raman lidar and the measured water vapor path from the microwave radiometer. Source: David Turner, University of Wisconsin, DOE ARM Program.

- Quantification of the potential changes in the cryosphere, including the effects of permafrost melting on regional hydrology and the carbon budget, and the consequences for mountain snowpacks, sea ice, and glaciers [beyond 4 years]. These efforts will complement those of the Climate Variability and Change and Carbon Cycle research elements in the area of cryosphere-atmosphere-ocean coupling and feedbacks.
- In collaboration with the Climate Variability and Change research element, sensitivity tests of global models to improve parameterizations of feedbacks and sub-grid-scale processes (land-cover change, land surface processes, precipitation, clouds, etc.) [beyond 4 years].
- Enhanced data sets for feedback studies, including water cycle variables, aerosols, vegetation, and other related feedback variables, generated from a combination of satellite and ground-based data [2-4 years]. These data sets will be critical for most CCSP research elements.
- Together with the Land-Use/Land-Cover Change and Carbon Cycle research elements, improved coupled land-atmosphere models and enhanced capability to assess the consequences of different land-use change scenarios [beyond 4 years].

> **Question 5.3**: What are the key uncertainties in seasonal-to-interannual predictions and long-term projections of water cycle variables, and what improvements are needed in global and regional models to reduce these uncertainties?

State of Knowledge

Improved seasonal predictions of water resource availability and their application can have major economic benefits. For example, in 1999, if the experimental spring runoff forecasts for the Green River had been used, improved water management decisions could have resulted in more efficient use of stored water and yielded more than $3.1 million in additional revenues from power production and irrigation. While precipitation forecasts on "weather" time scales have improved, current global and regional models demonstrate limited skill in predicting precipitation, soil moisture, and runoff on seasonal and longer time scales. Water managers indicate this skill level to be inadequate for their needs.

Seasonal-to-interannual predictability is a function of local and remote influences involving various ocean and land processes. Enhanced predictability can result from persistence of specific phenomena or slowly varying boundary conditions (soil moisture/groundwater, snow/ice, vegetation/land cover, and ocean and land surface temperatures) that persist over periods of weeks, months, or even years. More accurate initial surface fields for prediction models produced by recently developed land data assimilation systems provide a basis for reducing prediction errors. Understanding of the El Niño/La Niña cycle has provided some predictive skill, particularly with respect to seasonal outlooks for floods and droughts (see Figure 5-5); however, the memory effects of land conditions on the atmosphere are not well enough understood. Cloud and precipitation feedbacks and the interactions of the lower boundary layer (lower 500 meters of the atmosphere) with land and ocean surface conditions also are not well understood.

A critical prediction problem involves advance warning for major flood and drought events. The ability to reliably assess whether hydrologic extremes will increase as greenhouse gas concentrations rise is also important. Extreme events arise from a combination of large-scale circulation patterns that enhance atmospheric conditions conducive to flood or drought, regional patterns and feedbacks that accentuate the larger scale factors, and preconditioning of the system to increase the impacts of the flood or the drought event. Understanding the relative roles of remote and local factors in initiating, maintaining, and terminating extreme events will require the Water Cycle and the Climate Variability and Change research elements to work collaboratively on this topic.

Illustrative Research Questions

- How predictable are water cycle variables at different temporal and spatial scales over different regions of the Earth's surface (Chapter 4)?

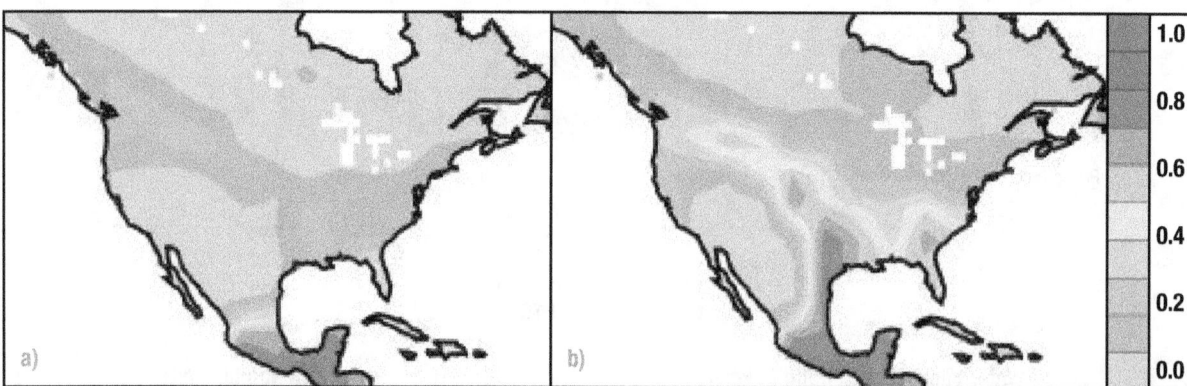

Figure 5-5: Predictability of precipitation in summer (June, July, and August) on seasonal time scales, through analysis of an ensemble of multi-decadal coupled land-atmospheric model simulations. In map (a), values close to 1 indicate areas where precipitation is strongly determined by sea surface temperature (SST) and therefore is predictable when SST is predictable; values close to 0 indicate where foreknowledge of SST may not lead to useful seasonal precipitation predictions. Map (b) shows the same information for SST and land surface moisture state. The addition of land surface information appears to improve predictability. Source: R.D. Koster, M. Suarez, and M. Heiser, NASA. For more information, see Annex C.

- For different model resolutions, how can key water cycle processes be better simulated, in order to enhance the capability of producing more accurate seasonal-to-interannual predictions of water cycle variables (Chapters 4 and 8)?
- How can the representation of water cycle processes in climate models be improved to reduce uncertainties in projections of climate change (Chapter 4)?
- What are the critical hydrological and atmospheric factors present in major flood and drought events that can be isolated, quantified, and incorporated into water cycle prediction methodologies (Chapter 4)?
- What model improvements are needed to assess the changes in seasonality, intensity, and variability of high-latitude freshwater fluxes (precipitation, evapotranspiration, runoff) and stores (soil moisture, snow/ice, permafrost) that may result from climate change, specifically in large basins covering a range of climate regions?
- What processes and model resolutions are needed to improve regional models used to downscale global predictions to local and watershed scales? Are there other downscaling techniques (e.g., statistical approaches) that can be equally effective (Chapter 4)?
- How can the uncertainty in the prediction of water cycle variables be characterized and communicated to water resource managers (Chapter 11)?

Research Needs

Advances in prediction capabilities will depend on improvements in model structure and initialization, data assimilation, and parameter representations. Predictability studies will be required to determine the regions, seasons, lead times, and processes most likely to provide additional predictive skill. Better understanding and improved model representations of less well-understood processes—such as the seasonal and longer term interactions of the atmosphere with vegetation, soils, oceans, and the cryosphere—are needed. The modeling of regional feedbacks leading to extreme events also requires a better understanding of land-atmosphere interactions, while the modeling of antecedent conditions requires hydrologic and biospheric models and monitoring programs that will account for the effects of prolonged rainfall, or lack thereof, in a given region. The goal of better predictions must be achieved by accurate representations of precipitation processes in climate models. The role of mountains in the annual water cycle also needs to be better understood. Data sets are needed for the calibration and validation of global coupled climate models and the development of regional downscaling and statistical forecasting techniques. In addition, model evaluation studies with enhanced data sets are needed to improve models and to characterize and reduce uncertainties.

Milestones, Products, and Payoffs

- Regional reanalysis providing a wide range of high-resolution, daily water cycle analysis products for a 25-year period, for use in analyzing features absent in global climate data assimilation products [less than 2 years].
- Observational data sets to initialize and test boundary layer and other components (including parameterizations) in mesoscale, regional, and global models [2-4 years].

- Results from field and modeling experiments to study the role of mountain environments on precipitation and runoff production [2-4 years].
- New drought monitoring and early warning products based on improved measurements of precipitation, soil moisture, and runoff, and data assimilation techniques to inform drought mitigation planning [2-4 years]. This initiative will be undertaken in collaboration with extreme event activities in the Climate Variability and Change and Ecosystems research elements.
- Metrics for representing the uncertainty in predictions of water cycle variables and measurably improved forecast products for water resource managers [2-4 years].
- In collaboration with the Climate Variability and Change research element, the capability for long-range prediction of drought and flood risks (seasonal-to-interannual time scales) [beyond 4 years]. The Water Cycle research element will focus on the role of terrestrial feedbacks to extreme events while the Climate Variability and Change research element will address the seasonal and longer term changes in remote influences such as sea surface temperatures.
- Downscaling techniques, such as improved regional climate models, that bridge the disparate spatial and temporal scales between global model outputs and atmospheric, land surface, and river basin processes, for improved evaluation of potential water resource impacts arising from climate variability and change [beyond 4 years].
- New space-based systems for measuring global precipitation will be developed and implemented to support the needs of the Climate Variability and Change, Carbon Cycle, and Land-Use/ Land-Cover Change research elements [beyond 4 years].

> **Question 5.4:** What are the consequences over a range of space and time scales of water cycle variability and change for human societies and ecosystems, and how do they interact with the Earth system to affect sediment transport and nutrient and biogeochemical cycles?

State of Knowledge

Variability and changes in the water cycle have been shown to have profound impacts on human societies (including human health) and ecosystems, but many of the linkages between these changes and their outcomes are not yet understood in the detail needed to inform policy and management responses. In addition, the strategies used for water management throughout the last century to adapt to climate variability have had impacts on water availability and water quality that must be identified and evaluated as part of the process of separating climate change effects from other forms of global change arising from factors such as industrialization and population growth. Furthermore, the ability to simulate variations in water availability and quality and their consequences for agriculture, wetlands, energy production and distribution, urban and industrial uses, and inland shipping, among others, should be further developed and integrated into a common modeling framework.

Many of the impacts of water variability arise because of its effects on sediment loadings and the transport of nutrients and sediments. As water cycles through the environment, it interacts strongly with other biogeochemical cycles—notably carbon, nitrogen, and other nutrients. Flowing water also erodes, transports, and deposits sediments in rivers, lakes, and oceans, altering water quality and affecting agricultural production and ecosystem functioning, among other socially relevant impacts. Yet our ability to quantify the role of flowing water as the primary agent for sediment transport that reshapes the Earth's surface, and for nutrient transport that feeds riparian (relating to rivers) habitats and degrades water bodies, is inadequate. Currently, we do not have the monitoring framework needed to generate a database to support research on these processes. The priority challenges are to quantify water flow and the various transport rates, biochemical transformations, and constituent concentrations and feedbacks whereby the water cycle alters media and ecosystems.

Illustrative Research Questions

- How does the water cycle interact through physical, chemical, biophysical, and microbiological processes with other Earth system components at the watershed scale (generally 50 to 20,000 km²) (Chapters 6, 7, and 8)?
- How do changes in climate, land cover, and non-point waste discharges alter water availability, river flows, water quality, and the transport of sediments, nutrients, and other chemicals, and how do these changes affect human and ecosystem health (Chapters 6 and 8)?
- How do surface and subsurface processes change the quantity and quality of water available for human and environmental uses (Chapters 6 and 8)?
- How might an intensification of the hydrological cycle enhance soil erosion and result in soil degradation, especially losses in soil carbon (Chapters 6 and 7)?
- How would ongoing systematic depletion of groundwater resources be influenced by climate change and how is this process affected by sea-level rise (Chapters 4, 6, and 8)?
- How might ongoing and potential future changes in hydrologic regimes, such as earlier melting of the snowpack and lake and river ice, affect the amount and timing of spring, summer, and fall flows, surface water temperatures, and their impacts on aquatic biota (Chapters 4 and 8)?
- How do variability and change in the water cycle affect riparian and estuarine environments in the United States (Chapters 6 and 8)?
- How might water cycle variability and change affect the spread of vector-mediated disease (Chapters 4, 6, and 8)?

Research Needs

Overall, there is a basic need to develop an integrated research vision (complete with hypotheses) for addressing multiple-process (hydrological, physical, chemical, and ecological) interactions between water and other Earth systems. Techniques that scale up processes active at watershed and sub-watershed scales to larger scales must be developed and tested. In addition, it is necessary to refine geophysical methods and the use of tracers, including isotopes, to determine subsurface paths, flow rates, and residence times, and to track pollution plumes. Experimental watersheds are needed to develop an understanding of these processes. Information on trends

in land use and land cover will be needed to assess consequences for water supply. It also will be essential to work closely with social and ecosystem scientists to develop understanding of impacts of water cycle variability, and social and biological responses and feedbacks arising from that variability.

Milestones, Products, and Payoffs

- Reliable, commensurate data sets at the watershed scale that scientists from various disciplines will use to examine critical water-Earth interactions for improved integrated watershed management [2-4 years].
- Development and application of better methods for monitoring subsurface waters for inventorying current and future water availability, including tracking aquifer storage and how it responds to water withdrawals and climate change [beyond 4 years].
- Development of data sets and analyses of spatial and temporal trends in stream flow and water quality (including sediments and chemicals) for a range of watershed conditions including watersheds with minimal effects from human activities, agricultural watersheds, sensitive ecosystems, and large river basins [2-4 years]. These activities will address the needs of the Ecosystems research element for baseline water cycle information.
- Evaluations of the effects of water cycle variability and change on trends in water-quality conditions [2-4 years]. Information from these evaluations is needed for studies by the Ecosystems research element.
- Improved modeling and remote-sensing methods for scaling up from individual pathways and mixing zones to the scale and complexity of watershed systems [beyond 4 years].
- Models that partition precipitation among surface and subsurface pathways, route flows, and quantify physical and chemical interactions for evaluating climate and pollution impacts [beyond 4 years].
- Contributions of data and advice to Human Contributions and Response studies, including use of improved epidemiological models, to examine the potential for major water-mediated infectious diseases [2-4 years].
- Scientific capability to assess the climate-related consequences of water use trends and water management practices [beyond 4 years]. This capability is needed to evaluate the effects of water management on the health of ecosystems, and together with the Climate Variability and Change research element to examine the combined effects and relative contributions of groundwater pumping and climate change on sea-level rise and coastal flooding.

> **Question 5.5**: How can global water cycle information be used to inform decision processes in the context of changing water resource conditions and policies?

State of Knowledge

Results of recent research on the water cycle can contribute to the capacities of decisionmakers in such sectors as water management, agriculture, urban planning, disaster management, energy, and transportation. Improved understanding of water cycle variability and trends at regional and watershed scales appears to benefit decisionmakers dealing with climate-sensitive issues (see Figure 5-6). Whether the water cycle is changing as a result of human activities or natural low-frequency variation, human societies will have to adapt, and focused scientific information must be provided to support their choices. The growing economic and social costs of extreme events indicate that there is need for improved responses to these disruptions, whether or not their frequency and intensity are changing. Currently, climate forecasts are often temporally or spatially too coarse to be of use for many water-dependent decision processes. In addition, factors such as regulatory inflexibility, institutional structures, and time pressures make it difficult to change established management and decision systems. Interactions between decisionmakers and research scientists are needed to make mutual adjustments as appropriate to match scientific information with decision processes. Efforts to eliminate the barriers between researchers and research users have been initiated and indicate that early collaboration and side-by-side demonstrations may be effective tools for speeding innovation.

Illustrative Research Questions

- How could climate variability and change potentially alter the effectiveness of current and future water management practices and their feedbacks on the climate (Chapter 9)?
- What are the consequences of demographic and land-use trends on groundwater and surface water supplies domestically and globally (Chapters 6 and 9)?
- What are effective means for transferring stochastic water cycle research products (e.g., hydro-climatological predictions, projections, and associated uncertainties) into the management,

planning, and design of water decision systems and infrastructure (Chapters 9 and 11)?
- How can advance knowledge of water availability be used to manage conflicting demands on domestic and transboundary water resources for water consumption, ecological functions, industrial uses, and transport (Chapter 9)?
- How can changes in the quality and quantity of water flowing within riparian and coastal environments arising from land management and policy decisions affect the provision of ecosystem services (Chapters 8 and 9)?
- How have water consumption patterns and trends changed as a result of major climatic events, technological innovation, and economic conditions, and how are they likely to change as a result of projected changes in the same factors (Chapters 9 and 11)?
- What are the trends in agricultural, industrial, municipal, and instream water uses, and what implications do they have for adaptation strategies (Chapters 6, 8, and 9)?
- What are the economic implications of strategies for the application of improved capabilities to predict seasonal water cycle variability and change (Chapters 9 and 11)?

Research Needs

For scientific information to have an impact, it will have to rely on refined and extended research on the role, entry points, and types of water cycle knowledge required for water management and policy decisionmaking processes. In order to make rapid progress it will be necessary to integrate data from a broad range of sources and disciplines. A basic requirement for achieving this goal is the

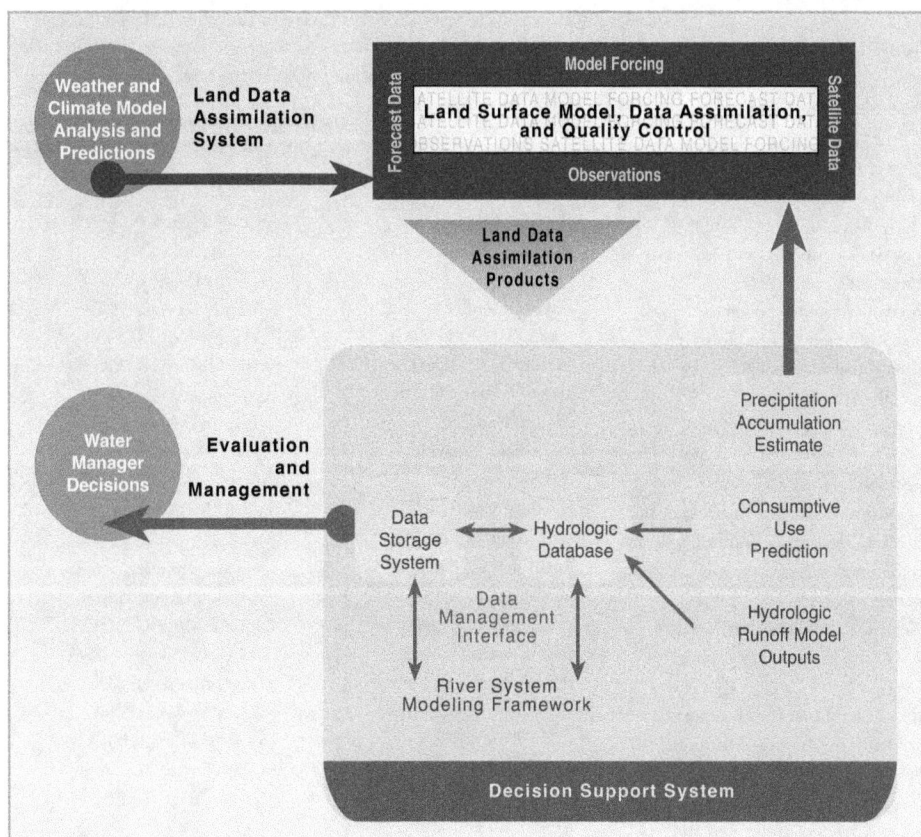

Figure 5-6: Schematic of the flow of data and information between land surface models, numerical forecasting products, and decision support systems used by water managers to operate reservoirs and river systems. Improved forecasts and data assimilation systems from water cycle research will lead to more equitable and sustainable management of precious water resources. Improved forecasts will lead to enhanced river system management, and increased water storage and efficient hydropower generation, while preserving flood control space.

development of frameworks for integrating the natural and social science information necessary for multiple-objective decisionmaking. Inputs should include research in remote sensing, uncertainty of predictions, data management for decision support, risk management, economic impact assessments, and water and environmental law, among many others.

An ability to assess the consequences of both historic and potential future water development paths is needed for assessing trends and variability in water resources. In addition, to determine patterns and trends, it will be necessary to inventory existing data sources and regional and sectoral studies, especially for data for which regional, national, and global repositories are rare or non-existent (e.g., water demand, diversion, use, and consumption). An integrated data system for collection, storage, and retrieval of these data would enhance national capacities for evaluating policy and decision options.

Milestones, Products, and Payoffs

A number of the following milestones, products, and payoffs will be developed in collaboration with the Decision Support Resources element.

- An experimental surface and subsurface moisture monitoring product for land resource management (e.g., fires or agriculture) [less than 2 years].
- An experimental online decision support tool designed to provide users with streamflow conditions and their accompanying probabilities in the Pacific Northwest arising from near- and far-term future climate predictions and projections [less than 2 years].
- Refinements of web-based tools that improve the communication and usability of climate/water forecasts [less than 2 years]. These developments will contribute to Human Contributions and Responses research element analyses of techniques for communicating and disseminating climate forecasts.
- Transfer to operations of capabilities to produce improved stochastic operational streamflow forecasts over a range of spatial and temporal scales (days, weeks, months, and seasons) to support water management decisions [2-4 years].
- Development of an ensemble forecast system to predict snowpack, streamflow, and reservoir storage over the western United States, based on information about global climate teleconnections, for lead times ranging from several weeks to 1 year for operational purposes, and from decades to a century for water resource planning [2-4 years].
- Demonstration of system simulation and forecasts using advanced watershed and river system management models and decision support systems, to facilitate acceptance and utilization of these advanced technologies for improved hydropower production and river system management [2-4 years].
- Assessment reports on the status and trends of water flows, water uses, and storage changes for use in analyses of water availability [2-4 years].
- Integrated models of total water consumption for incorporation into decision support tools that identify water-scarce regions and efficient water use strategies [beyond 4 years]. The Human Contributions and Responses research element will supply inputs on the effects of climate on human drivers of water demand.

National and International Partnerships

As indicated throughout this chapter, the Water Cycle research element will have strong ties to many of the other components of the CCSP. It will also contribute to water initiatives that may be forthcoming from the newly formed Subcommittee on Water Availability and Quality (SWAQ) and the hydrologic aspects of the U.S. Weather Research Program.

The Water Cycle research element has strong links to the World Climate Research Programme (WCRP) and its Global Energy and Water Cycle Experiment (GEWEX). In particular, the activities outlined in this chapter will use data sets from the Coordinated Enhanced Observing Period (CEOP) in model development. The Water Resources Applications Project (WRAP) will provide a framework for studies to assess the benefits of improved hydrological forecasts in decisionmaking. The Earth System Science Partnership's Global Water System Project will be a partner for studies of the feedbacks of water management practices and infrastructure to regional and global climate. Water cycle observational issues will form the basis for activities under the water cycle theme of the International Global Observing Strategy (IGOS) Partnership and related observational activities, namely the Global Climate Observing System (GCOS), the Global Ocean Observing System (GOOS), and the Global Terrestrial Observing System (GTOS). Improved coordination of terrestrial observations is a critical need of the Water Cycle research element that will be addressed at the international level in the IGOS Water Cycle theme report. The Water Cycle research element also has developed strong linkages with the Hydrology and Water Resources Programme of the World Meteorological Organization (WMO), the International Hydrology Programme of the United Nations Educational, Scientific and Cultural Organization (UNESCO), and the joint UNESCO/WMO Hydrology for Environment, Life, and Policy (HELP) initiative. Water Cycle efforts also will contribute to observational programs and research developed through bilateral treaties, particularly with countries such as Japan that have placed a priority on water cycle research, and with Canada through the International Joint Commission, the Great Lakes Commission, and the Great Lakes Environmental Research Laboratory.

CHAPTER 5 AUTHORS

Lead Authors

Rick Lawford, NOAA
Jared Entin, NASA
Susanna Eden, CCSPO
Wanda Ferrell, DOE
Harvey Hill, NOAA
Jin Huang, NOAA
L. Douglas James, NSF
William H. Kirby, USGS
David Matthews, DOI
Pamela L. Stephens, NSF
Sushel Unninayar, NASA

Contributors

Mike Dettinger, USGS
John Furlow, USEPA
David C. Goodrich, USDA
David Mathis USACE
Bryce Stokes, USDA
Mark A. Weltz, USDA
Jon Werner, USDA
Jordan West, USEPA

6 Land-Use/Land-Cover Change

CHAPTER

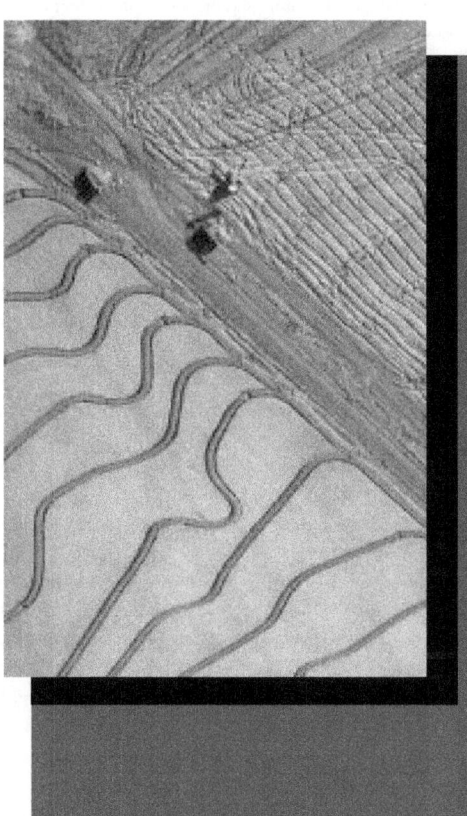

CHAPTER CONTENTS

Question 6.1: What tools or methods are needed to better characterize historic and current land-use and land-cover attributes and dynamics?

Question 6.2: What are the primary drivers of land-use and land-cover change?

Question 6.3: What will land-use and land-cover patterns and characteristics be 5 to 50 years into the future?

Question 6.4: How do climate variability and change affect land use and land cover, and what are the potential feedbacks of changes in land use and land cover to climate?

Question 6.5: What are the environmental, social, economic, and human health consequences of current and potential land-use and land-cover change over the next 5 to 50 years?

National and International Partnerships

Land-use and land-cover are linked to climate and weather in complex ways. Key links between changes in land cover and climate include the exchange of greenhouse gases (such as water vapor, carbon dioxide, methane, and nitrous oxide) between the land surface and the atmosphere, the radiation (both solar and longwave) balance of the land surface, the exchange of sensible heat between the land surface and the atmosphere, and the roughness of the land surface and its uptake of momentum from the atmosphere. Because of these strong links between land cover and climate, changes in land use and land cover can be important contributors to climate change and variability. Moreover, reconstructions of past land-cover changes and projections of possible future land-cover changes are needed to understand past climate changes and to project possible future climate changes; land-cover characteristics are important inputs to climate models. In addition, changes in land use and land cover, especially when coupled with climate variability and change, are likely to affect ecosystems and the many important goods and services that they provide to society. The National Research Council recently identified land-use dynamics as one of the grand challenges for environmental research (NRC, 2001b).

Determining the effects of land-use and land-cover change on the Earth system depends on an understanding of past land-use practices, current land-use and land-cover patterns, and projections of future land use and cover, as affected by human institutions, population size and distribution, economic development, technology, and other factors. The combination of climate and land-use change may have profound effects on the habitability of Earth in more significant ways than either acting alone. While land-use change is often a driver of environmental and climatic changes, a changing climate can in turn affect land use and land cover. Climate variability alters land-use practices differently in different parts of the world, highlighting differences in regional and national vulnerability and resilience.

The interaction between land use and climate variability and change is poorly understood and will require the development of new models linking the geophysics of climate with the socioeconomic drivers of land use. Providing a scientific understanding of the process of land-use change, the impacts of different land-use decisions, and the ways that decisions are affected by a changing climate and increasing climate variability are priority areas for research.

In addition to being a driver of Earth system processes affecting the climate, carbon cycle, and ecosystems, land-use and land-cover change is a global change in its own right, requiring its own research

Figure 6-1: Land use: strip cropping and woodlots in Leelanau County, Michigan. Source: USDA NRCS (photo by Lynn Betts, 2001).

---

foundation. Key issues to be addressed by this research element include the spatial and temporal dynamics of land-use change, the role of fragmentation and degradation, the role of multiple drivers, the role of institutions, and the interactions among drivers and types of land-use change.

This research element provides the scientific underpinning for land-use decisionmaking and projections of future land use, and has substantial benefits beyond climate change assessment and mitigation by supporting a wide array of issues important to users of this information. To meet multiple objectives, the land-use and land-cover change research element will address two overarching questions:

- What processes determine the temporal and spatial distributions of land cover and land use at local, regional, and global scales, and how and how well can land use and land cover be projected over time scales of 5-50 years?

- How may changes in land use, management, and cover affect local, regional, and global environmental and socioeconomic conditions, including economic welfare and human health, taking into consideration socioeconomic factors and potential technological change?

To address these overarching questions, a focused, integrated research agenda is required that includes process studies, systematic observations, modeling and prediction, retrospective studies, research on impacts, and regional science networks and assessments. In addition, research collaboration with other program elements will be necessary to gain detailed understanding of the direct impacts of land-use and land-cover change on climate, as well as the combined effects of land use and climate change on ecosystems, water, and carbon cycles. Answers to the overarching questions will require research focused on the five specific questions posed in this chapter.

Question 6.1: What tools or methods are needed to better characterize historic and current land use and land cover attributes and dynamics?

State of Knowledge

During the previous decade, significant progress was made in planning and launching satellites with instruments suited for Earth observation. In addition, a number of national- to global-scale experimental land-cover databases were developed that led to increased use of land cover in climate and carbon cycle models. Methodological advancements were also made. As a result, there is an improved capability for and strong reliance on remote-sensing and land-cover databases for multi-scale environmental studies. The research and development associated with this question involves the continuation of, and improvements in, data collection systems and data products. This will provide new information leading to regular updates of land-cover databases at scales relevant for issues ranging from local resource management to global-scale analyses.

While remote sensing provides quick and comparatively inexpensive information about land-cover changes over large areas, land-cover database improvements will require integration of data from ground-based networks (see Figure 6-2). These networks offer a wealth of historical data (often with data records extending back 50-100 years), and can provide detailed site information (e.g., primary production, species composition, habitat quality, wildlife population statistics, soil type, tillage and crop rotation history, and land-use classifications). Integrating ground-based and remote-sensing data collection systems provides an opportunity to vastly improve the speed and quality of land-use and land-cover data for use in applied research. Much of our understanding of land-use and land-cover change has built up from individual case studies, using both remote sensing and ground-based data, and we will continue to rely on case studies as a means to gain required knowledge.

Illustrative Research Questions

- What improvements need to be made to current observing systems and what programs need to be put in place to provide the necessary long-term data and information to support the study of land-use and land-cover change at the global, regional, and national scales?

- What are the methodological advances needed to improve land-use and land-cover change analyses, including strategies for integrating ground-based data, socioeconomic statistics (e.g., census information), and remotely sensed measurements?

- What are the historical and current patterns and attributes of land use and land cover at national to global scales that affect the carbon cycle, atmospheric processes, and ecosystem structure and function?

- What are the national and global rates, patterns, and characteristics of contemporary land-use and land-cover change?

- Where are the current areas of rapid land-use and land-cover change at national and global scales?

Research Needs

Evolving public and private land management questions call for new data and information and improved scientific bases for decisionmaking. They also require long-term continuity in data collection, and the acquisition of data at the global scale. Coordination and prioritization of the specific land-use and land-cover change data requirements with the Observing and Monitoring research element is imperative. Methods and procedures now exist to make routine global observations of land cover, but there is currently no operational program. With the current suite of satellite sensing systems and archived data sets available to the research community, studies at the large spatial scales needed to depict land-cover and management changes can begin.

While considerable progress has been made in mapping land-cover characteristics, the ability to accurately map the wide range of landscape attributes, including land use and biomass, will require a considerable research effort. Coordination of existing *in situ* data collection efforts, and the retrieval, analysis, and integration of historical data are needed to extend the usefulness of existing data and to enrich and standardize data attributes in future data collection and use. In addition, improvements in remotely sensed data quality and in automated algorithms for detection of local land-cover changes and their characteristics are needed. Data integration will be particularly important so that *in situ*, remotely sensed, and other forms of data can be merged and used to derive the needed land-use and land-cover information. As scientific demands and needs for

land-use and land-cover information change, parallel innovation in the resulting data products will be essential.

Milestones, Products, and Payoffs

- National land-cover database for the United States that includes attributes of land cover and vegetation canopy characteristics [less than 2 years]. This product is a required model input needed by the Carbon Cycle, Water Cycle, and Ecosystems research elements.
- Continued acquisition of global calibrated coarse-, moderate-, and high-resolution remotely sensed data [ongoing].
- Global moderate-resolution land-cover database with attributes required for parameterization of climate, carbon cycle, and ecosystem models [less than 2 years]. Parameter specifications must be coordinated with the Climate Variability and Change research element.
- Global maps of areas of rapid land-use and land-cover change and location and extent of fires [less than 2 years].
- Quantification of rates of regional, national, and global land-use and land-cover change [regional, less than 2 years; national, 2-4 years; global, beyond 4 years]. This product is a required model input needed by the Carbon Cycle, Water Cycle, and Ecosystems research elements.
- Integrated land-use and land-cover change detection strategies and operational prototypes of detection systems that enable accurate and cost-effective detection of local to global changes [less than 2 years].
- Global high-resolution satellite remotely sensed data and land-cover databases with attributes required for national to global scale applications [beyond 4 years].
- Global and national land-use history maps spanning the period from the industrial revolution to the present (300 years). This product is required by the Carbon Cycle and Ecosystems research elements [beyond 4 years].
- Operational global monitoring of land-use and land-cover conditions [beyond 4 years]. This product will be used by several other research elements including Carbon Cycle, Ecosystems, and Climate Variability and Change.
- Analysis of the global occurrence, extent, and impact of major disturbances (e.g., fire, insects, drought, and flooding) on land use and land cover [beyond 4 years]. Input on the frequency of drought and flooding will be sought from the Climate Variability and Change research element.

Question 6.2: What are the primary drivers of land-use and land-cover change?

State of Knowledge

The ability to forecast land-use and land-cover change and, ultimately, to predict the consequences of change, will depend on our ability to understand the past, current, and future drivers of land-use and land-cover change. These factors as well as other emerging social and political factors may have significant effects on future land use and cover. Patterns of land use, land-cover change, and land management are shaped by the interaction of economic, environmental, social, political, and technological forces on local to global scales (see, for example, Figure 6-3).

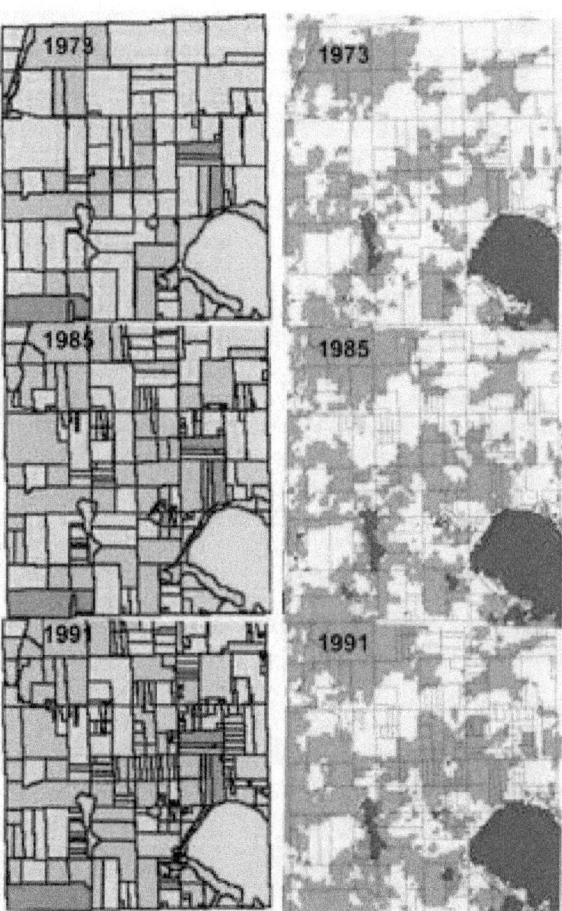

Figure 6-2: Forest cover increase and abandonment of marginal agricultural lands in Grand Traverse County, Michigan. The left image illustrates land-use change by parcel interpreted from aerial photographs. Green colors are forest, beige/yellow agriculture, and pink residential development. The right image is forest cover from Landsat satellite images. Green is forest and light yellow is not forest. Source: School of Natural Resources and Environment, University of Michigan. For more information, see Annex C.

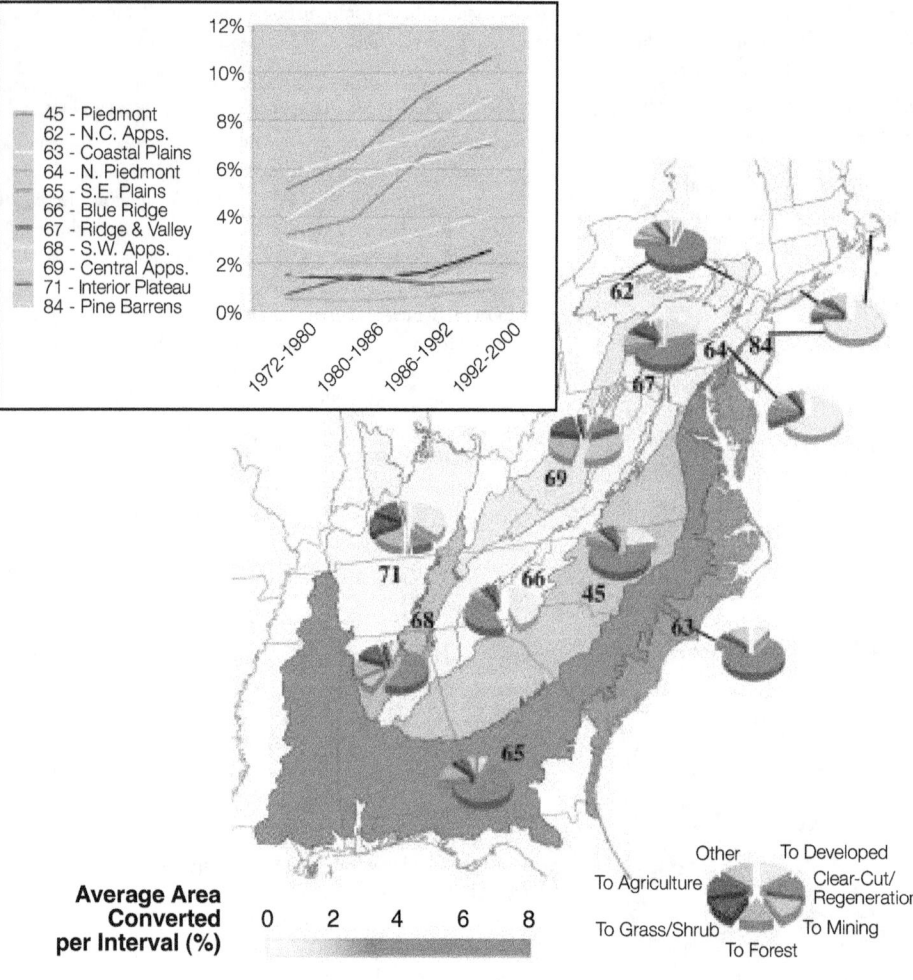

Figure 6-3: Land-cover change in eastern U.S. ecosystems, 1973-2000. An analysis of land-use and land-cover change in the eastern United States provides evidence of distinctive regional variation in the rates and characteristics of changes. The color of each ecoregion indicates the rate of change, while the pie charts indicate the type of change. In some areas, rapid, cyclic harvesting and replanting of forests was the main cause of change, while urbanization dominated in other areas. Source: USGS EROS Data Center. For more information, see Annex C.

An improved understanding of historical land-use and land-cover patterns provides a means to evaluate complex causes and responses in order to better project future trends of human activities and land-use and land-cover change. We must understand the primary modern and future drivers of land use and their interrelationship with land management decisions and resource policies to develop projections of future land-use and management decision outcomes under a range of economic, environmental, and social scenarios. This ability will allow better projections and hopefully minimize negative impacts, especially as related to climate change. This type of analysis will require the integration of various disciplines from the physical and social sciences.

Illustrative Research Questions

- What are the primary historic and contemporary natural and human drivers of land-use and land-cover change and what will they be in the future?
- What has been the historical relationship between land use and land management systems and how will the relationship change through the next few decades?
- What are the most significant drivers of land-use and land-cover change?
- How have and will the driving forces of change affect(ed) land use and cover at different scales (i.e., local, regional, and global)

and where are there opportunities for managing land-use change to minimize negative impacts and maximize positive outcomes?
- How, and to what extent, do extreme events (e.g., natural disasters, public health emergencies, and war) affect land-use and land-cover change and vice versa?
- How will environmental, institutional, political, technological, demographic, and economic processes determine the temporal and spatial distribution of land use and land cover over the next few decades?

Research Needs

An innovative approach is needed to quantify, understand, model, and project natural and human drivers of land-use and land-cover change. Research is needed to understand and project the interactions of economic, social, and environmental choices on land use and management policies and decisions. New techniques and tools that integrate understanding of human behavior, opportunities, consequences, and alternatives are needed for improved decisionmaking and policymaking. There must be close collaboration with the Human Contributions and Responses research element to understand the social and economic drivers that affect human choices.

Improvements are needed in process models of land-use and land-cover change dynamics in space and time, combining field-level case studies for analysis of processes and management systems, statistical studies for large regions, and empirical analyses using remote sensing at local scales. This process-level understanding of land-use and land-cover dynamics and interactions with socioeconomic and biophysical factors will aid the analysis of land-use and land-cover change across scales. Work is needed to understand the interactions between agents or causes of land-use change and climate change and variability.

Milestones, Products, and Payoffs

* Summary of the historical and contemporary regional driving forces of land-use and land-cover change [United States, less than 2 years; global, beyond 4 years]. This product is needed by the Ecosystems research element.
* Understand how primary drivers of land use and land management decisions are likely to change over the next few decades [United States, 2-4 years; global beyond 4 years].
* Quantify and project possible drivers of land-use change for a range of economic, environmental, and social values [beyond 4 years]. This product is needed by the Ecosystems and Carbon Cycle research elements.

> **Question 6.3**: What will land-use and land-cover patterns and characteristics be 5 to 50 years into the future?

State of Knowledge

To understand the historical, contemporary, and future linkages between land-use and land-cover change and its resulting effects on

biogeochemical cycles, climate, ecosystem health, and other systems, it will be necessary to make significant advances in documenting the rates and causes of land-use and land-cover change. Our current understanding of historic land-use and land-cover change is weak due to the anecdotal or very local nature of past research in this area. Future understanding of land-use and land-cover changes will be greatly improved through new systematic methods and study designs for land-use change research. To understand the forces of change that operate at different scales, it will be necessary to conduct studies that explicitly reveal the local and regional variations in land-use and land-cover changes. With this, the historical and contemporary data needed to develop models that project land use and land cover for specific intervals into the future will be produced.

Illustrative Research Questions

* What are the major feedbacks and interactions between climatic, socioeconomic, and ecological influences on changes in land use and land management?
* What spatial and temporal level of information and modeling are needed to project land use and land management and its impacts on the Earth system at regional, national, and global scales?
* Given specific climate, demographic, and socioeconomic projections, what is the current level of skill and what are the key sources of uncertainty and major sensitivities in projecting characteristics of land-use and land-cover change 5 to 50 years into the future?

Research Needs

A new suite of models that combine climatic, socioeconomic, and ecological data to model projected changes at scales that are relevant

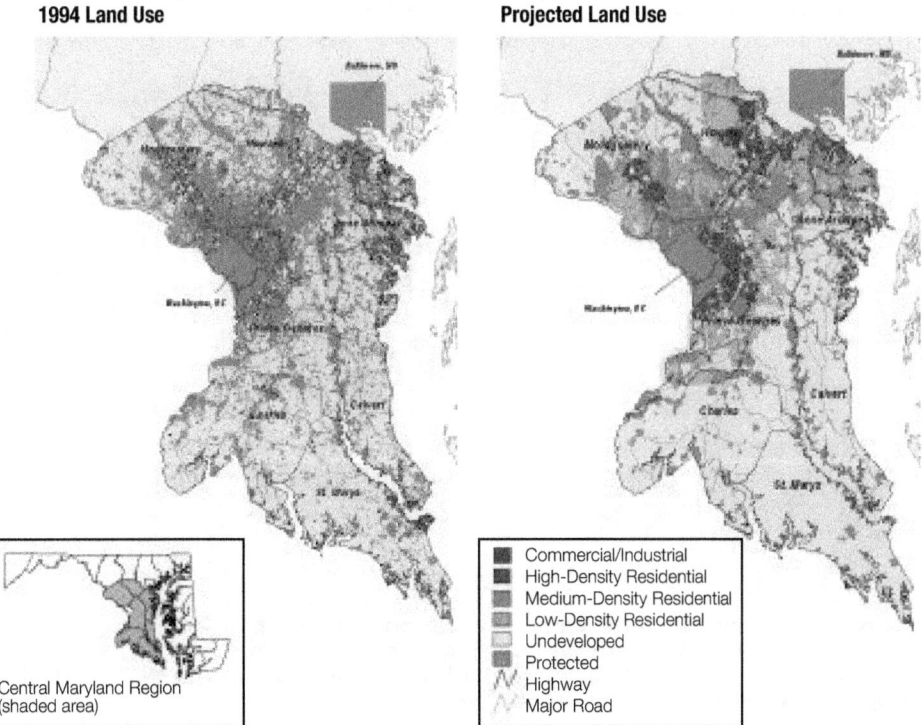

1994 Land Use

Projected Land Use

Commercial/Industrial
High-Density Residential
Medium-Density Residential
Low-Density Residential
Undeveloped
Protected
Highway
Major Road

Central Maryland Region
(shaded area)

Figure 6-4: The map on the left shows actual land use in the seven-county area of central Maryland in 1994, while the map on the right is the predicted distribution of land-use types based on a 'polycentric' city model. There is remarkable similarity between the two maps in commercial, high, and medium densities, but disagreement in the low-density residential category. Models with improved projections of this fragmented, low-density residential development—which consumes a disproportionate amount of open space and causes high public service costs—will better support decisionmaking regarding this type of development. Source: Nancy Bockstael, University of Maryland. For more information, see Annex C

to resource management and to those relevant for global assessments is needed. This need for predictive models calls for a better understanding of the drivers of land-use change, characterization/parameterization of land-use elements, and credible predictions of land cover and land use at annual to decadal time scales. Partnerships are needed with state and regional assessment and research efforts, to ensure comparability between national/global and state/regional models. Integration among the Carbon Cycle, Ecosystems, and Human Contributions and Responses research elements will be needed to develop and test models for generating scenarios of land-use and land-cover change, and for making projections of change that take into account the various influences of ecosystem functioning, carbon, water, and energy cycling as well as human-managed systems. Model validation will be a particularly challenging element of this research area. Simulation of past conditions will be a necessary strategy for testing the performance of models, placing more significance on the need to understand land-use and land-cover change in both historical and contemporary contexts.

Milestones, Products, and Payoffs

- Single-sector (e.g., urban, suburban, agriculture, forest, etc.) change models [prototypes for selected sectors, less than 2 years; operational models for selected sectors, 2-4 years; operational models for all sectors, beyond 4 years].
- Identification and integration of components of land-use and land-cover change models [regional, less than 2 years; national, 2-4 years; global beyond 4 years].
- Development of regional, national, and global land-use and land-cover change projection models, incorporating advances in understanding of drivers [regional, less than 2 years; national, 2-4 years; global beyond 4 years]. This product will be used by the Climate Variability and Change, Water Cycle, Carbon Cycle, Ecosystems, and Human Contributions and Responses research elements.

Question 6.4: How do climate variability and change affect land use and land cover, and what are the potential feedbacks of changes in land use and land cover to climate?

State of Knowledge

Land-use and land-cover change is linked in complex and interactive ways to global climate change, and the feedback between the two exists at multiple spatial and temporal scales. Changes in greenhouse gas emissions, albedo, and surface roughness are the primary mechanisms by which land-use and land-cover change affect climate. Climate variability and change, in turn, can affect the land cover of a given area and the ways in which land is used. Some of the impacts of these feedbacks are local while others have global ramifications. For example, trace gas emissions and removals by sinks depend strongly on land cover and land-use practices (see, for example, Figure 6-5), while the deposition of atmospheric constituents affects the potential rate and magnitude of terrestrial sinks.

Illustrative Research Questions

- What are the critical land uses and landscape variables that affect climate?
- What can we learn about the relationship between climate and land cover from studies of the past?
- How do climate variability and extreme events affect land use and land cover?
- How will climate and land use/cover influence each other in the future?

Research Needs

Simulation of climate-land use/cover feedbacks will require advancement of current understanding of multiple stress processes

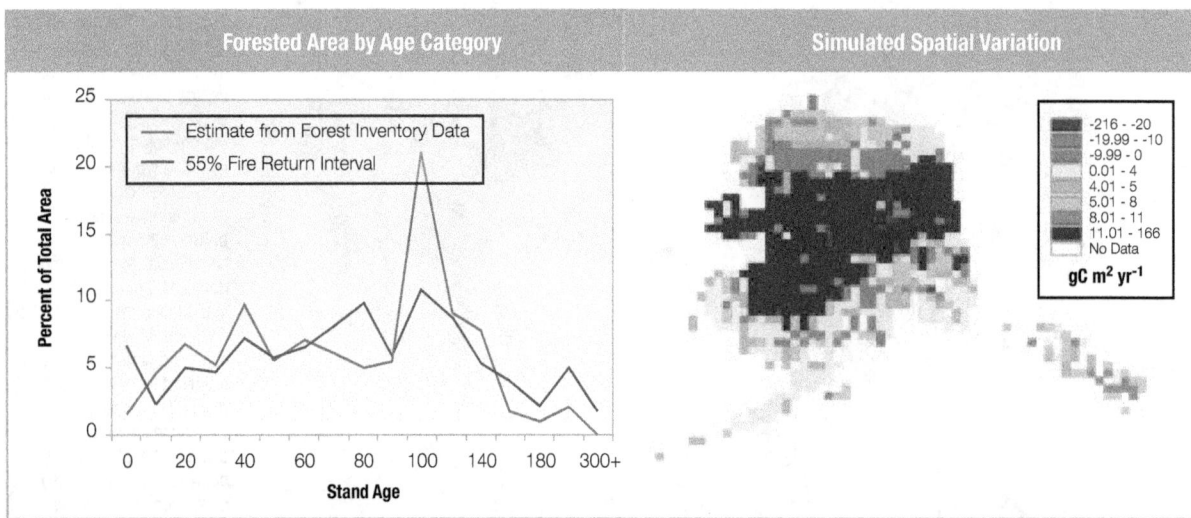

Figure 6-5: Increasing atmospheric carbon dioxide, climatic variation, and fire disturbance play substantial roles in the historical carbon dynamics of Alaska. Analyses of stand-age distribution in Alaska indicate that fire has likely become less frequent compared to the first half of the 20th century. Application of the Terrestrial Ecosystem Model indicates that regrowth under a less frequent fire regime leads to substantial carbon storage in the state between 1980 and 1989. Source: David McGuire and Dave Verbyla, University of Alaska, Fairbanks. For more information, see Annex C.

at local to global scales. We need to identify past changes in land use and land cover that are attributable to changes in climate, in order to project future changes in land use and land cover that could result from changes in climate. Validation of the interacting climate-land use effects for specific regions of the globe will be particularly challenging. Inputs will be needed from the Climate Variability and Change, Water Cycle, and Carbon Cycle research elements. International cooperation will be needed to optimize the currently existing and emerging observational networks.

Milestones, Products, and Payoffs

- Analysis of existing databases and theories about climate-related processes that affect land-use change, including uncertainty analysis [less than 2 years].
- Evaluation of how the type and distribution of land cover affects regional weather and climate patterns [less than 2 years]. Collaborative research with the Climate Variability and Change research element will be necessary.
- Sensitivity studies addressing how land-use and land-cover related changes in surface albedo, greenhouse gas fluxes, and particulates affect climate (such data will be given to the Climate Variability and Change research element) [2-4 years]. This investigation will involve collaboration with the Carbon Cycle and Climate Variability and Change research elements.
- Report on the effects of land-use and land-cover changes on local carbon dynamics. The report will discuss implications for sources and sinks of terrestrial carbon and the overall carbon budget and will be developed in collaboration with the Carbon Cycle and Climate Variability and Change research elements [2-4 years].
- Reports on the effects of land use and land cover on mitigation and management of greenhouse gases [beyond 4 years]. Data on greenhouse gas concentrations and history will be needed from the Carbon Cycle research element.
- Report on past trends in land cover or land use attributable to changes in climate (e.g., changes in forest type, changes in specific agricultural crops, or changes in the presence or absence of agriculture) [beyond 4 years]. Input will be needed from the Climate Variability and Change research element.
- Report on projected trends in land cover or land uses that are attributable to changes in climate (e.g., changes in forest type, changes in specific agricultural crops, or changes in the presence or absence of agriculture) [beyond 4 years].
- National and global models with a coupled climate-land use system [beyond 4 years]. This will be a collaborative task conducted with the Climate Variability and Change research element.

> **Question 6.5:** What are the environmental, social, economic, and human health consequences of current and potential land-use and land-cover change over the next 5 to 50 years?

State of Knowledge

There is clear evidence that changing land use and land cover has significant impacts on local environmental conditions and economic and social welfare. For example, the water cycle depends heavily on

vegetation, surface characteristics, soil properties, and water resources development by humans (e.g., dam construction, irrigation, channeling, and drainage of wetlands) which in turn affects water availability and quality. Land-use and land-cover change, climate variability and change, soil degradation, and other environmental changes all interact to affect natural resources through their effects on ecosystem structure and functioning. In turn, ecological systems may respond unexpectedly when exposed to two or more perturbations. The following research questions address the effects of changes in land use and land cover on other research elements (i.e., Ecosystems, Water Cycle, Carbon Cycle, Human Contributions and Responses).

Illustrative Research Questions

- How will different scenarios of land-use change stress or enhance the productivity of our natural resource base and the industries that depend on it, including agriculture and forestry?
- How will land-use and land-cover changes affect the form and functioning of ecosystems, including the ability to provide essential goods and services and levels of ecosystem biodiversity, and what are the ecological, economic, public health, and social benefits and costs of the changes?
- What are the impacts of future land-use and land-cover change on water quality and quantity (research will be undertaken with the Water Cycle research element)?
- Using focused case studies, how can landholders, land managers, and decisionmakers formulate land use and land management decisions and practices at various scales in order to mitigate negative impacts of, and take advantage of any new opportunities due to, climate change?

Research Needs

The other Climate Change Science Program research elements provide complementary information about the environmental and biophysical forces that influence potential land uses (e.g., atmospheric chemistry and processes, climate variability and change, water resources, nutrient flows, and ecological processes) and the anthropogenic pressures that will give rise to various land uses and processes (e.g., the Human Contributions and Responses research element). Development of coupled climate-land use/cover models that incorporate socioeconomic factors and ecosystem function should be accelerated. The challenge will be to use contemporary impacts of land-use and land-cover change to calibrate impacts on ecosystem goods and services; biogeochemical, water, and energy cycles; and climate processes. Research will require multidisciplinary cooperation to develop land-use and land-cover projections that address the necessary spatial and temporal scales, and include the necessary physical, biological, and social factors of interest, to ensure that projections of land use and land cover can be incorporated into models of impacts.

Milestones, Products, and Payoffs

- Report on the social, economic, and ecological impacts of urbanization on other land uses [less than 2 years]. This product will be used by the Ecosystems and Human Contributions and Responses research elements.
- Report on the social, economic, and ecological impacts of different scenarios of land-use change on agriculture, grazing, and

forestry [2-4 years]. This product will be used by the Ecosystems and Human Contributions and Responses research elements.

- Assessment of the impacts of different scenarios of urban and agricultural expansion on natural (terrestrial and aquatic) systems [2-4 years]. This product will be used by the Ecosystems research element.

- Reports on the relationship between land-use and land-cover change and human health [2-4 years]. This product will be used by the Human Contributions and Responses research element.

- Identification of the regions in the United States where land use and climate change may have the most significant implications for land management [2-4 years]. This product will be used by the Human Contributions and Responses research element.

- Report on the regional and national impacts of different scenarios of land use and land cover on water quality and quantity, conducted with the Water Cycle research element [regional, 2-4 years; national, beyond 4 years].

- Report on land management options associated with different climate change scenarios [beyond 4 years]. This product will be used by the Human Contributions and Responses research element.

National and International Partnerships

Nationally, several programs have identified land-use and land-cover change as part of their individual agency research agendas (e.g., the National Aeronautics and Space Administration, the U.S. Geological Survey, the National Science Foundation, the U.S. Environmental Protection Agency, and the U.S. Department of Agriculture) and have played an active role in developing this research element. It will be important as the program proceeds to engage multiple agencies and organizations working in this and related fields (e.g., the National Institutes of Health, the Department of Transportation, the Bureau of Land Management, and the U.S. Agency for International Development). In the next decade of global change research, it will be particularly important to include stakeholders (e.g., the National Governors Association, non-governmental organizations, and state and local land managers) in guiding this research element.

Global change research is strengthened through international partnerships. In the next 10 years, the establishment of international land-use and land-cover science programs will augment ongoing efforts such as the International Geosphere-Biosphere Programme to help bridge the gap between climate change researchers, land managers, and decisionmakers. For example, the Global Observation of Forest Cover and Global Observations of Land Cover Dynamics (GOFC-GOLD) is a new program and part of the Integrated Global Observing System (IGOS) for coordinating global land observations. GOFC-GOLD is implemented through regional networks of data providers and users to address a combination of global change and natural resource management questions, and engages local scientists with local and regional expertise and knowledge. Regional observational and monitoring networks and associated case studies are key to understanding phenomena at fine scales, and provide a test bed for models and a mechanism for comparative analysis.

Another example is the United Nations Land Cover Network—an emerging cooperative activity of the Food and Agriculture Organization (FAO) and the United Nations Environment Programme (UNEP) to develop monitoring and measurement of land-cover change in support of their global environmental outlooks and assessments (e.g., the Millennium Ecosystem Assessment). In addition to these activities, development agencies are attempting to address questions concerning the societal impacts of global change through new programs such as the U.S. Agency for International Development's Geographic Information and Sustainable Development program. Such programs can help in strengthening the scientific underpinning for the decisionmaking process.

To facilitate the U.S. science community achieving broad science and societal objectives in global land-use and land-cover research, international partnerships are being formed for regional studies of global importance. For example, during the last 10 years, studies of the Amazon have been conducted in the framework of the Large-Scale Biosphere-Atmosphere Experiment in Amazonia (LBA), a cooperative international project led by Brazil. Created through an international cooperative agreement, LBA has important institutional relations, including ties with over 40 Brazilian institutions, 25 institutions from various Amazonian countries, as well as institutions from the United States and eight European nations. The LBA project is expected to continue for at least 3 years.

During the next 3-5 years, the Northern Eurasia Earth Science Partnership Initiative (NEESPI) will produce a large-scale interdisciplinary program of research aimed at developing a better understanding of the interactions between ecosystems, the atmosphere, and human dynamics in northern Eurasia. This region, representing a quarter of the world's land masses, will be studied by U.S. scientists in partnership with scientists from northern Eurasia to enhance scientific knowledge and develop predictive capabilities to support informed decisionmaking with respect to land-use and land-cover changes.

CHAPTER 6 AUTHORS

Lead Authors
Tom Loveland, USGS
Garik Gutman, NASA
Marilyn Buford, USDA
Keya Chatterjee, NASA
Chris Justice, University of Maryland
Catriona Rogers, USEPA
Bryce Stokes, USDA
Julie Thomas, NPS

Contributors
Ken Andrasko, USEPA
Richard Aspinall, NSF
Virgil C. Baldwin, USDA
Matt Fladeland, NASA
Jeff Goebel, USDA
Mike Jawson, USDA

7 Carbon Cycle

Carbon is important as the basis for the food and fiber that sustain and shelter human populations, as the primary energy source that fuels economies, and as a major contributor to the planetary greenhouse effect and potential climate change. Carbon dioxide (CO_2) is the largest single forcing agent of climate change, and methane (CH_4) is also a significant contributor. Atmospheric concentrations of CO_2 and CH_4 have been increasing for about 2 centuries as a result of human activities and are now higher than they have been for over 400,000 years. Since 1750, CO_2 concentrations in the atmosphere have increased by 30% and CH_4 concentrations in the atmosphere have increased by 150%. Approximately three-quarters of present-day anthropogenic CO_2 emissions are due to fossil-fuel combustion (plus a small amount from cement production); land-use change accounts for the rest. The strengths of

CH_4 emission sources are uncertain due to the high variability in space and time of biospheric sources (IPCC, 2001a). Future atmospheric concentrations of these greenhouse gases will depend on trends and variability in natural and human-caused emissions and the capacity of terrestrial and marine sinks to absorb and retain carbon. The global carbon cycle is depicted in Figure 7-1.

Decisionmakers searching for options to stabilize or mitigate concentrations of greenhouse gases in the atmosphere are faced with two broad approaches for controlling atmospheric carbon concentrations: (1) reduction of carbon emissions at their source, such as through reducing fossil fuel use and cement production or changing land use and management (e.g., reducing deforestation); and/or (2) enhanced sequestration of carbon, either through enhancement of biospheric carbon storage or through engineering solutions to capture carbon and store it in repositories such as the

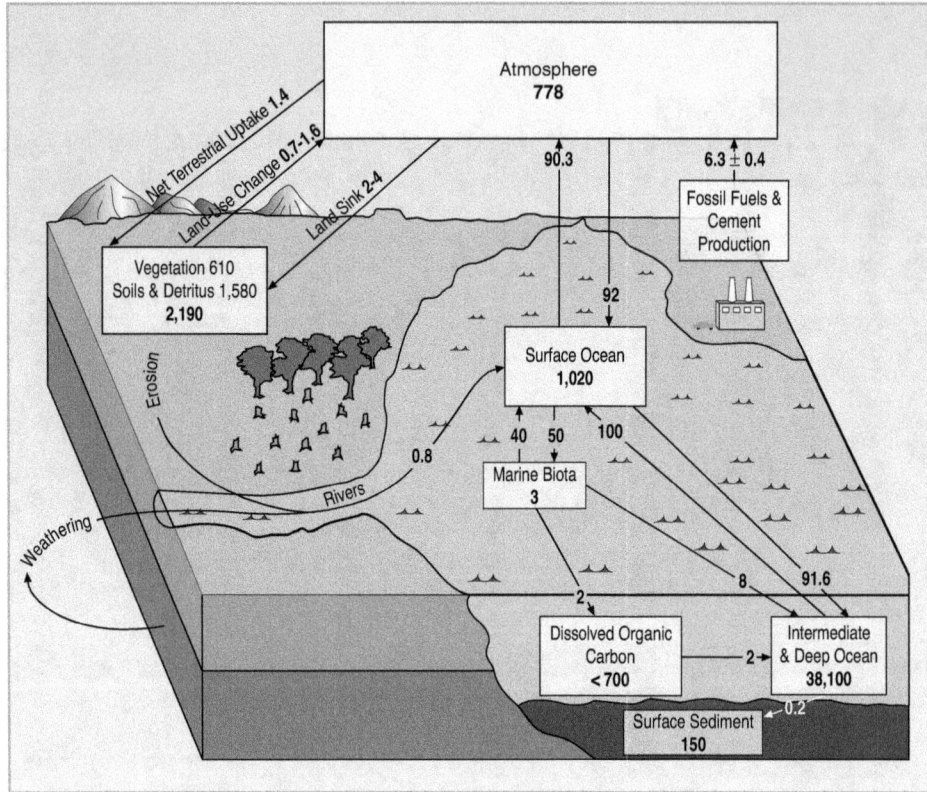

Figure 7-1: The global carbon cycle. Carbon is exchanged among the atmosphere, oceans, and land. This cycling of carbon is fundamental to regulating Earth's climate. In this static figure, the components are simplified and average values are presented. Storages [in petagrams carbon (PgC)] and fluxes (in PgC yr^{-1}) of carbon for the major components of the Earth's carbon cycle are estimated for the 1990s. For more information, see Annex C.

deep ocean or geologic formations. Enhancing carbon sequestration is of current interest as a near-term policy option to slow the rise in atmospheric CO_2 and provide more time to develop a wider range of viable mitigation and adaptation options. However, uncertainties remain about how much additional carbon storage can be achieved, the efficacy and longevity of carbon sequestration approaches, whether they will lead to unintended environmental consequences, and just how vulnerable or resilient the global carbon cycle is to such manipulations.

Successful carbon management strategies will require solid scientific information about the basic processes of the carbon cycle and an understanding of its long-term interactions with other components of the Earth system such as climate and the water and nitrogen cycles. Such strategies also will require an ability to account for all carbon stocks, fluxes, and changes and to distinguish the effects of human actions from those of natural system variability (see Figure 7-2). Because CO_2 is an essential ingredient for plant growth, it will be essential to address the direct effects of increasing atmospheric concentrations of CO_2 on terrestrial and marine ecosystem productivity. Breakthrough advances in techniques to observe and model the atmospheric, terrestrial, and oceanic components of the carbon cycle have readied the scientific community for a concerted research effort to identify, characterize, quantify, and project the major regional carbon sources and sinks—with North America as a near-term priority.

The overall goal for Climate Change Science Program (CCSP) carbon cycle research is to provide critical scientific information on the fate of carbon in the environment and how cycling of carbon might change in the future, including the role of and implications

for societal actions. In this decade, research on the global carbon cycle will focus on two overarching questions:

- How large and variable are the dynamic reservoirs and fluxes of carbon within the Earth system, and how might carbon cycling change and be managed in future years, decades, and centuries?
- What are our options for managing carbon sources and sinks to achieve an appropriate balance of risk, cost, and benefit to society?

A well-coordinated, multidisciplinary research strategy, bringing together a broad range of needed infrastructure, resources, and expertise from the public and private sectors, will be essential to answer these questions. A continuing dialogue with stakeholders, including resource managers, policymakers, and other decisionmakers, must be established and maintained to ensure that desired information is provided in a useful form (NRC, 1999c).

Specific research questions that will be addressed in support of the two overarching questions are covered in the following sections. These six carbon cycle questions focus on research issues of high priority and potential payoff for the next 10 years. They derive from the program goals recommended by the research community in *A U.S. Carbon Cycle Science Plan* (CCWG, 1999). Carbon cycling is an integrated Earth system process and no one of these questions can be addressed in isolation from the others—or without contributions from and interactions with the other research elements of the CCSP, the Climate Change Technology Program (CCTP), and the international scientific community. Many of the research activities, research needs, and milestones, products, and payoffs identified under each question will be relevant to more than one question, indicating a high degree of complementarity across questions.

Question 7.1: What are the magnitudes and distributions of North American carbon sources and sinks on seasonal to centennial time scales, and what are the processes controlling their dynamics?

State of Knowledge

There is compelling evidence of a current Northern Hemisphere extra-tropical terrestrial sink of 0.6-2.3 PgC yr^{-1} (IPCC, 2001a). Recent work suggests that this sink is a result of land-use change, including recovery of forest cleared for agriculture in the last century, and land management practices, such as fire suppression and reduced tillage of agricultural lands. Other studies suggest that elevated atmospheric CO_2 concentration, nitrogen deposition, changes in growing season duration, and changes in regional rainfall patterns also play a role. Atmospheric studies indicate that the net terrestrial sink varies significantly from year to year (see Figure 7-3).

Current estimates of regional distributions of carbon sources and sinks derived from atmospheric and oceanic data differ from forest inventory and terrestrial ecosystem model estimates, but there is growing confidence that these differences can be reconciled (IPCC, 2000a, 2001a). The Carbon Cycle science program will coordinate the observational, experimental, analytical, and data management activities needed to reconcile the discrepancies, to reduce the uncertainties, and to produce a consistent result for North America through the North American Carbon Program (NACP, 2002). When integrated with results from corresponding international research projects in Europe and Asia (Global Carbon Project, 2003), the results of the NACP will contribute to locating and accurately quantifying the Northern Hemisphere carbon sink.

Illustrative Research Questions

- What is the carbon balance of North America and adjacent ocean basins, and how is that balance changing over time? How large and variable are the sources and sinks, and what are the geographic patterns of carbon fluxes?
- What are the most important mechanisms, both natural and human-induced, that control North American carbon sources and sinks, and how will they change in the future?
- How much do North America and adjacent ocean basins contribute to the Northern Hemisphere carbon sink?

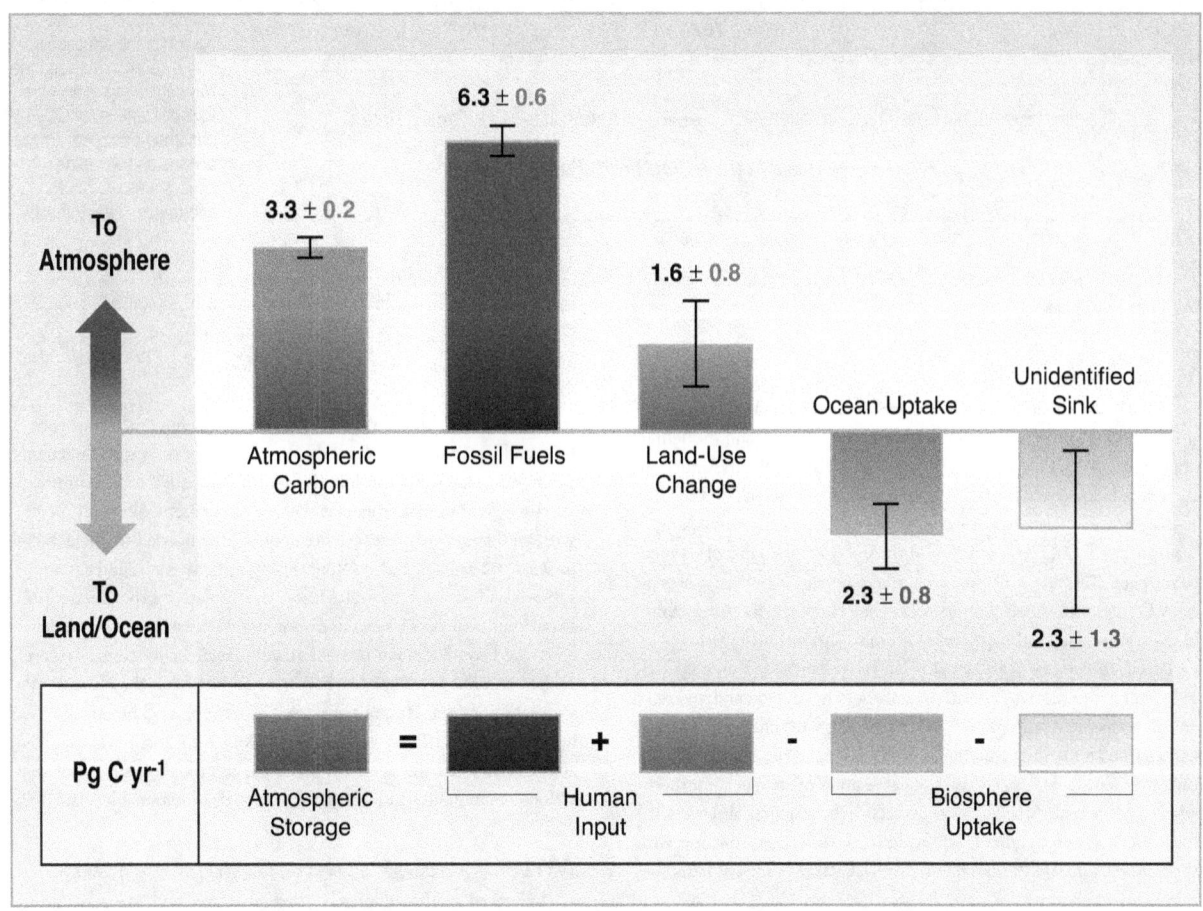

Figure 7-2: Average annual global budget of CO_2 and uncertainties for 1989 to 1998 expressed in PgC yr^{-1}. Error bounds correspond to a 90% confidence interval. The numbers reported here are from *Land Use, Land-Use Change, and Forestry*, a special report of the Intergovernmental Panel on Climate Change (IPCC, 2000a). There is compelling evidence that a large fraction of the "unidentified sink" may be accounted for by uptake in the temperate and/or boreal zones of the terrestrial Northern Hemisphere. Source: F. Hall, Office of Global Carbon Studies, NASA Goddard Space Flight Center.

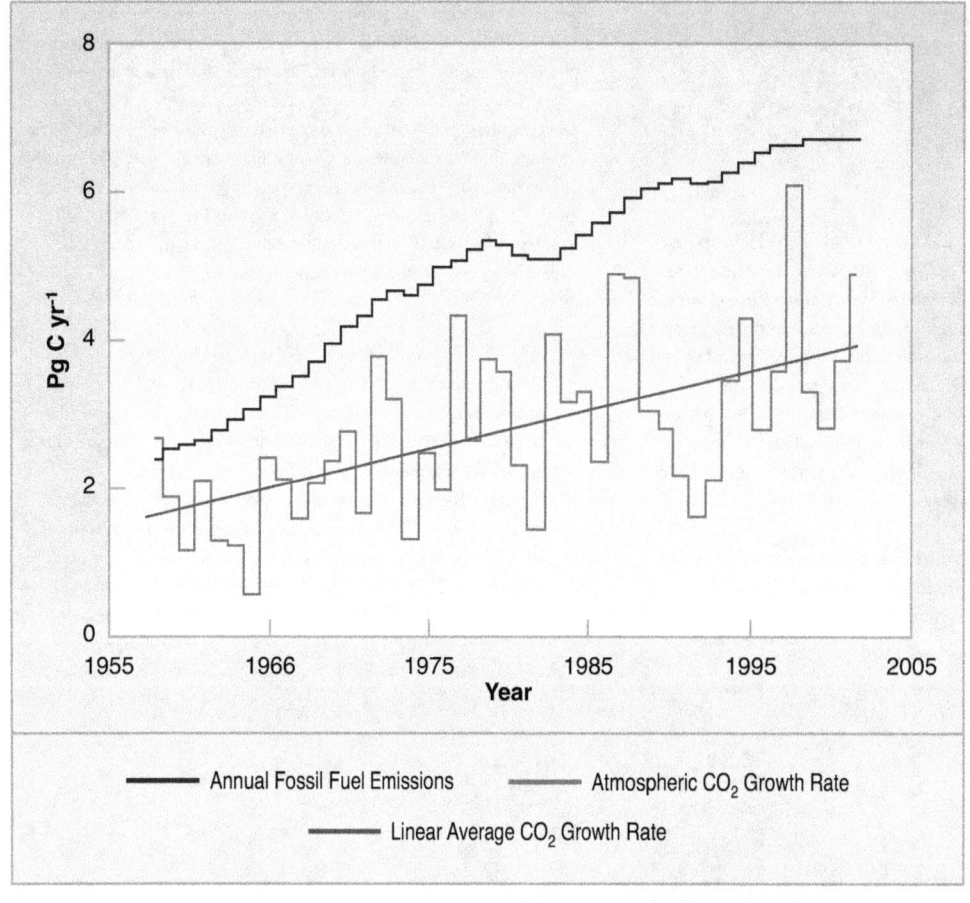

Figure 7-3: Rate of increase of atmospheric CO_2 and fossil-fuel emissions. The upper curve shows the annual global amount of carbon added to the atmosphere (in PgC yr^{-1}) in the form of CO_2 by burning coal, oil, and natural gas. The strongly varying curve shows the annual rate of increase of carbon in the atmosphere. The difference between the two curves represents the total net amount of CO_2 absorbed each year by the oceans and terrestrial ecosystems. For more information, see Annex C.

- Are there potential "surprises," where sources could increase or sinks disappear?

Research Needs

Continued and enhanced NACP research will require multidisciplinary investigations that merge observational, experimental, and modeling approaches at various scales to produce a consistent North American carbon balance (NACP, 2002). Observational needs include reliable measurements of atmospheric concentrations, vertical profiles, and transport of CO_2, CH_4, and related tracers (e.g., the ratio of oxygen to nitrogen (O_2/N_2), CO_2 isotopes); micrometeorological estimates of net CO_2 and CH_4 fluxes with accompanying biometric measurements at ecosystem and landscape scales; biomass and soil inventories of carbon in forests, crop and grazing lands, wetlands, and unmanaged ecosystems; monitoring of carbon transport by erosion and rivers; and measurements of coastal ocean carbon constituents, processes, and air-sea and land-ocean fluxes. There is a need for improved measurement technologies and standardization of analytical methods and reference materials. The work to secure the required observational networks and enhancements will need to be closely coordinated with the Climate Change Research Initiative (CCRI) plan for climate-quality observations.

A field program, with intensive campaigns, process studies, and remote sensing of productivity and land cover, will be conducted initially within selected regions of the United States, and subsequently expanded to include the entire continent. Innovative ecosystem, inverse, and data assimilation modeling approaches will be needed to enable comprehensive carbon accounting and improve understanding of North American carbon stocks and flows.

Joint activities with the Water Cycle research element to measure CO_2 and water vapor fluxes and the transport of carbon and nutrients through surface waters and wetlands will be needed to economize on costly field measurements and share expertise. Collaboration with the Ecosystems research element will be needed to characterize process controls on carbon flow through plants, soils, and aquatic systems and to conduct multi-factor manipulative experiments (i.e., experiments in which key environmental factors are varied under controlled conditions) in terrestrial and marine ecosystems. Research on human system influences, requiring inputs from the Human Contributions and Responses research element, will be equally important, especially in studies of emissions from fossil-fuel use; land, animal, and waste management practices; and the effects of adaptation actions implemented at local, state, regional, or national levels.

Milestones, Products, and Payoffs

- Improved methodologies for carbon source and sink accounting in agriculture and forestry [2 years]. This will be an important input for the Decision Support element.
- Quantitative measures of atmospheric CO_2 and CH_4 concentrations and related tracers in under-sampled locations [2-4 years].

- Landscape-scale estimates of carbon stocks in agricultural, forest, and range systems and unmanaged ecosystems from spatially resolved carbon inventory and remote-sensing data [beyond 4 years]. This will be an important input for the Ecosystems research element.

- Quantitative estimates of carbon fluxes from managed and unmanaged ecosystems in North America and surrounding oceans, with regional specificity [beyond 4 years]. Measurement sites will be coordinated with the Water Cycle, Ecosystems, and Land-Use/Land-Cover Change research elements.

- Carbon cycle models customized for North America [2-4 years], with improved physical controls and characterization of respiration [2 years] and employing the first carbon data assimilation approaches [2-4 years].

- Prototype *State of the Carbon Cycle* report focused on North America [2 years], followed by a more comprehensive report [beyond 4 years]. This will be a valuable input for the Decision Support element.

New data and models will provide enhanced capability for estimating the future capacity of carbon sources and sinks and will guide full carbon accounting on regional and continental scales. These results are a prerequisite for planning, implementing, and monitoring carbon management practices in North America. Decisionmakers will receive a series of increasingly comprehensive and informative reports about the status and trends of carbon emissions and sequestration in North America, and their contribution to the global carbon balance, for use in policy formulation and resource management.

> **Question 7.2:** What are the magnitudes and distributions of ocean carbon sources and sinks on seasonal to centennial time scales, and what are the processes controlling their dynamics?

State of Knowledge

The ocean is the largest of the dynamic carbon reservoirs on decadal to millennial time scales, and ocean processes have regulated the uptake, storage, and release of CO_2 to the atmosphere over past glacial-interglacial cycles. Globally, the ocean's present-day net uptake of carbon is approximately 2 PgC yr^{-1} (see Figure 7-4), accounting for the removal from the atmosphere of about 30% of fossil-fuel emissions (IPCC, 2001a). Marine carbon storage is jointly modulated by ocean circulation and biogeochemistry. Physical processes, primarily the ventilation of surface waters and mixing with intermediate and deep waters, have been largely responsible for regulating the historical uptake and storage of this anthropogenic carbon. However, knowledge is not yet sufficient to account for regional, seasonal, or interannual variations in ocean carbon uptake.

There is growing appreciation of the importance and complexity of factors governing the biological uptake of CO_2 and subsequent export of organic carbon to the deep sea (e.g., iron limitation, nitrogen fixation, calcification, aquatic community structure, subsurface re-mineralization). The discovery that iron is a limiting nutrient for major regions of the world's oceans has profound implications for understanding controls on ocean carbon uptake, as well as for evaluating carbon management options. The responses of air-sea CO_2 fluxes and marine ecosystems to daily, seasonal, and interannual variations in nutrient supply and climate are not well documented outside the equatorial Pacific.

Knowledge is also lacking on the magnitudes, locations, and mechanisms of surface carbon export and subsequent re-mineralization in the mesopelagic

---

BOX 7-1

GLOBAL CARBON CYCLE

FY04 CCRI Priority—North America's Carbon Balance

Climate Change Research Initiative research on the carbon cycle will focus on North America's carbon balance. This research will reduce uncertainties related to the buildup of CO_2 and CH_4 in the atmosphere and the fraction of fossil-fuel carbon being taken up by North America's ecosystems and adjacent oceans (NRC, 2001a). This work will be undertaken in the context of the U.S. Global Change Research Program's ongoing North American Carbon Program to quantify the magnitudes and distributions of terrestrial, oceanic, and atmospheric carbon sources and sinks, to understand the processes controlling their dynamics, and to produce a consistent analysis of North America's carbon budget that explains regional and sectoral contributions and year-to-year variations.

The CCRI will augment monitoring capabilities for atmospheric concentrations of CO_2, CH_4, and related tracers. The CCRI will invest in expanding and enhancing the AmeriFlux network, which measures net CO_2 exchange between terrestrial ecosystems and the atmosphere, documenting how much carbon is gained or lost on an annual basis. New experimental studies of carbon cycling processes in forests and soils and new ocean carbon surveys along the continental margins of North America will be conducted. Development of new *in situ* and remote-sensing technologies for measuring atmospheric CO_2 and CH_4 and carbon in plants, soils, and the ocean is being accelerated. New investments in diagnostic analyses and modeling will focus on developing innovative modeling frameworks and model-data fusion approaches that will bring together and ensure synergistic uses of diverse carbon data sets.

These CCRI investments will yield near-term information to be summarized in a first *State of the Carbon Cycle* report focused on North America that will provide: (1) an evaluation of our knowledge of carbon cycle dynamics relevant to the contributions of and impacts on the United States, and (2) scientific information for U.S. decision support focused on key issues for carbon management and policy.

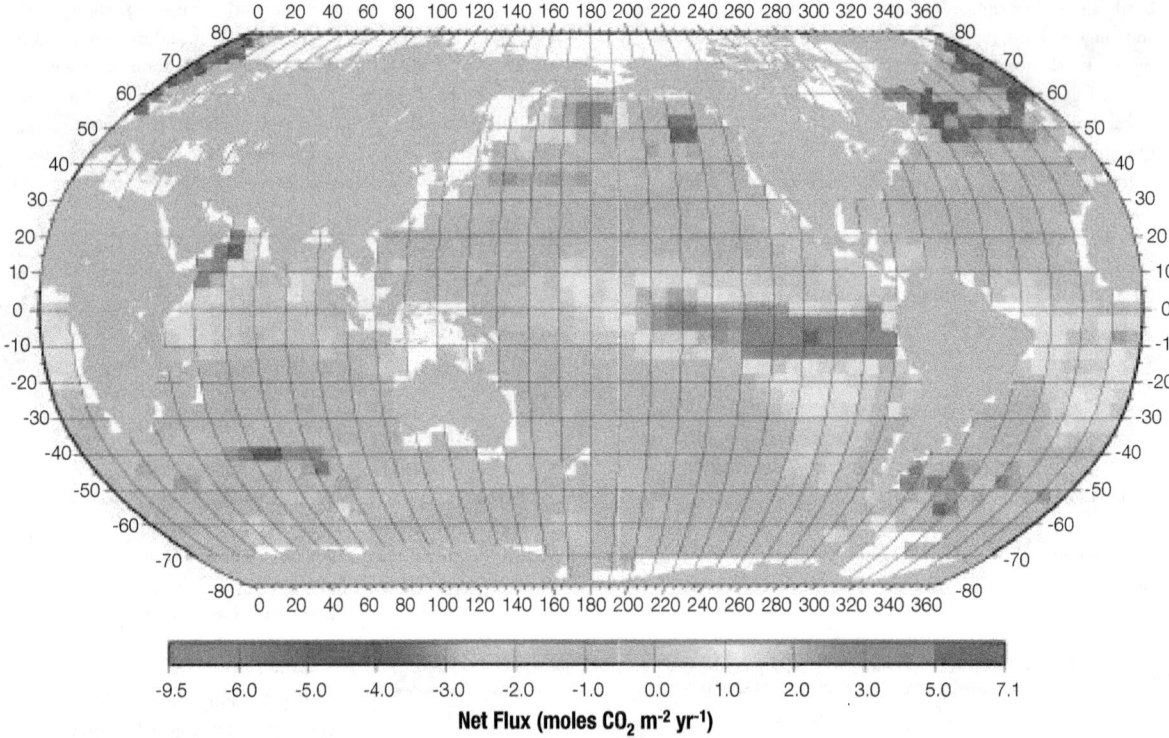

Figure 7-4: Map showing the climatological mean annual distribution of net sea-air CO_2 flux (in moles CO_2 m^{-2}yr^{-1}) over the global oceans. The blue-magenta areas indicate that the ocean is a sink for atmospheric CO_2, and the yellow-red areas, a source. The annual CO_2 uptake by the oceans has been estimated to be in the range of 1.5 to 2.1 PgC. Source: T. Takahashi, NOAA Lamont-Doherty Earth Observatory. For more information, see Annex C.

(100-1,000 m) zone, the deep ocean, and sediments. Past analyses have not fully accounted for carbon fluxes and dynamics on continental margins or in the coastal ocean, where sediment-water interactions are significant and terrestrial inputs and human-induced perturbations are large. The future behavior of the ocean carbon sink is most uncertain because of potentially large feedbacks among climate change, ocean circulation, marine ecosystems, and ocean carbon cycle processes (CCOIIG, 2003).

Illustrative Research Questions

- What is the global inventory and geographic distribution of anthropogenic CO_2 in the oceans?
- What is the magnitude, spatial pattern, and variability of air-sea CO_2 fluxes?
- How are the ocean inventories of carbon and related biogeochemical constituents changing over time?
- What biogeochemical, ecological, and physical processes control oceanic uptake and release of carbon, and how may these processes change in the future? What are the major feedback mechanisms and climate sensitivities for ocean carbon storage?
- What quantity of carbon is transported to the oceans from the land and associated freshwater systems, and how much of this carbon contributes to net storage?

Research Needs

CCSP carbon cycle research will need to continue and enhance global ocean observations (*in situ* and remotely sensed) to track the fate of carbon in the oceans, characterize fluxes of carbon and nutrients (e.g., iron, nitrogen, phosphorus), quantify transfers from the land and atmosphere to the ocean over large space and time scales, and understand the physical and biogeochemical processes that control ocean carbon fluxes now and in the future. Global carbon hydrographic surveys, surface water partial pressure of CO_2 (pCO_2) observations, time-series measurements, satellite remote sensing, and a North American coastal observing system are needed to understand the feedback links among the open ocean, ocean margins, the terrestrial environment, and the atmosphere (CCOIIG, 2003; GOOS, 2002; LSCOP, 2002; Ocean Theme Team, 2001).

Enabling activities will be needed in the areas of chemical reference standards, *in situ* sensor and autonomous platform technology development, remote-sensing algorithms and validation, data management and accessibility, and computational capacity for modeling. Research will be required to integrate and reconcile oceanic and atmospheric carbon uptake estimates through enhanced data synthesis and numerical modeling approaches including both forward and inverse models as well as data assimilation. Data from the Climate Variability and Change research element will be needed to support the development and implementation of models linking climate, ocean circulation, ocean carbon biogeochemistry, and marine ecosystem dynamics to assess more accurately the relationship of carbon sources and sinks to global climate change.

Multidisciplinary process studies will be needed to understand the response of ocean ecosystems to interannual climatic variability, biogeochemical cycling in the upper 1,000 meters of the ocean, continental margin carbon dynamics, and air-sea gas exchange.

Focused process studies in the North Atlantic, North Pacific, and along the continental margins of those basins, including inputs from rivers and estuaries, are needed in the next several years to provide independent constraints on quantification of the Northern Hemisphere carbon sink, enable designation of sources and source dynamics, and improve estimation of the retention capability of the sinks (NACP, 2002).

It is anticipated that in 5 to 10 years, an intensive program of research will be needed to resolve uncertainties about the sensitivity to climate change of the physical, chemical, and biological factors regulating carbon fluxes in the Southern Ocean. The specific objectives of this Southern Ocean Carbon Program will be defined in the next several years as the results of recent research in the Southern Ocean [i.e., Joint Global Ocean Flux Study (JGOFS) and World Ocean Circulation Experiment (WOCE)] are synthesized and become available, carbon cycle models are improved, and new enabling technologies are developed.

Milestones, Products, and Payoffs

- Quantification of temporal changes in the global ocean inventories of anthropogenic carbon and related biogeochemical constituents [2-4 years].
- Estimates of the interannual variability in the regional- and basin-scale air-sea CO_2 fluxes for the North Atlantic and North Pacific based on *in situ* measurements, remote sensing, and data assimilation [2-4 years].
- Greater understanding of the role of nutrients and trace metals, phytoplankton functional groups, primary productivity, and subsurface transport and dynamics in carbon export to the deep sea [2-4 years].
- Improved models of ocean biogeochemical processes based on linkages with ocean observations from repeat transects and time-series measurements [2-4 years].
- Observational and modeling constraints on ocean carbon dynamics, land-ocean exchange, and air-sea fluxes for the continental margin regions surrounding North America [beyond 4 years].
- Quantification of global air-sea fluxes of CO_2, lateral ocean carbon transport, delivery of carbon from the land to the ocean, particle fluxes and export rates, and the spatial distribution of carbon in the ocean on seasonal to interannual time scales using remote and *in situ* measurements and data assimilation models [beyond 4 years]. These products will be important inputs for the Climate Variability and Change and Modeling research elements.
- Models of ocean uptake of carbon that integrate biogeochemistry, ocean circulation, and marine ecosystem responses [beyond 4 years]. These models will be important inputs for the Climate Variability and Change and Modeling research elements.
- Estimates of the climate sensitivity and potential feedbacks to climate change of carbon cycling processes in the Southern Ocean [beyond 4 years].

This research will quantify the capacity of the oceans to absorb fossil-fuel CO_2 and sequester carbon through export to the deep sea. Scientific information will be provided on the potential for large feedbacks to the climate system.

Question 7.3: What are the effects on carbon sources and sinks of past, present, and future land-use change and resource management practices at local, regional, and global scales?

State of Knowledge

Historic and current land-use changes and resource management practices impact the overall carbon cycle. For example, there has been widespread reforestation since 1900 in the eastern United States following the movement of agricultural production toward the Midwest. Forest growth and conversion of forests to long-lived wood products increase the carbon stored in the forest products pool. Better land management practices (e.g., reduced soil tillage in cropping systems), increased agricultural productivity, and conversion from cropland to grassland can increase carbon storage in soil. However, changes in land use and management, such as clearing forests and grasslands and intensive tillage and harvest practices, release CO_2 to the atmosphere.

Currently, terrestrial ecosystems offset only a modest fraction of fossil-fuel emissions, but deliberate management options offer the potential to achieve additional carbon storage (IPCC, 2000a, 2001a). Utilizing terrestrial carbon sequestration as a near-term option for reducing the buildup of atmospheric greenhouse gases requires an improved understanding of the role of land use and management in the carbon cycle.

Illustrative Research Questions

- What are the roles of past and current land use and management in terrestrial carbon sources and sinks at local to continental scales?
- How do resource management practices and likely future changes in management, at local to continental scales, affect carbon stored in terrestrial ecosystems and durable products?
- How do social, political, and economic forces influence decisions regarding land use and resource management, and how might changes in these forces affect the carbon cycle?

Research Needs

Continued monitoring of carbon fluxes and storage (in soil, litter, root systems, vegetation, forest products, woody debris, and sediments) and their response to land-use changes, nutrient inputs, and resource management practices will be required to accurately quantify the role of land-cover and land-use change in the global carbon cycle. Maintenance and enhancement of the data collection and synthesis capabilities of national networks of long-term experimental sites in forests, rangelands, wetlands, agricultural lands, and other ecosystems are needed to provide an essential foundation of ecosystem monitoring data. Carbon cycle research will require close collaborations with national and international operational resource management and inventory programs to ensure the availability of reliable long-term observations of ecological processes, environmental changes and impacts, and treatment effects. Continued satellite land-cover data products and new remote-sensing estimates of aboveground biomass will be needed. Process studies linked with observations and long-term manipulative experiments will be required to identify cause-and-effect relationships and evaluate interactions with other biogeochemical cycles (e.g., the nitrogen cycle and effects on nitrous oxide emissions).

Research in this area will require inputs from the Land-Use/Land-Cover Change research element to document global patterns of land use and land cover and to understand changes in them, along with land management practices, as powerful drivers of terrestrial carbon sinks and sources. There is also an urgent need for improved understanding of the processes of land-use change and the impacts of environmental and resource management decisions. Models are needed to link ecosystem, management, policy, and socioeconomic factors to better project future changes in both carbon storage and flux and land use and development. Thus, collaborations with social scientists and inputs on the social and economic drivers of land-use change from the Land-Use/Land-Cover Change research element will be necessary.

Milestones, Products, and Payoffs

- Database of agricultural management effects on carbon emissions and sequestration in the United States [2-4 years].
- Evaluation of the effects of land-cover and land-use change and disturbance (e.g., fire, erosion) on carbon sources and sinks and the fate of carbon in selected ecosystems (e.g., in Amazonia [2-4 years], northern Eurasia [4 years], and the pan-tropics [beyond 4 years]). This product will be developed in collaboration with the Land-Use/Land-Cover Change research element.
- Quantification of the effects of different land-use changes and management practices on biomass and soil carbon storage and release and their costs [beyond 4 years]. These results will be an important input for the Decision Support element.
- Analysis of the effects of historical and contemporary land-use on carbon storage and release across environmental gradients [beyond 4 years]. This product will be developed in collaboration with the Land-Use/Land-Cover Change research element.
- Linked ecosystem, resource management, and human dimensions models that enable scientific evaluation of effects on carbon sequestration in a context of other management concerns such as market prices, land allocation decisions, and consumer and producer welfare [beyond 4 years]. These models will be developed in collaboration with the Human Contributions and Responses, Land-Use/Land-Cover Change, and Ecosystems research elements. They will provide important inputs for the Decision Support element.

Quantifying past and current effects of land-use change and resource management on the carbon cycle will enable policymakers and resource managers to predict how current activities will affect the carbon cycle at multiple scales and to develop alternative policies and practices for near-term carbon management.

> **Question 7.4:** How do global terrestrial, oceanic, and atmospheric carbon sources and sinks change on seasonal to centennial time scales, and how can this knowledge be integrated to quantify and explain annual global carbon budgets?

State of Knowledge

There is a growing realization that the carbon cycle can only be understood as an integrated global system. It is necessary to study individual components of the carbon cycle (e.g., North America, the world's oceans, ecosystems experiencing changes in land use), but additional attention must be devoted to integrating the results of such studies and placing them in a global context. Many of the most important advances over the last decade involved new combinations of data and models for components of the Earth's carbon cycle to provide valuable understanding of, and constraints on, carbon sources and sinks in other components. Estimates of regional ocean sources and sinks can now be used in combination with atmospheric data to constrain estimates of terrestrial carbon sinks. A major advance in the past decade has been the ability, enabled by new techniques for atmospheric measurement, to distinguish the roles of the ocean and land in the uptake and storage of atmospheric carbon (CCWG, 1999).

Inverse modeling techniques are approaching continental-scale resolution of sources and sinks, but with significant uncertainties due to sparse input data and the limitations of atmospheric transport models. For much of the world, CO_2 emissions are calculated from national reports of fossil-fuel usage; these country-level statistics are not adequately resolved spatially for many scientific studies and independent validation information is not available. Key processes dominating uptake and release of carbon can vary in different regions of the world, and can change in response to changes in natural and human forcings (IPCC, 2000a, 2001a). New remote-sensing observations have engendered a new appreciation for the significant spatial and temporal variability of primary productivity in Earth's ecosystems (FAO, 2002; Ocean Theme Team, 2001).

Illustrative Research Questions

- What are the current state of and trends in the global carbon cycle?
- What natural processes and human activities control carbon emissions and uptake around the world, and how are they changing?
- How will changes in climate, atmospheric CO_2 concentration, and human activity influence carbon sources and sinks both regionally and globally?

Research Needs

Sustained investments will be needed in the collection, standardization, management, analysis, and reporting of relevant global carbon monitoring and inventory data; in the elucidation of carbon cycling processes; and in the development of improved process models, interactive carbon-climate models, and, ultimately, Earth system models. New *in situ* and space-based observational capabilities will be needed (FAO, 2002; Ocean Theme Team, 2001). Measurements of atmospheric carbon isotopic signatures and new approaches for deriving or measuring fossil-fuel emissions at higher spatial resolution (i.e., at least 100 km) will be needed to better understand global fossil-fuel emissions. Process studies must focus on characterizing key controls as they vary around the world (e.g., land-use history, ecosystem disturbance regimes, nutrients, climatic variability and change, ocean circulation, resource and land management actions) and on explaining changes in the growth rates of atmospheric CO_2, CH_4, and other greenhouse gases (the effects of black carbon aerosols on the global radiation balance are addressed by the Atmospheric Composition research element).

Improving models will require development of innovative new assimilation and modeling techniques and rigorous testing, evaluation, and periodic intercomparison. Climatic and hydrological data products (e.g., precipitation, soil moisture, surface temperature, sea surface winds, salinity) from the Water Cycle and Climate Variability and Change research elements will be required as inputs for these models. Activities to secure enhanced measurements of atmospheric CO_2, CH_4, other important carbon-containing greenhouse gases, carbon monoxide, and related tracers will need to be coordinated with the Atmospheric Composition research element. Joint improvement of land-atmosphere exchange and atmospheric transport models with the Water Cycle, Atmospheric Composition, and Ecosystems research elements will be required. The Carbon Cycle science program will need to collaborate with all CCSP research elements to assemble, merge, and integrate carbon, biogeochemical, physical, and socioeconomic information for comprehensive reporting on the state of the global carbon cycle. Continued international cooperation will be necessary to integrate results and ensure widespread utility.

Milestones, Products, and Payoffs

- U.S. contributions to an international carbon observing system, including measurements of carbon storage and fluxes, complementary environmental data, and assessment of the current quality of measurements [ongoing; less than 2 years for enhancements]. Some of these will be the systematic climate quality observations needed for the CCRI.
- Global, synoptic data products from satellite remote sensing documenting changes in terrestrial and marine primary productivity, biomass, vegetation structure, land cover, and atmospheric column CO_2 [all but CO_2, ongoing; CO_2, beyond 4 years].
- Global maps of carbon stocks derived from model-based analysis of actual land cover [1-km resolution, 2 years; 30-m resolution, beyond 4 years].
- Identification and quantification of the processes controlling soil carbon storage and loss and global CO_2 exchange among the land, ocean, and atmosphere [2-4 years]. Aspects related to intercontinental transport of carbon-containing trace gases will be addressed in collaboration with the Atmospheric Composition research element.
- Identification of the processes controlling carbon sources and sinks through multi-factor manipulative experiments, studies of disturbance, and integration of decision sciences and risk management studies [beyond 4 years].
- Interactive global climate-carbon cycle models that explore the coupling and feedbacks between the physical and biogeochemical systems [2-4 years].
- First *State of the Global Carbon Cycle* report and balanced global carbon budget [beyond 4 years]. These products will be important inputs for the Decision Support element.

Policymakers and resource managers will be provided with consistent, integrated, and quantitative information on global carbon sources and sinks that can be used in worldwide carbon accounting and for evaluating and verifying carbon management activities. Improved global carbon models and understanding of key process controls on

carbon uptake and emissions, including regional variations, will be made available to improve applied climate models and decision support systems.

Question 7.5: What will be the future atmospheric concentrations of carbon dioxide, methane, and other carbon-containing greenhouse gases, and how will terrestrial and marine carbon sources and sinks change in the future?

State of Knowledge

Geological and paleoclimatic records indicate that major changes in carbon cycle dynamics have occurred in the past. These changes have been attributed to a variety of feedbacks, non-linear responses, threshold effects, or rare events. For example, there is evidence that huge, near-instantaneous releases of CH_4, very likely from clathrate (CH_4 hydrate) deposits, have affected the climate system in the past. Changes in the ocean's thermohaline circulation may have caused large changes in ocean CO_2 uptake. Failure to consider such possibilities in model projections could result in large over- or under-estimates of future atmospheric carbon concentrations, with consequent implications for policy scenarios (IPCC, 2001a).

Understanding of how carbon cycling may change in response to conditions significantly different from those of the present can be achieved through three complementary approaches: (1) Paleoclimatic and paleoecological information is used to refine our understanding of the processes controlling carbon cycling under past conditions; (2) analogs of future states are employed in manipulative experiments to observe and quantify the behavior of the system under new combinations of conditions; and (3) models (both inverse and forward) are used to simulate future system behavior based on a set of assumed initial conditions and hypothesized system interactions.

Several different types of carbon models are available, but most lack complete integration of all components, interactive coupling, ability to portray rare events or abrupt transitions, and/or full validation. While no one of these models is ideal, as a group they are becoming quite useful for exploring global change scenarios and bounding potential future CO_2 conditions and responses of ecosystems (IPCC, 2001a). Current models are less useful for projecting future CH_4 conditions, primarily because sufficient measurements and process understanding are lacking.

Illustrative Research Questions

- How will the distribution, strength, and dynamics of global carbon sources and sinks change in the next few decades and centuries?
- What are the processes that control the responses of terrestrial and marine carbon sources and sinks to future increases in CO_2, changes in climate, and inherent natural variability?
- How can we best represent carbon cycle processes in models to produce realistic projections of atmospheric concentrations?
- What are the important land use-climate-ocean-carbon cycle interactions and feedbacks? Which of these are most sensitive to climate change and/or have the potential to lead to anomalous responses?

Research Needs

Accurate projections of future atmospheric CO_2 and CH_4 levels are essential for calculating radiative forcings in models that project changes in climate and their impact on the sustainability of natural resources and human populations (NRC, 2001a). Advances will require a combination of observations, manipulative experiments, and synthesis via models enabled by increases in computational capabilities. Paleoecological and paleoclimatic studies will be needed to provide insight into the magnitude and mechanisms of past changes and the potential for abrupt changes in atmospheric levels of CH_4 or CO_2. Manipulative experiments will be needed to understand physiological acclimation to enhanced CO_2 levels, effects on processes that influence feedbacks to the climate system (e.g., nitrous oxide production, evapotranspiration, melting of permafrost), and the integrated effects of changes in multiple, interacting environmental factors.

Modeling activities will need to focus on incorporating improved process understanding into carbon cycle models, developing new generations of terrestrial and ocean carbon exchange models, and developing Earth system models with a dynamic coupling between carbon cycle processes, human activities, and the climate system. Collaborations with the Ecosystems research element on processes of ecosystem change and nitrogen cycling and with the Atmospheric Composition research element on future atmospheric greenhouse gas composition will be essential. Modeling of future carbon conditions will require inputs on future human actions and responses (including changes in energy consumption, land use and management, technology utilization, and adaptation and mitigation practices) from the Human Contributions and Responses and Land-Use/Land-Cover Change research elements and from the CCTP.

Milestones, Products, and Payoffs

- Synthesis of whole ecosystem response to increasing atmospheric CO_2 concentrations, and changes in temperature, precipitation, and other factors (e.g., iron fertilization for the ocean, nitrogen fertilization for terrestrial ecosystems) based on multi-factor experimental manipulation studies [2-4 years]. This synthesis will be developed in collaboration with the Ecosystems research element.
- Advanced carbon cycle models that incorporate improved parameterizations based on data from manipulative experiments, soil carbon transformation studies, and paleoclimatic and paleoecological studies [beyond 4 years].
- Carbon models that include the long-term effects of land use [2-4 years].
- Advanced carbon models able to simulate interannual variability at ecosystem, landscape, and ocean basin scales for selected areas [2-4 years].
- Analysis of global CH_4 dynamics, with the potential for reduced uncertainties, based on a new synthesis of observational data and improved models that address radiative forcing and the potential for abrupt change [beyond 4 years]. This product will be developed in collaboration with the Atmospheric Composition research element.
- Evaluation of the potential for dramatic changes in carbon storage and fluxes due to changes in climate, atmospheric composition, ecosystem disturbance, ocean circulation, and land-use change,

and characterization of potential feedbacks to the climate system [beyond 4 years]. This information will provide important inputs to the Climate Variability and Change and Modeling research elements.

- Improved projections of climate change forcings (i.e., atmospheric CO_2 and CH_4 concentrations) and quantification of dynamic feedbacks among the carbon cycle, human actions, and the climate system, with better estimates of errors and sources of uncertainty, from prognostic models [beyond 4 years]. This information will provide important inputs to the Climate Variability and Change, Modeling, and Land-Use/Land-Cover Change research elements and will be of use to the Ecosystems research element for designing and interpreting ecosystem experiments.

New understanding of the controls on carbon cycle processes will be provided to improve parameterizations and/or mechanistic portrayals in climate models. Projections of future atmospheric concentrations of CO_2 and CH_4 will be made available for use in applied climate models and analysis of impacts on ecosystems. Both will aid in improving model projections of future climate change.

> **Question 7.6:** How will the Earth system, and its different components, respond to various options for managing carbon in the environment, and what scientific information is needed for evaluating these options?

State of Knowledge

Questions about the effectiveness of carbon sequestration, the longevity of storage, the economic consequences of reducing emissions, technological options, resultant impacts on natural and human systems, and the overall costs and economic viability of carbon management approaches create an imperative for better scientific information to inform decisionmaking. Current interest in carbon sequestration centers on land management practices that enhance the storage of carbon in soils and biomass (see Figure 7-5), fertilization of the ocean via iron inputs that enhance biological uptake of carbon, and direct CO_2 injection into the deep sea or geological formations. Presently, there is limited scientific information to evaluate the full range of impacts of these various carbon management strategies. Little is known about the long-term efficacy of new management practices for enhancing carbon sequestration or reducing emissions or how such practices will affect components of the Earth system (NRC, 1999a,c). Basic research is needed to assess new management practices, their feasibility and effectiveness in keeping carbon out of the atmosphere on centennial time scales, and their potential environmental consequences or benefits. This research element is tightly linked to the CCTP, which focuses on engineered technologies, carbon offsets, and economic systems.

Illustrative Research Questions

- What is the scientific basis for mitigation strategies involving management of carbon on the land and in the ocean, and how can we enhance and manage long-lived carbon sinks to sequester carbon?
- What are potential magnitudes, mechanisms, and longevity of carbon sequestration by terrestrial and marine systems?

- How will elevated atmospheric CO_2, climatic variability and change, and other environmental factors and changes (such as air, water, and land pollution; natural disturbances; and human activities) affect carbon cycle management approaches?
- What scientific and socioeconomic criteria should be used to evaluate the sensitivity of the carbon cycle and the vulnerability and sustainability of carbon management approaches?

Research Needs

Field studies, manipulative experiments, and model investigations will be needed to evaluate the effectiveness of designed management approaches (e.g., emissions reductions, sequestration through manipulation of biological systems, engineered sequestration) to manipulate carbon in the ocean, land, and atmosphere and to assess their impacts on natural and human systems. Approaches for establishing a baseline against which to measure change will be required as will be methods for distinguishing natural system variability from the effects of particular human actions. New monitoring techniques and strategies to measure the short- and long-term efficacy of carbon management activities will be needed, including reliable reference materials that are available to the international community. Experiments and process studies also will be needed to evaluate the likelihood of unintended environmental consequences resulting from enhanced carbon sequestration or emissions reductions practices as well as the full cost of such approaches.

Research on the scientific underpinning for carbon management will require coordination with the Ecosystems research element and

the CCTP as well as public and private programs responsible for developing and/or implementing carbon management. It is expected that, whenever possible, field and experimental studies will be conducted in collaboration with existing carbon management projects. Inputs from the social sciences will be needed to characterize societal actions and responses. Models will be needed to incorporate understanding of basic processes into evaluation of natural and enhanced mechanisms of carbon sequestration and to assess the economics of direct CO_2 injection in the ocean and carbon management practices in the agricultural and forestry sectors. Research is needed to support assessments of carbon sequestration and emissions reduction potentials, comprehensive accounting of carbon stocks and fluxes, decisionmaking processes that involve multiple land management scenarios, approaches for calculating net carbon emissions intensity, and verification of the reports required by international agreements.

Milestones, Products, and Payoffs

- Evaluation of the biophysical potential of U.S. ecosystems to sequester carbon [selected regions, 2 years; United States, 4 years] and assessment of carbon sequestration management practices in crops and grazing systems [2-4 years].
- Monitoring techniques and strategies to improve quantitative measurement of the efficacy of carbon management activities: from new *in situ* and existing satellite capabilities [2-4 years] and from new satellite capabilities [beyond 4 years]. This will be an important input for the Decision Support element.
- Identification of the effects of enhanced nutrient availability on carbon uptake in the oceans and on land and of elevated CO_2 on

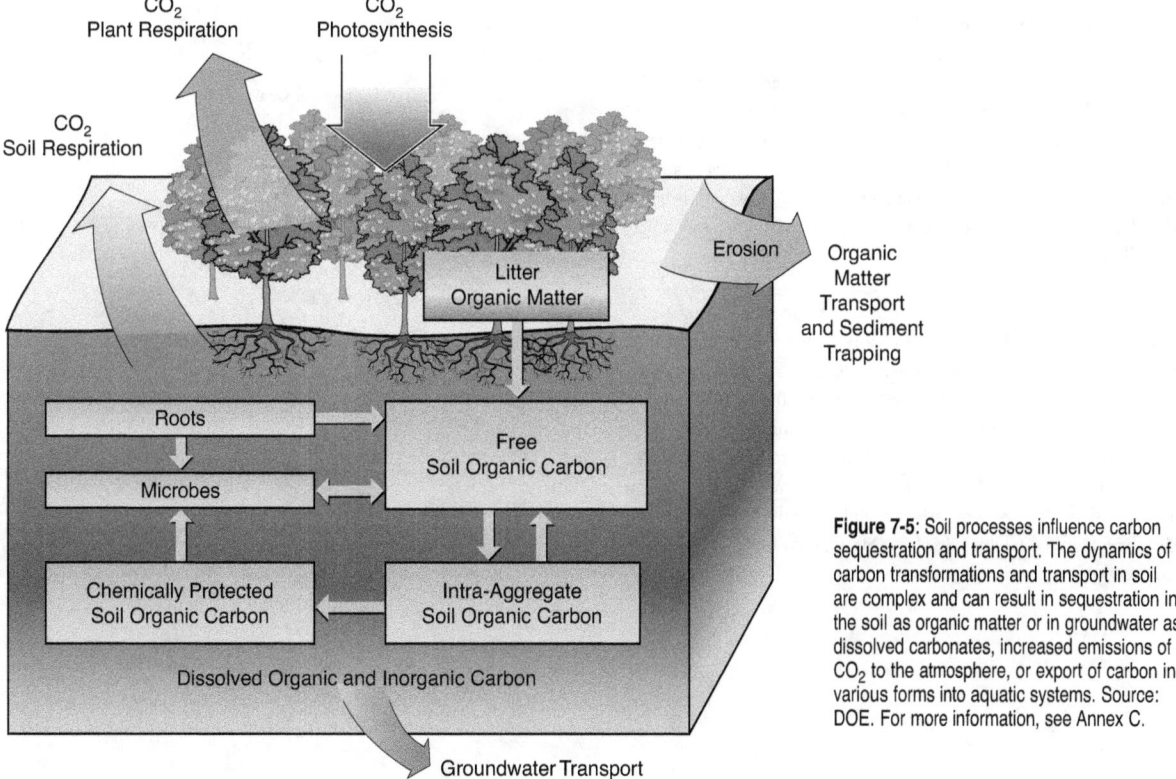

Figure 7-5: Soil processes influence carbon sequestration and transport. The dynamics of carbon transformations and transport in soil are complex and can result in sequestration in the soil as organic matter or in groundwater as dissolved carbonates, increased emissions of CO_2 to the atmosphere, or export of carbon in various forms into aquatic systems. Source: DOE. For more information, see Annex C.

physiological processes and carbon allocation [beyond 4 years]. The terrestrial product will be developed in collaboration with the Ecosystems research element.

- Evaluation of the environmental effects of mitigation options that involve reduction or prevention of greenhouse gas emissions [beyond 4 years]. This product will be developed in collaboration with the Ecosystems research element.
- Analysis of options for science-based carbon management decisions and deployment by landowners [beyond 4 years].

This research will provide the scientific foundation to inform decisions and strategies for managing carbon stocks and enhancing carbon sinks in terrestrial and oceanic systems. It will provide scientific information for evaluating the effects of actions that change emissions. Firm quantitative estimates of key carbon cycle properties (e.g., rate, magnitude, and longevity) will provide fundamental information for projecting carbon sequestration capacity, for calculating net emissions, for full carbon accounting, and for evaluating compliance with international agreements.

National and International Partnerships

Carbon cycle research will depend upon contributions from national programs and projects managed outside of the CCSP and CCTP as well as significant investments by state, tribal, and local governments and the private sector. Priorities for cooperative partnerships are: (1) coordination with long-term ecological research and monitoring programs; (2) collaboration with agricultural, forest, and soil inventories; (3) coordination with the National Oceanographic Partnership Program (NOPP); and (4) collaboration with carbon sequestration and emissions reduction projects supported by governments, non-governmental organizations, and industry. Partnerships with established data repositories and information systems will be needed to ensure access to required data. Cooperation with programs that provide national computational infrastructure will be essential for advances in carbon cycle and climate system modeling. These partnerships will be accomplished, when feasible, through joint planning and coordinated implementation with all participants.

International cooperation will be necessary to integrate scientific results from around the world and ensure widespread utility of *State of the Carbon Cycle* reports and model projections. Close cooperation with Canada and Mexico under the NACP will be important to the success of the program. The United States will depend upon observational networks and field-based carbon programs in Europe (e.g., CarboEurope) and Asia to contribute to an overall understanding of the Northern Hemisphere carbon sink. Partnerships are anticipated with the Integrated Global Observing Strategy-Partnership (IGOS-P) and the global observing system programs [the Global Terrestrial

Observing System (GTOS), the Global Ocean Observing System (GOOS), and the Global Climate Observing System (GCOS)] in securing the global observations needed to support carbon cycle modeling and projections of future atmospheric concentrations of CO_2 and CH_4 (FAO, 2002; GOOS, 2002; Ocean Theme Team, 2001). Coordination to establish common standards and measurement protocols and means of interrelating measurements will be a priority.

Interactions with, and contributions to, the joint Global Carbon Project (Global Carbon Project, 2003) of the International Geosphere-Biosphere Programme (IGBP), the International Human Dimensions Programme (IHDP), and the World Climate Research Programme (WCRP) will be essential for scientific interpretation of results from around the world and global-scale integration. CCSP carbon cycle research will contribute to bilateral activities of the Administration by promoting cooperative research on carbon cycle science. Results of CCSP carbon cycle research are anticipated to contribute substantially to future international assessments, especially Intergovernmental Panel on Climate Change (IPCC) assessments and the Millennium Ecosystem Assessment (MA Secretariat, 2002).

CHAPTER 7 AUTHORS

Lead Authors
Diane E. Wickland, NASA
Roger Dahlman, DOE
Jessica Orrego, CCSPO
Richard A. Birdsey, USDA
Paula Bontempi, NASA
Marilyn Buford, USDA
Nancy Cavallaro, USDA
Sue Conard, USDA
Rachael Craig, NSF
Michael Jawson, USDA
Anna Palmisano, DOE
Don Rice, NSF
Ed Sheffner, NASA
David Shultz, USGS
Bryce Stokes, USDA
Kathy Tedesco, NOAA
Charles Trees, NASA

Contributors
Enriqueta Barrera, NSF
Cliff Hickman, USDA
Carol Jones, USDA
Steven Shafer, USDA
Michael Verkouteren, NIST

CHAPTER 8 Ecosystems

CHAPTER CONTENTS

Question 8.1: What are the most important feedbacks between ecological systems and global change (especially climate), and what are their quantitative relationships?

Question 8.2: What are the potential consequences of global change for ecological systems?

Question 8.3: What are the options for sustaining and improving ecological systems and related goods and services, given projected global changes?

National and International Partnerships

Ecosystems shape our societies and nations by providing essential renewable resources and other benefits. They sustain human life by providing the goods and services it depends on, including food, fiber, shelter, energy, biodiversity, clean air and water, recycling of elements, and cultural, spiritual, and aesthetic returns. Ecosystems also affect the climate system by exchanging large amounts of energy, momentum, and greenhouse gases with the atmosphere. The goal of the Ecosystems research element of the U.S. Climate Change Science Program (CCSP) is to understand and be able to project the potential effects of global change on ecosystems, the goods and services ecosystems provide, and ecosystem links to the climate system (see Box 8-1 and Figure 8-1).

Global change is altering the structure and functioning of ecosystems, which in turn affects availability of ecological resources and benefits, changes the magnitude of some feedbacks between ecosystems and the climate system, and could affect economic systems that depend on ecosystems. Research during the last decade focused on the vulnerability of ecosystems to global change and contributed to assessments of the potential effects of global change on ecological systems at multiple scales.

We now know that effects of environmental changes and variability may be manifested in complex, indirect, and

BOX 8-1

TWO KEY DEFINITIONS

Ecosystem
A community (i.e., an assemblage of populations of plants, animals, fungi, and microorganisms that live in an environment and interact with one another, forming together a distinctive living system with its own composition, structure, environmental relations, development, and function) and its environment treated together as a functional system of complementary relationships and transfer and circulation of energy and matter.

Ecosystem Goods and Services
Through numerous biological, chemical, and physical processes, ecosystems provide both goods and services. Goods include food, feed, fiber, fuel, pharmaceutical products, and wildlife. Services include maintenance of hydrologic cycles, cleansing of water and air, regulation of climate and weather, storage and cycling of nutrients, provision of habitat, and provision of beauty and inspiration. Many goods pass through markets, but services rarely do.

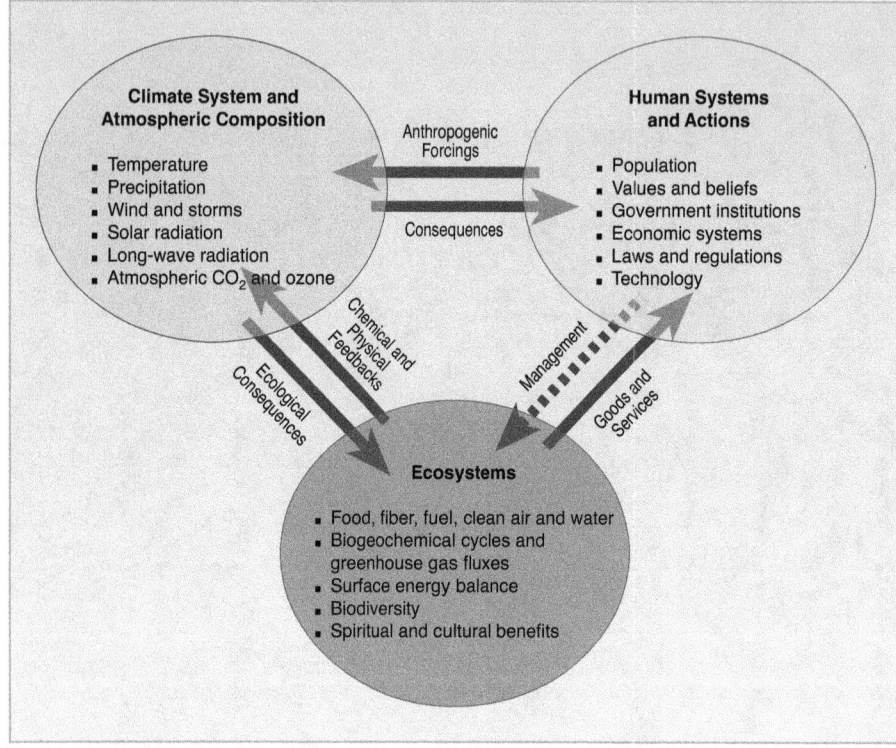

Figure 8-1: Key linkages and feedbacks between ecological systems, human systems (societies), and the climate system and atmospheric composition. Spatial scales are implicit and span from local to global. Humans manage some ecosystems intensively and others lightly. All ecosystems are affected to some degree, positively or negatively, by major global changes.

conflicting ways. For example, warming may enhance tree growth by extending growing season length (in temperate and cool regions), but pathogens better able to survive the winter because of higher temperatures may decrease forest productivity and increase vulnerability of forests to disturbances such as fire. Subtle changes in winds over the ocean can affect currents which in turn may alter the ranges and population sizes of fish species and increase or decrease fish catches. Whether environmental changes are anthropogenic or natural in origin, human societies face substantial challenges in ensuring that ecosystems sustain the goods and services on which we depend for our quality of life and survival itself.

During the next 10 years, research on ecosystems will focus on two overarching questions:
- How do natural and human-induced changes in the environment affect the structure and functioning of ecosystems at a range of spatial and temporal scales, including those processes that can in turn influence regional and global climate?
- What options does society have to ensure that ecosystem goods and services will be sustained or enhanced in the face of potential regional and global environmental changes?

Research should be focused on building the scientific foundation needed for an enhanced capability to forecast effects of multiple environmental changes (such as concurrent changes in climate, atmospheric composition, land use, pollution, invasive species, and resource management practices) on ecosystems, and for developing products for decision support in managing ecosystems. Near-term priorities will be placed on economically important ecosystems (e.g., Figure 8-2) and special studies relevant to regions where abrupt environmental changes or threshold responses by ecosystems may occur. Investigations will emphasize changes in ecosystem

structure and functioning and changes in the frequency and intensity of disturbance processes anticipated to have significant consequences for society during the next 50 years, including altered productivity, changes in biodiversity and species invasions (including pests and pathogens), and changes in carbon, nitrogen, and water cycles.

Ensuring the desired provision of ecosystem goods and services will require an understanding of interactions among basic ecosystem processes and developing approaches to reduce the vulnerabilities to, or take advantage of opportunities that arise because of, global and climatic changes. Scientific research can contribute to this societal goal by addressing three questions that focus on linkages and feedbacks between ecosystems and drivers of global change, important consequences of global change for ecological systems, and societal options for sustaining and enhancing ecosystem goods and services as environmental conditions change. This research will produce critical knowledge and provide a forecasting capability that will continuously improve decisionmaking for resource management and policy development.

Question 8.1: What are the most important feedbacks between ecological systems and global change (especially climate), and what are their quantitative relationships?

State of Knowledge
Biological, chemical, and physical processes occurring in ecosystems affect and are affected by weather and climate in many ways. For example, ecosystems (and the organisms they contain) exchange large amounts of greenhouse gases with the atmosphere, including

water vapor, carbon dioxide (CO_2), methane (CH_4), and nitrous oxide (N_2O). Moreover, the reflection (or absorption) of solar radiation by ecosystems is important to the temperature of Earth's surface. Linkages among the physical, chemical, and biological components of ecosystems are important on short (minutes to days) and long (years to millennia) time scales, as well as local to global spatial scales.

Global change has the potential to alter ecosystem structure (e.g., amount of leaf area, plant height, or species composition) and ecosystem functioning (e.g., rates of evapotranspiration, carbon assimilation, and biogeochemical cycling), and those potential changes in ecosystems might enhance or reduce global change through numerous feedback mechanisms (see Box 8-2). In addition to its direct linkages with ecological systems, global change could alter human actions that affect the structure, functioning, and spatial distribution of ecosystems, which in turn could alter important feedbacks from ecological systems to climate.

The most important feedbacks, either positive or negative, are likely to involve:
* Altered ecosystem/atmosphere exchanges of greenhouse gases

* Altered releases of aerosols from ecosystems (including black carbon and sulfur resulting from controlled and uncontrolled ecosystem burning)
* Altered releases of volatile organic compounds from ecosystems
* Changes in surface albedo resulting from changes in ecosystems
* Changes in the fraction of absorbed solar radiation that drives evapotranspiration compared to directly heating the plants and soils in terrestrial ecosystems
* Long-term changes in ecosystem structure or shifts in the geographic distribution and extent of major ecosystem types.

Illustrative Research Questions
* How might changes in temperature and precipitation affect net ecosystem exchanges (or timing or geographic distribution of those exchanges) of greenhouse gases and aerosols?
* How might changes in climate and atmospheric composition, in combination with other factors such as land-use/cover changes, affect ecosystem albedo, evapotranspiration, and nutrient cycling?
* How might changes in regional air quality (including chemicals and aerosols released from industrial sources or ecosystem disturbances such as wildfires and crop residue burning), in combination with climatic variability and change, affect ecosystem albedo and exchange of greenhouse gases?
* How might changes in ecosystems alter Earth's radiation balance, freshwater cycle, and carbon cycle, and could any such alterations contribute to abrupt climate change?
* How might human activities affect the release or uptake of greenhouse gases by ecosystems?

Figure 8-2: Landscapes and ecosystems are managed (some intensively, some lightly) by humans to produce grain, timber, and cattle, among many other goods and services desired by societies. For example, according to the United Nations Food and Agriculture Organization, 96% of the protein and 99% of the energy (usable calories) in the U.S. food supply come from terrestrial ecosystems (the remainder is derived from the ocean). Globally, about 94% of protein and 99% of energy in the human food supply come from terrestrial ecosystems. Photo sources: Tim McCabe (left and right) and Jeff Vanuga (bottom), USDA NRCS.

BOX 8-2

FEEDBACKS

A feedback from an ecosystem to climate or atmospheric composition occurs when a change in climate or atmospheric composition causes a change in the ecosystem that in turn alters the rate of the "original" change in climate or atmospheric composition. A *positive feedback* intensifies the original change whereas a *negative feedback* slows the original change (but does not change its sign). A positive feedback could occur, for example, if warming and drying (caused by increasing atmospheric CO_2 concentration) of high-latitude terrestrial ecosystems containing large amounts of carbon in plants and soils (e.g., tundra and peatland) resulted in greater ecosystem respiration, and this increased the rate of atmospheric CO_2 increase, which then accelerated the warming and drying. A negative feedback might occur, for example, if increasing atmospheric CO_2 concentration increased primary production in aquatic and/or terrestrial ecosystems, and that increased production resulted in greater carbon storage on land and in waters. This could slow the increase in radiative forcing from greenhouse gases in the atmosphere.

Research Needs

Research needs include improved experimental facilities and capabilities for making measurements in those facilities, ecosystem models, and ecosystem observing capabilities (and their related measurements) at multiple scales (to scale up from point observations with remotely sensed data). Initial efforts will be directed at enhancing existing capabilities and improving use of existing data streams. Studies should include identification of early indicators of changes in ecosystems that may be important as feedbacks to climate and atmospheric composition. Specific research needs include:

- Field and controlled-environment experimental facilities and long-term ecological observing systems at multiple locations to quantify ecosystem-environment interactions (focusing on ecosystem greenhouse gas and energy exchanges) to better parameterize, calibrate, and evaluate models of land-ocean-atmosphere chemistry feedbacks. Primary linkages are to the Carbon Cycle and Water Cycle research elements to share data and experimental sites and facilities.
- Spatially explicit ecosystem models capable of representing complex interactions between diverse ecosystems and their physical and chemical environments.
- Models that link remote sensing of land surface albedo to changes in the spatial distribution of ecosystems and exchanges of mass, energy, and momentum for implementation in climate models. It is anticipated that these models will be developed in collaboration with the Water Cycle and Carbon Cycle research elements. A primary linkage is to the Land-Use/Land-Cover Change research element to provide model-based projections of future land cover.
- Social science research to explore human factors in ecosystem-climate linkages and feedbacks. The Human Contributions and Responses element must supply information on the magnitude and significance of the primary human drivers of global change.

Milestones, Products, and Payoffs

- Reports presenting a synthesis of current knowledge of observed and potential (modeled) feedbacks between ecosystems and climatic change to aid understanding of such feedbacks and identify knowledge gaps for research planning [2-4 years]; *Arctic Climate Impact Assessment* [2 years].
- Definition of the initial requirements for ecosystem observations to quantify feedbacks to climate and atmospheric composition, to enhance existing observing systems, and to guide development of new observing capabilities [2-4 years]. This will provide key input to the Observing and Monitoring component of CCSP research.
- Quantification of important feedbacks from ecological systems to climate and atmospheric composition to improve the accuracy of climate projections [beyond 4 years]. This product will be needed by the Climate Variability and Change research element to ensure inclusion of appropriate ecological components in future climate models.

Figure 8-3: When stressed, corals frequently expel their symbiotic algae en mass, leaving coral bereft of pigmentation and appearing nearly transparent on the animal's white skeleton, a phenomenon referred to as coral bleaching. This image of bleached coral colonies was obtained during the January-March 2002 coral bleaching event in Great Barrier Reef, Australia, the worst bleaching event on record for this reef. Bleaching events reported prior to the 1980s were generally attributed to local phenomena (e.g., major storm events, sedimentation, or pollution); but, since then, a direct relationship between bleaching events and elevated ocean temperature (see Figure 8-4) was found. Source: Ray Berkelmans, Australian Institute of Marine Science.

Question 8.2: What are the potential consequences of global change for ecological systems?

State of Knowledge

Many research programs that support long-term observations [e.g., forest productivity, ultraviolet-B (UV-B) radiation received by ecosystems, greenhouse gas concentrations and fluxes, atmospheric nitrogen deposition, nutrient loading, fisheries, and the spread of invasive species] have unambiguously established that large-scale ecological changes are occurring, and there is considerable evidence that some of those changes are the result of ecological responses to climatic variability and change. For example, recent warming has been indicated as potentially linked to longer growing seasons (i.e., period of leaf display) in temperate and boreal terrestrial ecosystems, grass species decline, changes in aquatic biodiversity, and coral bleaching (IPCC, 2001b) (see Figures 8-3 and 8-4). Natural modes of climatic variability (e.g., El Niño-Southern Oscillation, North Atlantic Oscillation, and Pacific Decadal Variability) are known to affect plankton and fisheries, such as those yielding sardine, anchovies, and salmon. Soil-borne plant pathogens and parasitic nematodes have been found to move northward (in the Northern Hemisphere) with increased surface temperature. Because survival and spread of pathogens and their vectors (carriers) depend on climate and weather, climatic change and increased natural climatic variability would be expected to affect disease-causing organisms that could alter the ecological status of fauna and flora. These and other observations and expectations have come from both experiments and *in situ* monitoring.

Most ecosystems are now subject to multiple environmental changes. The dynamics and interactions of those changes and the

consequences for ecological systems are poorly understood (NRC, 1999a). Recent reviews (e.g., IPCC, 2001b) summarized the range of observed and potential consequences of combinations of changes in climate, atmospheric composition, and local drivers (e.g., invasive species, pollution, and physical habitat modification) on ecological systems. For example, in aquatic systems, alterations in wind speeds and precipitation patterns, in combination with increased air temperature, would affect water column stratification and circulation, resulting in changes in the rates of nutrient supply and productivity at all trophic levels. For terrestrial ecosystems, a large knowledge base of effects of a change in a single environmental parameter exists, but effects of multiple changes on most ecosystem processes are uncertain (see, e.g., Table 8-1). Nonetheless, we know, for example, that interactions among changes in temperature, precipitation, and fire regimes can influence vulnerability to invasive species in terrestrial ecosystems. We also know that elevated atmospheric CO_2 concentration can sometimes eliminate the negative effects of elevated tropospheric ozone (O_3) concentration and warming on crop yields.

Illustrative Research Questions

- How might the combination of increasing CO_2 concentration, increasing tropospheric O_3 concentration, and warming affect yield of major U.S. crops?
- What are the effects of changes in atmospheric CO_2 concentration, precipitation, and temperature on the structure and functioning of boreal forests?
- What are the effects of increased UV-B radiation, increased rates of sea-level rise, temperature changes, and elevated concentration of CO_2 on biodiversity, structure, and functioning of coastal ecosystems?

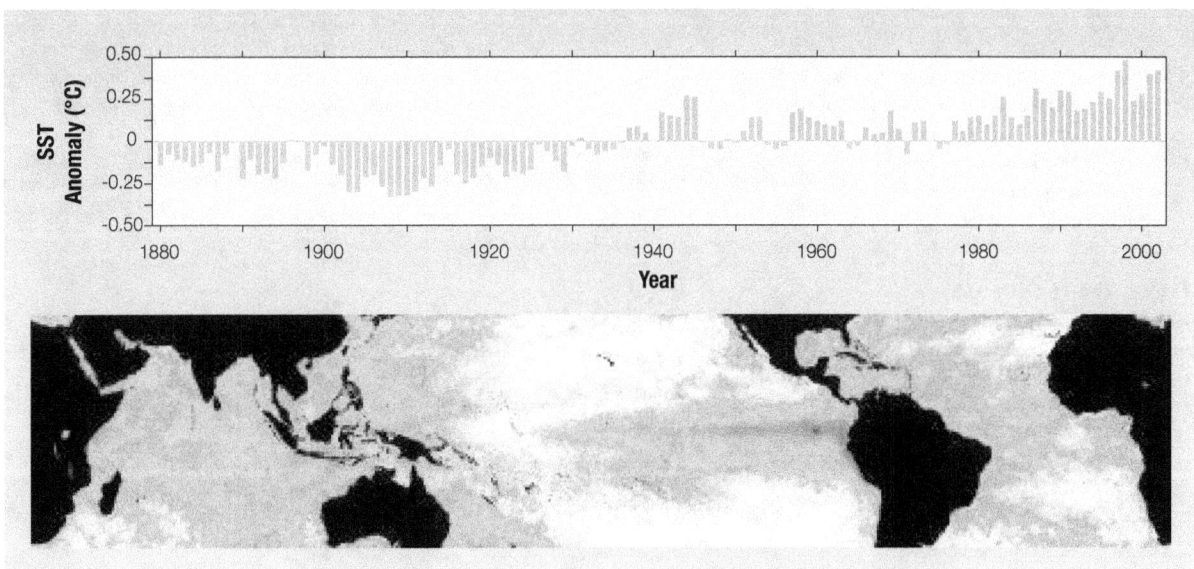

Figure 8-4: *Top*: Annual global sea surface temperature (SST) anomalies relative to the 1880-2002 mean based on *in situ* and satellite measurements. The warmest annual global SST occurred in 1998. *Bottom*: Incidences of coral bleaching were influenced by unprecedented SST anomalies in 1998, due to a severe El Niño event as shown by this satellite retrospective annual composite monthly mean coral bleaching "HotSpot" chart for 1998. A coral bleaching HotSpot is defined as an SST anomaly above a coral-bleaching threshold SST climatology. HotSpot charts illustrate the magnitude and spatial distribution of thermal stresses that may contribute to coral bleaching. This chart was derived from the NOAA/NASA 9-km satellite AVHRR (Advanced Very High-Resolution Radiometer) Oceans Pathfinder SST data set, the most refined available. Source: NOAA National Climatic Data Center.

Figure 8-5: Scientific user facility for the study of effects of simultaneous changes in concentrations of CO_2 and O_3 on the structure and functioning of northern hardwood tree stands. This facility is located near Rhinelander, Wisconsin, and is supported by DOE, USDA, NSF, and others. Source: North Central Station, U.S. Forest Service.

- Does climatic variability and change modify effects of other changes (e.g., pollution, invasive species, and changes in land, water, and resource use) on ecosystems?
- How will basin-scale changes in physical forcing mechanisms affect the productivity, distribution, and abundance of plankton, fish, seabird, and marine mammal populations in coastal marine ecosystems?
- How do changes in climate and weather (both its variability and extremes) affect the ecology and epidemiology of infectious pathogens, dissemination by their vectors, and susceptibility of the humans, animals, and plants that are their hosts?
- How rapidly might ecosystems, or individual species, move poleward and to higher elevations in response to regional warming?
- What are the effects of increasing atmospheric CO_2 concentration, warming, and sea-level rise on wetland plant contributions to soil elevation and shoreline stability?
- How will changes in the hydrologic cycle affect aquatic, riverine, and inland wetland ecosystems?

Research Needs

Identifying and quantifying the rates and consequences of global change for ecological systems is essential for appropriately evaluating options for responding to such changes. Determining the most important and societally relevant effects of global change on ecosystems will require collaboration among physical, biological, and social scientists and an improved understanding of complex interactions between natural and human disturbances and climatic variability. Near-term research priorities should include both ecosystems of special importance to society (e.g., major crops, commercial forests, and

parks and preserves) and regions where abrupt changes or threshold responses may occur, such as high-altitude and high-latitude ecosystems and transitional zones between ecosystems (e.g., forest-grassland, agriculture-native prairie, ocean boundary currents, coastal zones, and/or rural-urban interfaces). Field studies should, where appropriate, share experimental facilities and research sites (see, e.g., Figure 8-5) with the Carbon Cycle and Water Cycle research elements. High-resolution ocean shoreline topographic data will be needed to adequately project effects of sea-level rise on coastal ecosystems.

Specific research needs include:

- Experiments to study the interactive effects of climatic variability and change, elevated atmospheric CO_2 concentration, nutrient/pollution deposition, increased UV-B radiation, invasive species, and land use on key species and intact ecosystems. Understanding the effects of warming, increasing CO_2 concentration, and

TABLE 8-1	
Documented effects of environmental changes on net primary production (NPP) of terrestrial ecosystems	
Environmental Change Factor	**Observed Effect on NPP**
Increasing atmospheric CO_2 concentration	Stimulation
Increasing tropospheric O_3 concentration	Inhibition
Increasing temperature	Stimulation *or* inhibition[a]
Changes in regional hydrologic cycles	Inhibition *or* stimulation[a]
Increasing atmospheric nitrogen deposition	Stimulation *or* no effect[a]
Increasing land-surface UV-B radiation	No effect *or* inhibition[a]
Combinations of the above	Highly uncertain for most combinations and ecosystems

[a] Depends on other factors and circumstances. For example, warming might stimulate NPP in a presently cool region, but inhibit NPP in a presently warm region.

changing precipitation on the structure and functioning of ecosystems will require improved projections of the rate of change of climate and atmospheric CO_2 concentration. These improved projections will be provided by the Climate Variability and Change and Carbon Cycle research elements, and will be needed to both design experiments and interpret experimental results.

- Quantification of biomass, species composition, and community structure of terrestrial and aquatic ecosystems in relation to disturbance patterns, through observations, modeling, and process studies. Information on disturbance patterns will require inputs from the Land-Use/Land-Cover Change research element and information on biomass (carbon) pools will require inputs from the Carbon Cycle research element.

- Experiments and models that can identify threshold responses of ecosystems and species to potential climatic variability and change.

- Studies to connect paleontological, historical, contemporary, and future changes and rates of change in ecosystem structure and functioning. Long-term data sets and projections of climate and the spatial distribution and intensity of various land uses will be required from the Climate Variability and Change and Land-Use/Land-Cover Change research elements.

- Maintenance and enhancement of long-term observations to track changes in seasonal cycles of productivity, species distributions and abundances, and ecosystem structure. Improved spatial, spectral, and especially temporal resolution in observing systems from the Observing and Monitoring research element is needed to better understand and project ecological processes to parameterize models and verify model projections.

- Determination of organismal rates of adaptation and the magnitude of subsequent effects on community structure in relation to rates of global change.

- Investigations of the link between biodiversity and ecosystem functions and resulting services.

- Studies of effects of changes in "upstream" ecosystems on receiving-water ecosystems. Primary linkages are to the Water Cycle and Land-Use/Land-Cover Change research elements.

Milestones, Products, and Payoffs

- Data from field experiments quantifying aboveground and belowground effects of elevated CO_2 concentration in combination with elevated O_3 concentration on the structure and functioning of agricultural [less than 2 years], forest [2-4 years], and aquatic [beyond 4 years] ecosystems. Some of the data will be obtained in collaboration with the Carbon Cycle research element.

- Reports describing the potential consequences of global and climatic changes on selected arctic, alpine, wetland, riverine, and estuarine and marine ecosystems; selected forest and rangeland ecosystems; selected desert ecosystems; and the Great Lakes based on available research findings, to alert decisionmakers to potential consequences for these ecosystems [2-4 years]. Results will be important input to the Human Contributions and Responses research element.

- Field experiments (and user facilities) in place to study responses of ecosystems (including any changes in nutrient cycling) to combinations of elevated CO_2 concentration, warming, and altered hydrology, with data collection underway. Such facilities

will be essential for evaluating ecosystem models used to assess effects of climatic variability and change on ecosystem goods and services (and therefore input to the Human Contributions and Responses research element), as well as feedbacks to the climate system and atmospheric composition (and therefore input to the Climate Variability and Change and Atmospheric Composition research elements). Where appropriate, these research facilities will be developed in collaboration with the Carbon Cycle research element [2-4 years].

- Synthesis of known effects of increasing CO_2, warming, and other factors (e.g., increasing tropospheric O_3) on terrestrial ecosystems based on multi-factor experiments [2-4 years]. This synthesis will be developed with the Carbon Cycle research element.

- A new suite of indicators of coastal and aquatic ecosystem change and health based on output from ecosystem models, long-term observations, and process studies [2-4 years].

- Definition of the initial requirements for observing systems to monitor the health of ecosystems, to serve as an early warning system for unanticipated ecosystem changes, and to verify approaches for modeling and forecasting ecosystem changes [2-4 years]. This will be an important input to the Observing and Monitoring component of the program.

- Development of data and predictive models determining the sensitivity of selected organisms and their assemblages to changes in UV-B radiation and other environmental variables relative to observations of UV-B radiation in terrestrial, aquatic, and wetland habitats [beyond 4 years].

- Development of data and predictive models determining the sensitivity of selected organisms and their assemblages to contaminants and other environmental variables in terrestrial, aquatic, and wetland habitats [beyond 4 years].

- Spatially explicit ecosystem models at regional to global scales, based on data from remote-sensing records and experimental manipulations focused on effects of interactions among global change variables, to improve our understanding of contemporary and historical changes in ecosystem structure and functioning [beyond 4 years].

- Enhanced understanding of potential consequences of major global changes on key ecological systems [beyond 4 years].

> **Question 8.3:** What are the options for sustaining and improving ecological systems and related goods and services, given projected global changes?

State of Knowledge

Experiments and observations have demonstrated linkages between climate and ecological processes, indicating that future changes in climate could alter the flow of ecosystem goods and services (IPCC, 2001b). Several specific mitigation and adaptation measures have been identified and evaluated, including integrated land and water management; genetic selection of plants and livestock; multiple cropping systems; multiple use of freshwater and terrestrial ecosystems; programs for protection of key habitats, landscapes, and/or species; intervention programs [e.g., captive breeding and (re)introduction programs]; more efficient use of natural resources;

and institutional and infrastructure improvements (e.g., market responses, crop insurance, and water flow and supply management) (IPCC, 2001c).

It is clear that management practices can affect climate-related ecosystem goods and services. For example, management can influence the emission of greenhouse gases and aerosols from ecosystems; the rate at which ecosystems gain or lose carbon, nitrogen, phosphorus, and other elements as well as the total amount of those elements stored; the radiation balance of ecosystems (i.e., land surface albedo); and the production of goods valued by humans. While some management strategies have been studied, society's knowledge and ability to manage the broad array of ecosystem goods and services in the context of increasing and potentially conflicting demands (e.g., increasing food and fiber production while storing more carbon in soils and reducing CH_4 emissions) is very limited.

Illustrative Research Questions

- How can aquatic ecosystems be managed to balance the production and sustenance of ecosystem services across multiple demands (e.g., management of rivers to supply freshwater for drinking, irrigation, recreation, hydropower, and fish), considering potential effects of interacting environmental changes?
- How can terrestrial ecosystems such as rangelands, forests, woodlands, and croplands be managed (e.g., maintaining wildlife corridors) to balance the production and sustenance of ecosystem goods and services across multiple demands (e.g., food, fiber, fuel, fodder, recreation, biodiversity, biogeochemical cycles, tourism, and flood control), considering the future effects of interacting environmental changes?

- What options exist for society to preserve genetic diversity; respond to species migrations, invasions, and/or declines; and manage changing disease incidence and severity in the face of global change?
- How can coral reefs be managed for tourism, erosion protection, refugia for commercially and recreationally important species, and biodiversity, considering potential global changes?
- How can coastal and estuarine ecosystems be managed to sustain their productivity and use in the face of existing stresses (e.g., pollution, invasive species, and extreme natural events) and potential global changes?
- What options exist for responding to abrupt changes in ecological systems?
- What are the effects of management practices on global and regional environments (e.g., atmospheric chemistry, water supply, and water quality), nitrogen cycling, and the health, productivity, and resilience of ecosystems?

Research Needs

Much ecosystem management for the foreseeable future will proceed with imperfect knowledge about the effects of multiple global change processes and about fundamental aspects of ecosystem structure and functioning. Routine monitoring (see, e.g., Figure 8-6), scientific evaluation, and feedback from managers could enable adaptive shifts in management strategies as knowledge about an ecological system grows, and at the same time will provide important opportunities for scientists to test hypotheses about ecosystem responses to environmental change. Substantial improvements in modeling capabilities are also needed to develop and deploy effective options to maintain and enhance the supply of critical goods and services and to evaluate alternative management options under changing environmental conditions. Modeling alternative management

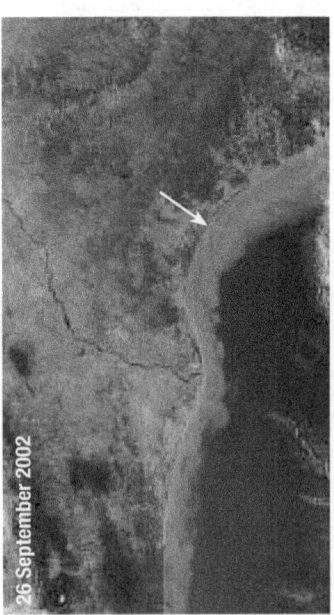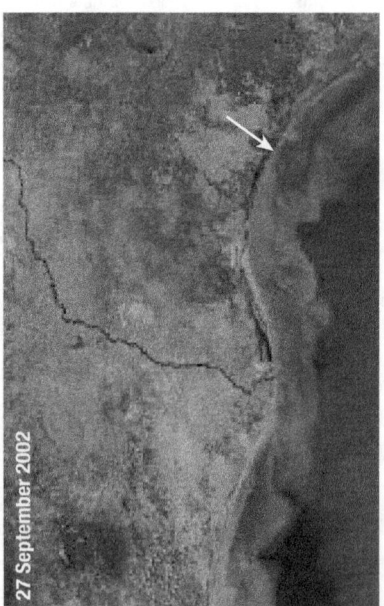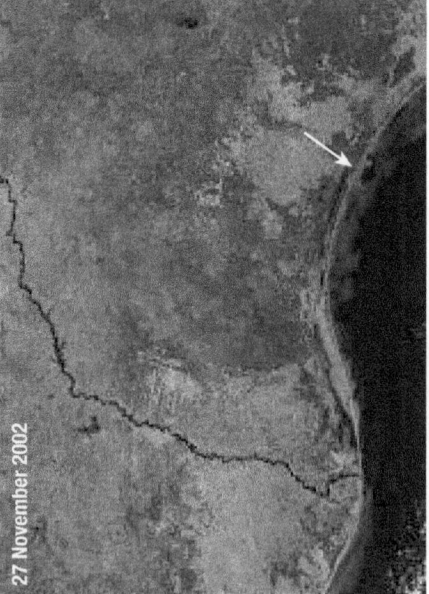

Figure 8-6: Brightly colored waters in the Gulf of Mexico indicate the presence of sediment, detritus, and blooms of marine plants called phytoplankton (noted by arrow). The blooms may be caused, or enhanced, by changes in land management "upstream" and/or changes in regional climate. By late November, this bloom appears to have subsided. Images are true-color Moderate-Resolution Imaging Spectroradiometer (MODIS) products. This type of remote-sensing technology is essential to monitoring and quantifying ecosystem and landscape states and changes. Source: NASA MODIS Ocean Team.

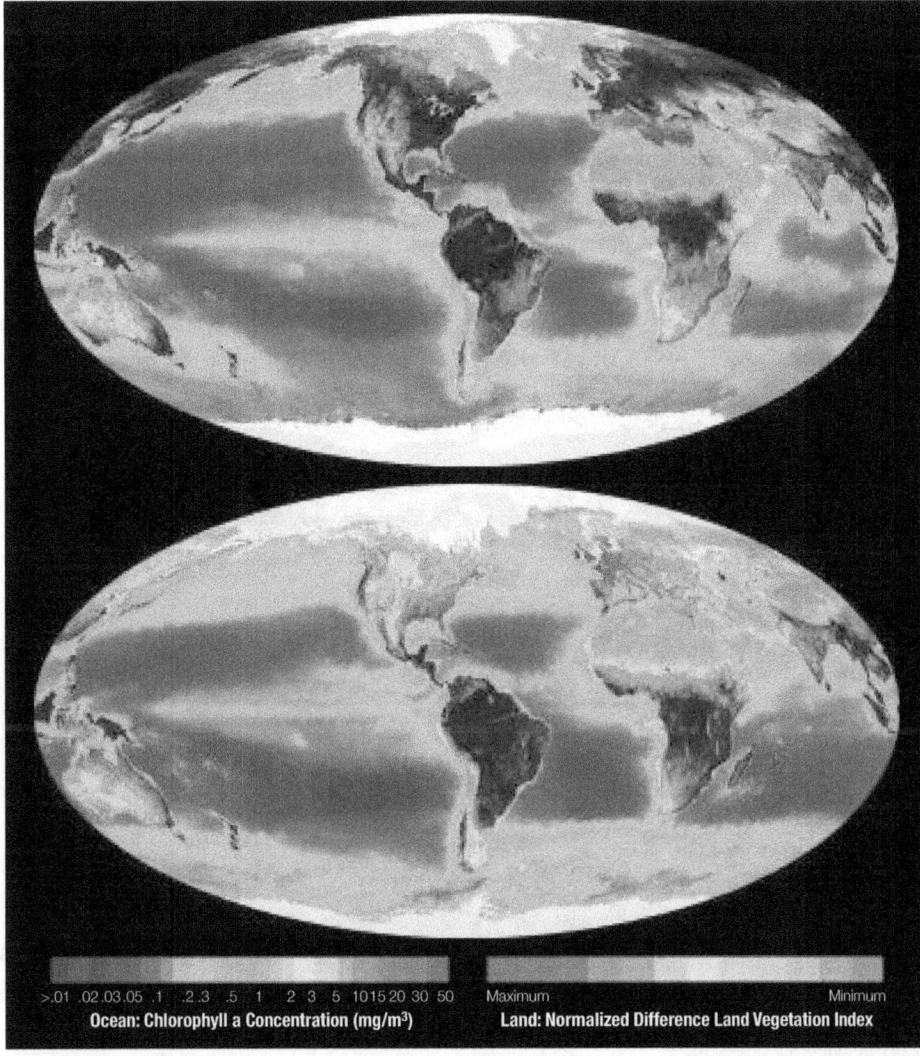

Figure 8-7: Summer (top) and winter (bottom) composite of global ocean chlorophyll *a* concentration (a surrogate for phytoplankton biomass) and terrestrial vegetation "greenness" (a measure of potential productivity) from September 1997 to December 2001. Some responses to global change may be evident on extremely large spatial scales, requiring global-level observing systems and international collaborations for detecting and interpreting changes. When evaluated over time, composite photographs such as this may reveal global-scale changes in the spatial distribution, structure, and functioning of ecosystems. Source: SeaWiFS Project, NASA/Goddard Space Flight Center and ORBIMAGE.

options will require evaluation of the influence of societal demands on ecosystems.

Specific research needs include:
- Improve understanding of causal mechanisms that drive complex changes in ecological systems, including robust indicators and likely rates of change, to develop predictive management tools such as ecological forecasting models using socioeconomic data as an input.
- Identify ecological systems susceptible to abrupt environmental changes with potentially significant (positive or negative) impacts on goods and services in order to develop adequate mitigation and adaptation responses.
- Evaluate the use of ecological information and projections in decisionmaking.
- Apply information on water quantity, quality, and delivery from the Water Cycle research element and frequency of extreme events from the Climate Variability and Change research element to evaluate ecosystem performance and management options.
- Develop and evaluate local- to regional-scale ecosystem-climate models.

- Assess the direct and indirect ecological effects and economic costs of management practices through regular monitoring, evaluation, and experimentation with a goal of enabling adaptive shifts in management.
- Explore obstacles to the implementation of ecosystem management strategies.
- Understand consequences of harvest practices in marine fisheries on changes in the age structure of the harvested populations, the structure and productivity of fisheries ecosystems, and responses of fisheries to global change.

Milestones, Products, and Payoffs
- For forests, agricultural systems, rangelands, wetlands, fisheries, and coral reefs, conduct preliminary comparisons of the effectiveness of selected management practices in selected regions focusing on greenhouse gas exchange, health, productivity, and biodiversity of the targeted ecosystems and their goods and services under changing environmental conditions [2-4 years].
- Initiation of development of decision support tools relevant to regions where abrupt or threshold ecological responses may

occur, especially high-altitude and high-latitude ecosystems and transitional zones between ecosystems (i.e., ecotones) such as forest-grassland, agriculture-native prairie, riparian and coastal zones, and rural-urban interfaces [2-4 years].

• Data sets and spatially explicit models for examining effects of management and policy decisions on a wide range of ecosystems to predict the efficacy and tradeoffs of management strategies at varying scales [beyond 4 years]. A subset of these products will be developed in collaboration with the Carbon Cycle research element.

National and International Partnerships

Interagency and international facilities and mechanisms must be in place to process, archive, and distribute the data collected and to generate relevant products. Given the nature of global change, research must span large spatial scales (from small experimental plots to global satellite image mosaics; see, e.g., Figure 8-7) and long time scales (paleontological data from ice cores, tree rings, and fossil pollen to near-real-time forecast models), and monitor a wide range of variables important for characterizing the state of ecosystems. National and international observing systems at multiple spatial scales are needed to develop a consistent record of environmental change over time. Data from such observing systems would provide inputs to models and allow evaluation and improvement of model performance. The resulting large collections of ecological and environmental data will necessitate large databases and new approaches to data integration and analysis and will require new and enhanced national and international partnerships.

Future experimental and observing systems may rely on networks of terrestrial and aquatic ecosystem observatories within particular biomes or larger ecoregions. They should link together efficiently and build on existing networks of field stations, experimental forests and ranges, environmental and resource monitoring programs, and long-term ecological research sites sponsored by many governmental and nongovernmental organizations, some of which have lengthy records (many in non-machine-readable forms) of ecological and environmental data.

Scientists conducting research under the Ecosystems research element of the CCSP will participate in the planning of international collaboration activities, including those sponsored wholly or in part by the International Geosphere-Biosphere Programme (IGBP), such as the Global Climate and Terrestrial Ecosystems (GCTE) project, the Global Environmental Change and Food Systems (GECaFS) project, the Land-Ocean Interactions in the Coastal Zone (LOICZ) project, the Surface Ocean-Lower Atmosphere Study (SOLAS), the Global Ocean Ecosystem Dynamics (GLOBEC) program, the Global Ecology and Oceanography of Harmful Algal Bloom (GEOHAB) program, and the Biospheric Aspects of the Hydrological Cycle (BAHC) project. Also important are the Global Ocean Observing System (GOOS), the Global Terrestrial Observing System (GTOS), the Global Climate Observing System (GCOS), the Millennium Ecosystem Assessment, and the International Long-Term Ecological Research (ILTER) network.

CHAPTER 8 AUTHORS

Lead Authors
Susan Herrod Julius, USEPA
Steve Shafer, USDA
Jeff Amthor, DOE
Paula Bontempi, NASA
Marilyn Buford, USDA
John Calder, NOAA
Stan Coloff, USGS
Susan Conard, USDA
Sara Mirabilio, NOAA
Knute Nadelhoffer, NSF
Bill Peterson, NOAA
Don Scavia, NOAA
Bryce Stokes, USDA
Woody Turner, NASA

Contributors
Larry Adams, USDA
Nancy Cavallaro, USDA
Pat Megonigal, Smithsonian Institute
Jessica Orrego, CCSPO
Catriona Rogers, USEPA
Chuck Trees, NASA

Human Contributions and Responses to Environmental Change

CHAPTER CONTENTS

Question 9.1: What are the magnitudes, interrelationships, and significance of the primary human drivers of and their potential impact on global environmental change?

Question 9.2: What are the current and potential future impacts of global environmental variability and change on human welfare, what factors influence the capacity of human societies to respond to change, and how can resilience be increased and vulnerability reduced?

Question 9.3: How can the methods and capabilities for societal decisionmaking under conditions of complexity and uncertainty about global environmental variability and change be enhanced?

Question 9.4: What are the potential human health effects of global environmental change, and what climate, socioeconomic, and environmental information is needed to assess the cumulative risk to health from these effects?

National and International Partnerships

Human activities play an important part in virtually all natural systems and are forces for change in the environment at local, regional, and even global scales. Social, economic, and cultural systems are changing in a world that is more populated, urban, and interconnected than ever. Such large-scale changes increase the resilience of some groups while increasing the vulnerability of others. A more integrated understanding of the complex interactions of human societies and the Earth system is essential if we are to identify vulnerable systems and pursue options that take advantage of opportunities and enhance resilience. Basic social science research into human-environment interactions provides a foundation for applied analyses and modeling of human behavior at its interface with global environmental change.

The need for understanding human contributions and responses—sometimes referred to as "human dimensions" of global change—motivates the research questions in this chapter and elsewhere in this plan. Human dimensions research includes studies of potential technological, social, economic, and cultural drivers of global change, and how

these and other aspects of human systems may affect adaptation and the consequences of change for society. Much of this research is "cross-cutting"—integral to explorations of causes and impacts of changes in atmospheric composition, climate, the water cycle, the carbon cycle, ecosystems, land use and land cover, and other global systems. Research on human contributions and responses integrates information from different research elements to establish baseline characterizations of man acting in and reacting to his environment. The complex interactions of multiple environmental stressors on human activities must be examined. It is widely acknowledged that human dimensions research has special challenges associated with the cross-disciplinary nature of its topics and with the mix of qualitative and quantitative data and analyses employed in its pursuit.

Across the range of human dimensions research there is a particularly strong need for the integration of social, economic, and health data with environmental data. Such integration requires data from physical, biological, social, and health disciplines on compatible temporal and spatial scales, to support the synthesis of data for research and to support decisionmaking. There is an especially critical need for geo-referenced data.

A broad research agenda for human contributions and responses has been identified in a series of national and international reports, including the assessment reports of the Intergovernmental Panel on Climate Change (IPCC, 2001a,b,c,d) and a series of focused reports and monographs from the National Research Council (NRC, 1999a,e, 2001c,e). The NRC report *Climate Change Science: An Analysis of Some Key Questions* concluded that: "In order to address the consequences of climate change and better serve the Nation's decisionmakers, the research enterprise dealing with environmental change and environment-society interactions must be enhanced." Such an enterprise should include, "…support of interdisciplinary research that couples physical, chemical, biological, and human systems" (NRC, 2001a). This chapter draws from these reports and from priority areas identified by the research community through federal research programs.

Two overarching questions for research on the human contributions and responses to global change are:

- How do humans and human societies drive changes in the global environment?
- How do humans prepare for and respond to global environmental change?

These questions frame the human dimensions research outlined in the four key questions that follow.

> **Question 9.1**: What are the magnitudes, interrelationships, and significance of primary human drivers of and their potential impact on global environmental change?

State of Knowledge

Human drivers of global environmental change include consumption of energy and natural resources, technological and economic choices, culture, and institutions. The effects of these drivers are seen in population growth and movement, changes in consumption, de- or reforestation, land-use change, and toleration or regulation of pollution. The IPCC (IPCC, 2000 a,b, 2001a,b,c,d), the NRC (NRC, 1999a, 2001a), and additional U.S. studies have summarized social science research on these drivers in the specific context of climate change, and the International Human Dimensions Programme has contributed to this body of knowledge. For example, research has pointed to population changes (including an aging population in the United States with rapid growth of human settlements, especially in the South, West, and coastal areas) that have affected consumption patterns and other drivers of global environmental change.

Research into the human drivers of global change has focused on changes in land use and energy use. But there is also a growing body of work on fundamental socioeconomic processes that drive human use of the environment (e.g., changes in population densities, advances in technology, the emergence of new institutional structures). Furthermore, current research on sustainability emphasizes the roles played by societies in driving global environmental change. Recent research has also improved our understanding of many of the factors that affect environmentally significant consumption at the household level. Important advances have been made in understanding the effects of economic transformation—for instance, how the growth of the service sector in urban areas contributes both to social wealth and vulnerability of human settlements. Similarly, research on technological change has helped to identify trends in innovation, efficiency, and expanded living standards, and their implications for natural and depletable resources.

Illustrative Research Questions

Research questions related to human drivers span a range of topics, including: What are the key processes and trends associated with population growth and demographic change, management of natural resources (including land and water), the development of advanced technologies, and trade and global economic activity? How can improved understanding of these issues be used to improve scenarios and projections of global change? Who are the principal actors, both individuals and institutions, and what are the key factors, such as households, markets, property and land tenure, and government policies and practices? How can researchers develop appropriate scenarios and link them to decisionmaking frameworks? How can stakeholder involvement be used to help determine the research agenda? In addition, questions specific to population and technological change and the role of trade and economic activity include:

- **Population growth and demographic change**
  - How do population growth, composition, distribution, and dynamics (fertility, mortality, migration, and household change) affect the sustainability of energy and land use, economic activity, land cover, the climate system, and other global environmental systems?
  - How is the relationship between population dynamics and environmental change affected by the scale at which population-environmental linkages are measured (e.g., the plot, the community, the state, the region, the nation)?
  - How do people use information and form perceptions about potential or actual global environmental changes, along with other social, economic, and political considerations, to make decisions about production, consumption (including use of natural resources), and mobility (including migration)?
  - What are the roles played by institutions in structuring the activities that drive global environmental change?
- **Technological change**
  - What are the drivers, especially institutional factors that induce technological innovation and adoption of new technologies? What influences the transfer of technology from region to region or country to country? How does technological innovation and transfer impact systemic environmental change and figure in adaptation and mitigation strategies?
  - How can research that identifies viable technology options be modeled as policy options?
  - What can be projected about the effectiveness, cost, and environmental and health effects of alternative energy and mitigation technologies, including sequestration options?
  - How can this research contribute to efforts to develop mitigation technology options by, for example, placing values on such items as temporary carbon storage and the availability of limited resources such as land and water?

- **Trade and global economic activity**
  - What influences the movement of goods and services domestically and from one country to another? How does movement of goods and services impact global environmental change?
  - How do operational and technological changes affect economic productivity and energy use?

Research Needs

Key needs have been identified, including:

- Development of a connection to decision support capabilities by improving the scientific information that helps to inform the policy process.
- Development of more coherent and plausible scenarios with projections of social, economic, and technology variables. Here, linkages to outputs from the Climate Change Technology Program (CCTP) will be made.
- Development of integrated assessment models with the ability to better analyze the effects (social, economic, and health) of measures directed at reducing greenhouse gas emissions and that include non-market submodels for the analysis of quantitative and qualitative data related to human health and well-being.
- Development of integrated assessment models that introduce new energy and carbon sequestration technologies (including technologies under consideration in the CCTP) and incorporate new knowledge about innovation and diffusion.
- Development of the capability to study the economic and trade effects of various mitigation options that differ in complex ways, both within and among countries, including broad policy approaches (e.g., emissions targets, technology subsidies, voluntary national goals) and means of implementation (e.g., voluntary programs, incentives, taxes, cap and trade systems, and quantity constraints).
- Assessment of the full costs and benefits (including productivity impacts) of environmental policy and technology choices (mitigation and adaptation) that affect human well-being at different scales, including the individual or household level.

Milestones, Products, and Payoffs

- Research in this area is expected to improve our understanding of how human societies drive global environmental change. Inputs on potential future human drivers of change are required for Ecosystems, Water Cycle, and Carbon Cycle research. Available information will be reassessed for its relevance and contribution to interdisciplinary studies of human-induced environmental change [less than 2 years].
- Scenarios will be strengthened by an improved understanding of the interdependence among economic growth; population growth, composition, distribution, and dynamics (including migration); energy consumption in different sectors (e.g., electric power generation, transportation, residential heating and cooling); advancements in technologies; and pollutant emissions [less than 4 years] (a benefit to the Carbon Cycle research element).
- Evaluations will be developed of the economic opportunities to reduce greenhouse gas emissions or increase sequestration in the agricultural and forestry sectors [2-4 years].

- Structured methods will be developed to define the connections and tradeoffs among economic development, technological change, and human well-being at multiple scales and at the intersection of complex institutional arrangements [beyond 4 years].

Human Contributions and Responses products will provide needed inputs to the Carbon Cycle and Land-Use/Land-Cover Change research elements related to changes in energy consumption, technology utilization, and adaptation policies (see also Question 9.2).

> **Question 9.2:** What are the current and potential future impacts of global environmental variability and change on human welfare, what factors influence the capacity of human societies to respond to change, and how can resilience be increased and vulnerability reduced?

State of Knowledge

For the purpose of this question, "global environmental variability and change" includes climate variability and change and related sea-level rise. These environmental changes need to be analyzed in the context of other natural and social system stresses, such as land-use and land-cover change, population changes and migrations, and global economic restructuring. There has been significant progress in analyzing and modeling regional vulnerabilities and possibilities for adaptation, including in the context of multiple stresses. Progress has been made in understanding how society adapts to seasonal climate variability and, by extension, how it may adapt to potential longer term climate change (IPCC, 2001b).

The state of global change impact and adaptation research varies, depending on the nature of the impact, the scale of the analysis, and the region of the world. For most types of impacts, this field of inquiry has advanced from modeling direct impacts on natural and human systems (e.g., crops, forests, water flows, coastal infrastructure) to analyses of how people might alter specific activities in reaction to changing climate, and for several types of impacts, anticipatory responses have been investigated as well. For example, with respect to sea-level rise, the direct impacts and possible responses are fairly well established for the United States. However, a high priority for research concerns the environmental impacts of adaptive responses in the future.

On a global scale, considerable gaps exist in understanding, modeling, and quantifying the sensitivity and vulnerability of human systems to global change and measuring the capacity of human systems to adapt. For instance, little is known about the effectiveness of applying adaptation experiences with past and current climate variability and extreme events to the realm of climate change adaptation; nor about how this information could be used to improve estimates of the feasibility, effectiveness, and costs and benefits of adaptation to long-term change. Gaps also exist in understanding differences in adaptive capacity across regions of the world and different socioeconomic groups (IPCC, 2001b). Also less well known are the roles that institutional change and consumption patterns in the future will play in the capacity of society to prepare for and respond to global changes.

Illustrative Research Questions

- What factors determine the vulnerability of human systems to climate variability and change, and how can vulnerability be reduced?
- What factors determine the vulnerability of natural systems to the adaptive measures that people may implement in response to global change?
- How are climate variability, trends in climate, and sea-level rise likely to affect resource management (e.g., water, fish, agriculture, forestry, transportation, energy supplies), urban planning, coastal zone management, and the effectiveness of federal environmental and infrastructure programs?
- What are the economic and social costs and benefits of current climate variability and longer term climate change and what are the market and non-market tradeoffs, feasibility, and effectiveness of potential adaptation and/or mitigation options?
- To what extent will consumption patterns (e.g., per capita water consumption) and/or land-use changes influence the vulnerability of human systems to the impacts of climate variability and change?
- How may methods be refined to accurately assess the combined impacts of the full range of potential climate change, water quality and availability, land use, sea-level rise, and ecosystems on human welfare?

Research Needs

Research needs include empirical studies and model-based simulation studies of the influence of social and economic factors on vulnerability and adaptive capacity in households, organizations, and communities; assessments and economic analyses of the potential impacts of climate variability and change (including using products from the Ecosystems research element); retrospective analyses of the consequences of surprising shifts in climate and the ability of society to respond to negative impacts and potential opportunities; and studies analyzing the factors that affect adaptive capacity in the context of multiple social and natural system stresses (climate change, land-use change, population change and movements, sea-level rise, changes in political institutions, technology gains, and economic restructuring).

Much of this research will need to be place-based analysis at regional and local scales in order to capture the complexities of the human-environment interface and the adaptive strategies of individuals, industries, institutions, and communities (requiring connection to the place-based research planned through the Decision Support element). Comparative studies at different locations and in different socioeconomic contexts are critical. Longitudinal data sets need to be developed, as do data sets that track adaptation strategies across time (linkage to the Observing and Monitoring working group is key).

More extensive research crossing social science disciplines as well as research integrating social and natural system components is needed for improved understanding and modeling of impacts and adaptation and their feedback to possible mitigation efforts. Integrating across these connections is complex and will require methods for integrating qualitative and quantitative data and analyses as well as improvements in linking component models. Use of qualitative and quantitative approaches is critical if we are to make progress at a range of scales. Specifically, attention needs to be paid to the associated costs and benefits of adaptation strategies, strategies for mitigating the impacts of global change on different economic sectors and people in different locations and economic brackets, market and non-market valuation of positive and negative impacts, the possibility of new economic instruments for responding to global change, and the role of public and private institutions and public policies in influencing adaptive capacity (IPCC, 2001b). Research could include input from studies of mitigation and adaptation measures undertaken by the Ecosystems research element.

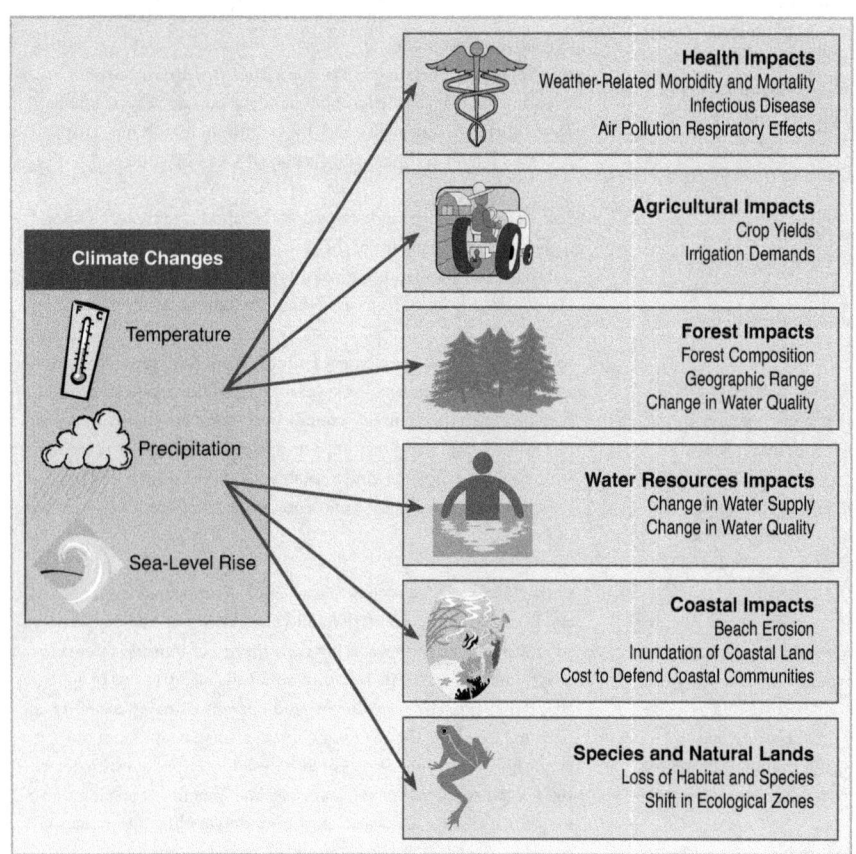

Figure 9-1: Potential climate variability and change impacts.

Milestones, Products, and Payoffs

Research on these questions can be expected to improve analytical methods and models of how climate variability and change and sea-level rise combined with socioeconomic changes are likely to affect decisionmaking in water management, agriculture, forest management, transportation infrastructure, urban areas, coastal areas, public health, and other climate-sensitive sectors in the United States and developing countries. The research will enable decisionmakers in both the public and private sectors to make more enlightened choices regarding the nature and timing of actions to undertake in response to the challenges and opportunities associated with anticipated climate variability, potential climate change, and sea-level rise.

Some expected products/milestones include:

- Improved characterization and understanding of vulnerability and adaptation based on analyses of societal adjustment to climate variability and seasonal-to-interannual forecasts [2-4 years].

- Identification of adaptation strategies effective for managing the impacts of seasonal and year-to-year climate variability that could prove useful for adaptation to projected longer term climate change [beyond 4 years].

- Elevation maps depicting areas vulnerable to sea-level rise and planning maps depicting how state and local governments could respond to sea-level rise [less than 2 years].

- Assessments of the potential economic impacts of climate change on the producers and consumers of food and fiber products [2-4 years].

- Estimates of the value of the ancillary benefits (e.g., enhanced wildlife habitat and improved water quality) that could result from implementing various mitigation activities within the forestry sector [2-4 years].

- Assessments of how coastal environmental programs can be improved to adapt to sea-level rise while enhancing economic growth [2-4 years].

> **Question 9.3**: How can the methods and capabilities for societal decisionmaking under conditions of complexity and uncertainty about global environmental variability and change be enhanced?

State of Knowledge

Decisionmaking is rife with uncertainties including risks of irreversible and/or non-linear changes that may be met with insufficient or excessive responses whose consequences may cascade across generations. The difficulties associated with characterizing and explaining uncertainty have become increasingly salient given the interest of policymakers in addressing global environmental change. Uncertainties arise from a number of factors, including problems with data, problems with models, lack of knowledge of important underlying relationships, imprecise representation of uncertainty, statistical variation and measurement error, and subjective judgment (IPCC, 2001b,d).

BOX 9-1

HUMAN CONTRIBUTIONS AND RESPONSES TO ENVIRONMENTAL CHANGE

FY04 CCRI Priority - Decisionmaking Under Uncertainty

The Climate Change Research Initiative (CCRI) will leverage existing U.S. Global Change Research Program (USGCRP) efforts to provide structured information to inform national, regional, and local discussions about possible global change causes, impacts, and mitigation and adaptation strategies.

The CCRI will provide continuing support for a set of interdisciplinary centers focusing on *Decisionmaking Under Uncertainty* associated with climate change and variability. These centers, which should be established in FY04 following a special competition, will conduct fundamental research on decisionmaking associated with climate change and

variability. The centers are expected to advance basic understanding about decision processes dealing with issues such as inter-temporal choice, risk perception, hazards and disaster reduction, opportunities, trade-offs, equity, framing, and probabilistic reasoning associated with uncertainty.

The centers will develop tools that people, organizations, and governments can use to better understand the risks and uncertainties associated with climate variability and change and the options they have to address them. In order to do this, they will develop and disseminate tangible products for researchers, decisionmakers and other relevant stakeholders and make them readily accessible through a range of media.

Illustrative Research Questions

How can methods or approaches be improved:

- For representing, analyzing, describing, and communicating uncertainties and for evaluating and addressing scientific disagreements about the nature and extent of risks?

- For understanding the costs (market and non-market) and opportunities (societal, organizational, and individual) associated with global climatic variability and change?

- For representing how individuals, organizations, and societies make choices regarding risks whose consequences are long-term and uncertain?

- For evaluating and comparing the effectiveness of different approaches to modeling decisionmaking?

- For understanding how public and private decisionmaking impacts human health and the natural environment?

- For designing processes that combine scientific analyses with the policy deliberations and judgments of decisionmakers?
- For disseminating, communicating, and evaluating climate forecast information for use by decisionmakers?

Research Needs

Associated research needs include analysis of decision processes to identify what information on global environmental variability and change is most useful and at what stage in the decision process that the information is needed. This work will be done in collaboration with the Decision Support Resources development described in Chapter 11.

Milestones, Products, and Payoffs

- Research on these questions will enable the development of assessments of the kind of knowledge and information needed by different decisionmakers and stakeholders in order to enhance decisionmaking associated with climate change, and will produce decision support resources [beyond 4 years].
- Research centers will be expected to facilitate interactions among researchers and relevant decisionmakers and stakeholders [less than 2 years]; provide educational opportunities for U.S. students and faculty [less than 2 years]; increase understanding of the types of information needed by decisionmakers [less than 2 years]; develop tools that people, organizations, and governments can use to better understand the risks associated with climate variability and change and the options they have to address those risks [2-4 years]; and increase basic understanding of decisionmaking processes associated with climate change and variability [2-4 years].
- In addition to the advancement of basic understanding and modeling of decisionmaking, the program expects to develop improved modeling frameworks that better link general circulation, ecological, and economic models of the agricultural and forestry sectors [2-4 years].
- Recommendations will be developed for producing, communicating, and disseminating climate information and its associated uncertainties to resource managers (e.g., farmers, forest landowners, drought policy planners, water utilities) and urban planners at local to national levels. Integration with Decision Support efforts will be important as will drawing on communication and dissemination tools developed by the Water Cycle research element [2-4 years].

> **Question 9.4**: What are the potential human health effects of global environmental change, and what climate, socioeconomic, and environmental information is needed to assess the cumulative risk to health from these effects?

State of Knowledge

It is well established that human health is linked to environmental conditions, and that changes in the natural environment may have subtle, or dramatic, effects on health. Timely knowledge of these effects may support our public health infrastructure in devising and implementing strategies to compensate or respond to these effects.

Over the past decade, several research and agenda-setting exercises have called for continued and expanded research and development of methods in this area (WAG, 1997; NRC, 1999a,d, 2001c; EHP, 2001; IPCC, 2001b). Given the complex interactions among physical, biological, and human systems, this research must be highly interdisciplinary, well integrated, and span the breadth from fundamental research to operations. A multi-agency interdisciplinary research effort to examine the linkages across these sectors is in place with research focusing on global and developing country impacts, and on the effects of simultaneous environmental and economic shifts on human health and well-being.

Federally supported research has thus far provided information on a broad range of health effects of global change, including the adverse effects of ozone, atmospheric particles and aeroallergens, ultraviolet (UV) radiation, vector- and water-borne diseases, and heat-related illnesses (see Figure 9-2). Research continues to improve understanding of the potential impact of climate variability on certain infectious diseases, and researchers are developing and evaluating tools and information products for anticipating and managing any such impacts that capitalize on the enormous protections afforded by wealth and the public health infrastructure. However, many questions remain unanswered.

Illustrative Research Questions

- What are the impacts of changes in water quantity and quality, temperature, ecosystems, land use, and climate on infectious disease and what is the relative importance of these impacts compared to other socioeconomic and technological factors?
- What are the impacts of atmospheric and climatic changes on the health effects associated with ambient air quality and UV radiation?
- What are the health effects and effective preparedness and response strategies associated with temperature extremes and with extreme weather events?
- What are the best methods for assessing known and potential climate-related health impacts and for developing and evaluating useful tools and information products to enhance public health and support decisionmaking?
- How can we improve the capacity of public health and societal infrastructure to prevent, detect, and effectively respond to health impacts that may be associated with climate change?
- How can the incorporation of health impacts and trends into climate change scenarios improve tools for decisionmaking at various time scales?

Research Needs

Research needs include:

- Work on improved understanding of the health effects of UV radiation, including exposure across regions and populations, risk awareness, and early detection.
- Initiation of a temporally and spatially compatible long-term field study, empirical analysis, and integrated modeling effort of the physical, biological, and social factors affecting the potential impact of climate variability and change on public health issues of national importance.
- Research on the climatic effects of temperature on air quality, particularly in urban heat islands and other regional settings, and the potential health consequences.

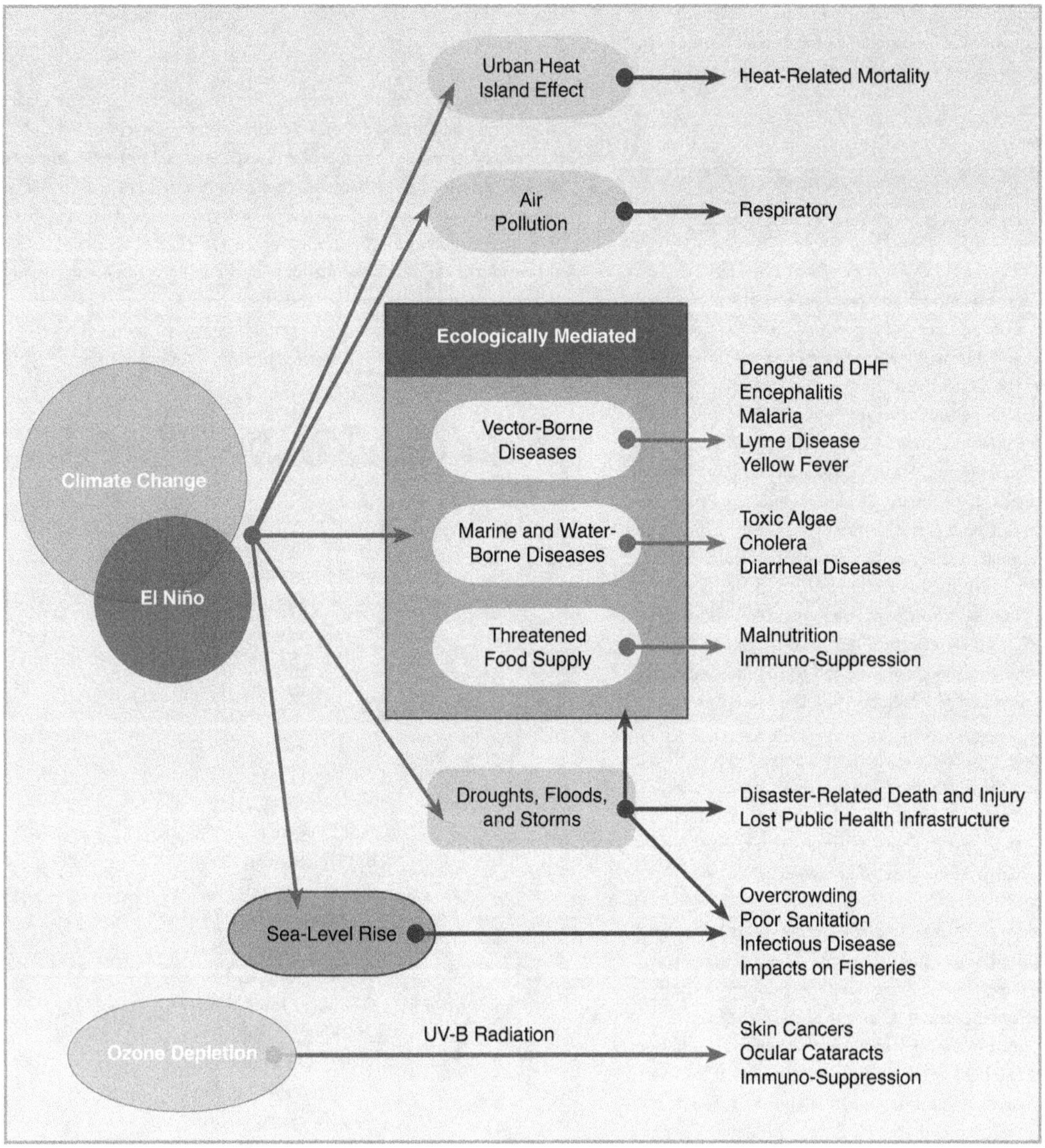

Urban Heat Island Effect → Heat-Related Mortality	
Air Pollution → Respiratory	
Ecologically Mediated	
Vector-Borne Diseases →	Dengue and DHF Encephalitis Malaria Lyme Disease Yellow Fever
Marine and Water-Borne Diseases →	Toxic Algae Cholera Diarrheal Diseases
Threatened Food Supply →	Malnutrition Immuno-Suppression
Climate Change / El Niño	
Droughts, Floods, and Storms →	Disaster-Related Death and Injury Lost Public Health Infrastructure
Sea-Level Rise →	Overcrowding Poor Sanitation Infectious Disease Impacts on Fisheries
Ozone Depletion — UV-B Radiation →	Skin Cancers Ocular Cataracts Immuno-Suppression

Figure 9-2: Possible pathways of public health impacts from climate change. For more information, see Annex C.

- Research on the effect of seasonal-to-interannual climate variation on public health, especially at a regional scale, and the integration of this information into decisionmaking processes.
- Research on preventing and reducing the adverse health impacts of extreme weather events.
- Research on prevention and control of infectious diseases that might increase in incidence as a result of climate change.
- Research on the regional control and treatment of vector- and water-borne diseases (this work should be linked to the Ecosystems research element).
- Economic analysis of the prevention, control, and treatment strategies for potential public health impacts associated with climate variability and change.
- Studies on the costs and benefits to public health of mitigation strategies for greenhouse gas emissions.

A parallel need exists to develop additional appropriate tools and methods for assessing and adapting to potential health outcomes,

and for evaluating the impact of research, the effectiveness of Earth science information and products, the methods for communicating that information, and the systematic identification of knowledge gaps and feedback to the research communities.

Milestones, Products, and Payoffs

Products from this area include operational tools, research to support innovative institutional arrangements and processes, and research results that may be used by decisionmakers. Expected milestones, products, and payoffs include:

- Additional tools for preventing and managing the public health threat of infectious diseases [2-4 years].
- Assessments of the potential health effects of combined exposures to climatic and other environmental factors (e.g., air pollution, and including input from the Atmospheric Composition research element) [beyond 4 years].
- A multi-agency joint award for competitive grants to support research on climate variability and health [2-4 years and beyond].

- The next phase of health sector assessments to understand the potential consequences of global change for human health in the United States, especially for at-risk demographic and geographic subpopulations [2-4 years].

National and International Partnerships

The study of human contributions and responses to global change within the Climate Change Science Program (CCSP) has ties to a number of national and international programs beyond those represented among the CCSP member agencies, including the International Human Dimensions Programme, the Intergovernmental Panel on Climate Change, the World Health Organization, the Pan American Health Organization, Environment Canada, Health Canada, the Climate Change Technology Program, the National Research Council, the Centers for Disease Control and Prevention, the National Institute of Environmental Health Sciences, the National Institute for Child Health and Human Development, the National Institute of Allergy and Infectious Diseases, the U.S. Census Bureau, the Bureau of Labor Statistics, and other federal agencies and programs, and to the International Research Institute for Climate Prediction and the Inter-American Institute for Global Change Research (see Chapter 15). Collaborations between the federal agencies involved in global change research and the abovementioned organizations include co-sponsorship of scientific workshops and conferences, efforts to set scientific agendas in research areas of mutual interest, and collaborative assessments of the state of knowledge.

Furthermore, numerous collaborative research projects between scientists in the United States and other countries are underway. As examples, U.S. scientists collaborate with developing country scientists to analyze coping strategies and the use of climate information in the face of year-to-year climate variability. In addition, U.S. researchers in the field of economics and other areas associated with creating decisionmaking frameworks collaborate through institutions such as Stanford University's Energy Modeling Forum (EMF). For example, an annual EMF meeting of specialists in integrated assessment modeling and related disciplines, such as climate science, biology, and health, has generated a great many

successful research partnerships across both countries and disciplines.

The scientific community has called for strengthening international cooperation and coordination related to human contributions and responses research, particularly in the areas of the potential impacts of climate on human welfare and resource management, vulnerability assessments, and adaptation research. Progress depends on advances in these areas, as well as in improvements in climate modeling, observations, and our understanding of the integrated climate system and associated socioeconomic and environmental responses. Cooperation should include the collection and archiving of social and economic data, as well as exchanging methodologies and research insights.

CHAPTER 9 AUTHORS

Lead Authors
Janet Gamble, USEPA
Caitlin Simpson, NOAA

Contributors
Mitch Baer, DOE
Tom Baerwald, NSF
Nancy Beller-Simms, UCAR
Rebecca Clark, NIH
Cheryl Eavey, NSF
Mary Gant, HHS
Cliff Hickman, USFS
Bill Hohenstein, USDA
John Houghton, DOE
Carol Jones, USDA
David Kirtland, USGS
Elizabeth Malone, PNNL
Melinda Moore, HHS
Claudia Nierenberg, NOAA
Robert O'Connor, NSF
Warren Piver, NIH
Joel Scheraga, USEPA
Jim Titus, USEPA
Juli Trtanj, UCAR

10 Modeling Strategy

CHAPTER CONTENTS

Goal 1: Improve the scientific basis of climate and climate impacts models.

Goal 2: Provide the infrastructure and capacity necessary to support a scientifically rigorous and responsive U.S. climate modeling activity.

Goal 3: Coordinate and accelerate climate modeling activities and provide relevant decision support information on a timely basis.

This chapter introduces goals and objectives intended to provide accelerated scientific improvements in climate and climate impact models responsive to the needs of the Climate Change Science Program's (CCSP) scientific research and decision support activities.

Models are essential tools for synthesizing observations, theory, and experimental results to investigate how the Earth system works and how it is affected by human activities. Models can be used in both a retrospective sense, to test the accuracy of modeled changes in Earth system forcing and response by comparing model results with observations of past change, and in a prognostic sense, for calculating the response of the Earth system to projected future forcing. Comprehensive climate system models provide the primary quantitative means to integrate scientific understanding of the many components of the climate system and, thus, are the principal tools available for making quantitative projections.

The CCSP modeling strategy consists of three goals: Improve the scientific basis of climate and climate impacts models; provide the infrastructure and capacity necessary to support a scientifically rigorous and responsive U.S. climate modeling activity; and coordinate and accelerate climate modeling activities and provide relevant decision support information on a timely basis. In order to achieve these goals, three modeling arenas will be implemented: (1) diverse and disparate research activities that represent new process understanding in models; (2) assimilation and integration efforts that employ new types of observations and tools and next-generation understanding and model coupling; and (3) "high-end" climate models run for scenarios required in periodic scientific assessments or to achieve higher resolutions.

The CCSP strategy envisions two complementary streams of climate modeling activities. The first is principally a research activity, which will maintain strong ties to the global change and computational science research communities to rapidly incorporate new knowledge into a comprehensive climate and Earth system modeling capability. Closely associated with the research activity, but distinct from it, will be the sustained and timely delivery of predictive model products that are required for assessments and other decision support needs. CCSP will ensure that a productive partnership is maintained between product-driven modeling activities and the discovery-driven modeling research program that will underpin its credibility and future success.

In his 11 June 2001 speech, the President asked his Administration to work to "develop state-of-the-art climate modeling that will help us better understand the causes and impacts of climate change." In response to this directive, the program is addressing the following overarching question:

How can we most effectively accelerate the development, testing, and application of the best possible scientifically based climate and climate impact models to serve scientific research and decision support needs?

Based on recommendations in National Research Council (NRC) reports on U.S. climate modeling (NRC, 1999b, 2001d), the CCSP agencies initiated new activities to strengthen the national climate modeling infrastructure. These activities will accelerate the delivery of improved model products that are especially important for making climate simulations, predictions, and projections more usable and applicable to the broader research, assessment, and policy communities (see Annex D for definitions of climate "prediction" and "projection"). The new activities form the basis of a longer term solution that will maintain the pace and progress of the basic research, while simultaneously creating a path for the rapid exploitation of new knowledge in model development, testing, and applications.

> **Goal 1**: Improve the scientific basis of climate and climate impacts models.

Virtually all comprehensive climate models project a warmer Earth, an intensified hydrologic cycle, and rising sea level as consequences of increasing atmospheric concentrations of greenhouse gases. However, projections of the details about the magnitude, timing, and specific regional impacts and consequences are variable (IPCC, 2001a,b). The program is placing the highest priority on research aimed at addressing known modeling deficiencies (see Chapter 4, Question 4.1). The following objectives in pursuit of Goal 1 describe CCSP's long-term approach.

Objective 1.1: Accelerate research on climate forcing, responses, and feedbacks aimed at improving methods for quantifying and reducing uncertainties in the current generation of prediction and projection models

The climate system responds in complex ways to changes in forcing that may be natural (e.g., variations in the magnitude of solar radiation reaching the top of the atmosphere) or human-induced (e.g., changing atmospheric concentrations of greenhouse gases). Several of the program's science elements will provide climate modelers with the best scientific estimates of past and expected future climate forcing factors—for example, the Climate Variability and Change research element for solar variability; the Atmospheric Composition, Carbon Cycle, and Human Contributions and Responses research elements for radiatively active trace gases and aerosols; and the Land-Use/Land-Cover Change and Ecosystems research elements for land surface cover changes and energy exchanges.

The direct response of the climate to a change in forcing may be either diminished or amplified by feedback processes within the climate system itself. For example, warmer upper oceans will result in increased evaporation and, thus, increased concentrations of atmospheric water vapor—itself a strong greenhouse gas—a positive or amplifying feedback. Increased water vapor will alter cloudiness, which may be either a positive or a negative feedback, depending on the cloud height and type. Climate-induced changes at the land surface (e.g., through more intense and higher frequency droughts) may in turn feed back on the climate itself, for example, through changes in soil moisture, vegetation, radiative characteristics, and surface-atmosphere exchanges of water vapor.

Near-Term Priorities

- Because of the highly interdisciplinary and complex nature of climate processes in general, understanding and modeling feedbacks is a challenging research task. The program will give high priority to research conducted under the research elements aimed at understanding and modeling the most important known feedback processes (see Chapter 4, Question 4.1), with the goal of better quantifying and reducing uncertainties in climate predictions and projections.
- CCSP-supported climate modeling centers will work closely with scientists to use observations and research advances to improve modeling capability and provide more useful products for decision support. The knowledge transfer will be enabled and accelerated by **C**limate **P**rocess and Modeling **T**eams (CPTs), a new paradigm for CCSP climate modeling and applications research (CPTs are discussed later in this chapter and in Box 4-1).

Enhanced understanding and improved representation in models of the processes that influence climate will improve confidence in model forecasts and projections. Reductions in uncertainty will be measured by the degree to which differences between the major climate models as well as differences between observations and relevant model fields are reduced.

Objective 1.2: Develop the next generation of global climate models through the addition of more complete representations of coupled interactive atmospheric chemistry, terrestrial and marine ecosystems, biogeochemical cycling, and middle atmospheric processes

Past emphasis has been on the development and testing of physical aspects of coupled atmospheric and ocean general circulation models (GCMs). This occurred primarily because climate models have their roots in numerical weather prediction models (atmospheric GCMs with some ocean coupling that treat primarily physical processes), and the available observations were those derived for the application of weather prediction models. Significant advances have been achieved on the physical aspects of climate modeling, but much more research is required (see Chapter 4).

In the 1990s, motivated by unresolved questions about long-term climate change, modeling efforts expanded to include additional components of the climate system, such as chemistry and biology, that are important to longer term climate processes. Here too, much has been accomplished, but much remains to be done. In parallel with continued research into physical climate processes and modeling, the program will enhance efforts to more fully develop the chemical and biological components of climate models, including their human dimensions, in the context of a coupled interactive system, and also expand the atmospheric domain to include middle-atmosphere processes. This priority is motivated by the need to provide answers to pressing questions about long-term change and variability that may result from human-induced climate forcing involving chemical, biological, and human-induced processes.

Near-Term Priorities

- In the near term, work will concentrate on improved representations of aerosols, elements of the carbon cycle,

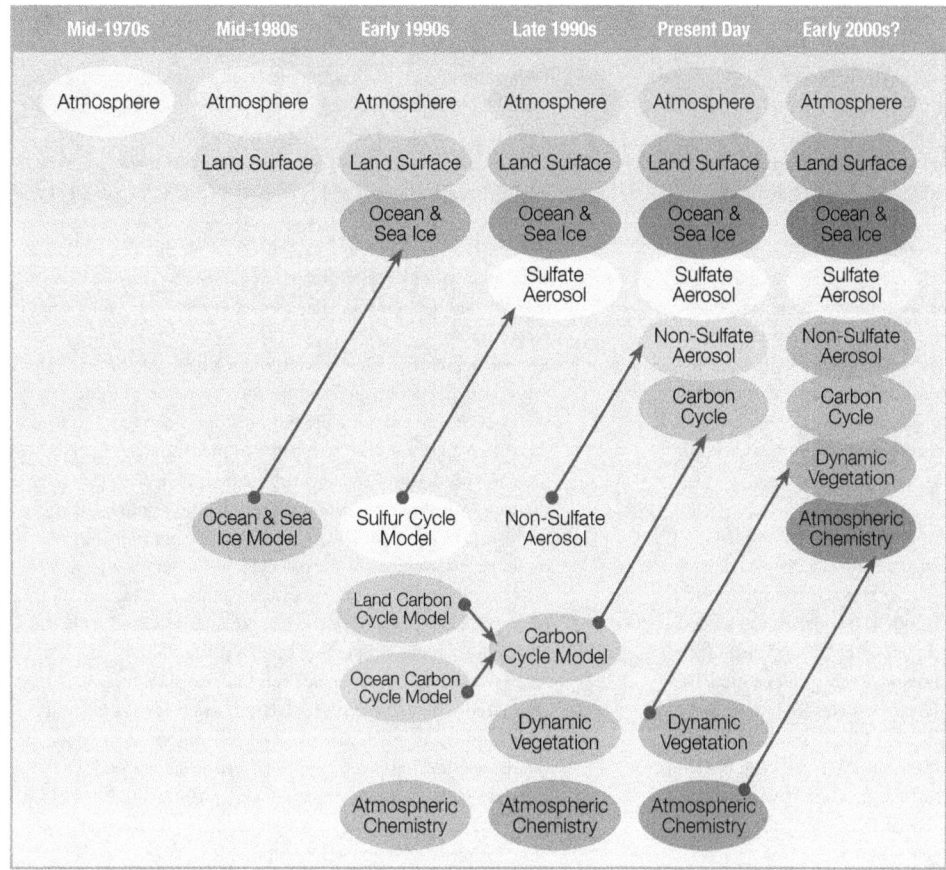

Mid-1970s	Mid-1980s	Early 1990s	Late 1990s	Present Day	Early 2000s?

Figure 10-1: The development of climate models over the last 25 years showing how the different components are first developed separately and later coupled into comprehensive climate models. Source: IPCC (2001a).

interactive land surface-atmospheric processes, and middle-atmosphere dynamics and chemistry. This work will integrate research advances by the CCSP research elements (see Chapters 3-9). The products will be next-generation climate system models with enhanced capabilities to more comprehensively model the interactive physical, chemical, and biological components of the climate system. This work will continue over the longer term, leading to fully interactive Earth system models. For a more detailed description of research planned for one key next-generation climate system model, see the Community Climate System Model website, <http://www.ccsm.ucar.edu/management/sciplan2004-2008>.

The Role of the Middle Atmosphere in Climate
- The representation of the stratosphere will be improved in climate models, including feedbacks between stratospheric dynamics and stratospheric ozone, and between stratospheric dynamics and water vapor (see Chapter 4).
- CCSP will examine whether variations in solar irradiance can play a significant role in the natural variability of the climate system. Models will be developed that extend vertically through the mesosphere and include interactions between ultraviolet radiation, ozone chemistry, and atmospheric circulation and transports (see Chapters 3 and 4).

Interaction of Aerosols, Chemistry, Ecosystems, and Hydrology
- CCSP will develop the capability to model a fully interactive aerosol system within climate models, in order to examine the

multi-faceted roles of aerosols in the climate system (see Chapters 3, 4 and 5).

Atmospheric Composition
- CCSP climate system models will be developed to include both tropospheric and stratospheric chemistry, chemical processes related to interactions at the Earth's surface, and interactions with hydrologic processes, in order to adequately represent the sources, sinks, and transformation processes of those molecules that are important for climate because of their ability to absorb and/or emit radiation and whose concentration and properties must be adequately simulated in climate models (see Chapter 3.)

Biogeochemistry and Ecosystems
- CCSP will develop the modeling tools and validation data sets for incorporating and assessing historical and future land use; the dynamics of managed forest, rangeland, agricultural, coastal, and ocean systems; and deliberate carbon sequestration activities (see Chapters 6-9).

High-Resolution Climate Model
- At present, there are two complementary approaches to modeling climate change and climate impacts at regional and sector scales. One approach uses a variety of "downscaling" techniques, ranging from nested mesoscale (regional scale) models to adaptable global grids (see Objective 1.6). The second approach is to increase the resolution of the global models themselves, throughout the entire global domain. The former approach is

complicated by several unresolved issues, ranging from the effects of lateral boundary processes to conservation principles. The second approach is not practical without very large increases in computing capability. The CCSP strategy is to continue to support research in and applications of regional-scale climate models and other downscaling methods. On a smaller scale, the program will support pilot projects for next-generation very high-resolution global climate model development, in anticipation of continuing advances in computational technologies.

Climate models and observations are intimately connected, as described under Objectives 1.3-1.5 below. Models must be evaluated and constrained by observations, which also serve to initialize models used for prediction. Models provide a dynamically consistent framework into which diverse climate observations can be assimilated to produce "value-added" data sets of gridded, continuous time series of hybrid field observations and modeled data.

Objective 1.3: Foster model analysis and testing through model diagnostics and intercomparison activities, including comparison with observations
Given the complexity of climate models, it is difficult to ascertain why a particular model performs "better" or "worse" than others in any given situation. Generally speaking, a measure of a model's quality is its ability to simulate the current climate [global averages, annual cycle, major modes such as the El Niño-Southern Oscillation (ENSO) and North Atlantic Oscillation (NAO), etc.] and the climate of the past centuries (19th and 20th) as measured by available observations. Therefore, it is essential to carefully identify and document model deficiencies, such as systematic biases ("bias" refers to the tendency of a model's prediction to drift toward the model's climatology, which may be at variance with the real world). Furthermore, thorough diagnostics of model deficiencies (e.g., intercomparison of various models, comparison of models to observations, analysis of physical mechanisms using simplified or conceptual models, and carrying out model sensitivity experiments) are essential to identify sources of model errors, which then provide the basis for model improvement.

Near-Term Priorities
- Various projects aimed at fostering intercomparison among climate models or their components are underway. The program will continue to support these efforts in model intercomparisons, with an emphasis on the diagnostics for the sources of deficiencies common to many climate models, such as the "double Intertropical Convergence Zone (ITCZ)" problem in the tropical Pacific Ocean, which is closely associated with the modeling of tropical deep convection and cloud feedbacks. A key approach is to bring together the expertise and interests of observation specialists, process modelers, diagnosticians, and climate modeling centers to tackle the problem from various angles in a coordinated manner (e.g., the CPT approach).
- Given the status of observations, it is not obvious which model(s) are "better" (see Chapter 12, Goal 4). In practice, when assessing how models simulate the present climate, most of the comparisons are made using "reanalysis data" provided by one of several groups. A wealth of satellite data is already available, with much more coming available. The linkage between satellite data and the needs of modeling groups that has been made very

effectively in some modeling organizations (e.g., NASA's Data Assimilation Office, NOAA's National Centers for Environmental Prediction) should be replicated by other CCSP modeling groups. The program will address the role of both the modeling groups and the remote-sensing teams (see Chapter 12, Goal 6). Those modeling groups that have not made use of satellite data in the past will be encouraged to utilize it as part of their activity. Remote-sensing projects will be better integrated into the modeling community, with the scientific research team associated with each instrument encouraged to provide user-friendly data sets for modeling as part of their data collection responsibility.
- For paleoclimate studies, there is often a large gap between the actual data being obtained and the interpretation of those data in terms of the climate variables output from each model. When there is an apparent mismatch between model output and climate interpretation, it is often unclear whether the interpretation of the paleoclimate proxy data is responsible. To minimize this problem, the program will encourage modelers to include direct representations of the paleoclimate proxy data (e.g., water isotopes), so that comparisons can then be made with the proxy data independent of any climate interpretation. This will also help to improve the value of proxy data.
- At times, model performance can look satisfactory when compared with one reanalysis data set, and less so compared with another. As in the case of GCM differences, the reasons for the reanalysis differences need to be better understood. The program will encourage improved understanding through close cooperation between the reanalysis modeling data centers and observational and empirical scientific research (e.g., via a CPT). This effort will require international cooperation, and possibly some joint funding of such comparisons through international entities, such as the World Meteorological Organization (see also Objective 1.5).

Some key examples of global climate model intercomparison projects include the Atmospheric Model Intercomparison Project (AMIP, <www-pcmdi.llnl.gov/amip/>), the Coupled Model Intercomparison Project (CMIP, <www-pcmdi.llnl.gov/cmip/>), and the Paleoclimate Model Intercomparison Project (PMIP, <www-lsce.cea.fr/pmip/>). AMIP was initiated in 1989, for the purpose of identifying and documenting the differences among the various atmospheric models and observations, to provide a basis for model improvement. Nearly all atmospheric models in use today have been calibrated using the AMIP experimental protocol. AMIP is closely integrated with CMIP, and provides a calibration for understanding how coupled models respond to increasing levels of greenhouse gases. Model-model and model-measurement intercomparisons have been carried out for a number of component models that will ultimately need to be incorporated into global climate models. For instance, multi-dimensional stratospheric composition models have been intercompared with both each other and with observations in a series of "Modeling and Measurement" papers.

CMIP, like AMIP, is an activity of the World Climate Research Programme (WCRP). It was initiated in 1995 with the goal of collecting and intercomparing simulations from global coupled climate models—that is, models that operate over the complete global three-dimensional domain of the climate system, with components typically consisting of atmosphere, ocean, sea ice, and land surface. Virtually every global coupled model group worldwide is participating in CMIP, including groups from Australia, Canada, China, France, Japan, Germany, South Korea, Russia, the United Kingdom, and the United States.

PMIP, launched in 1994 and endorsed by the International Geosphere-Biosphere Programme and WCRP, is an international project involving members of all the major modeling groups worldwide. The aims of the PMIP project are to improve understanding of the mechanisms of climate change by examining such changes in the past, and to evaluate the ability of climate models to reproduce paleoclimate conditions radically different from present-day climate.

Objective 1.4: Improve short-term climate predictions through model initialization with enhanced observational data

Over the past several decades, progress in improving numerical weather forecasting has been achieved primarily through the continuing increase in model resolution and improvement in initializing (specifying the values of the model's variables at the start of the forecast) the prediction models. Initialization has been improved through the increase in available observations, particularly from remote-sensing platforms, and by advances in data assimilation techniques.

Improving climate prediction can follow a similar path. Although seasonal climate forecasting using coupled climate models is still at an early stage of development, the most significant impact on forecasts in the past decade has come through the use of ocean data assimilation to assimilate *in situ* observations for initializing the coupled models for ENSO forecasts.

Near-Term Priorities

- To improve short-term climate forecasts, CCSP modeling will incorporate new and improved technologies in data assimilation, such as coupled ocean-atmosphere data assimilation and land data assimilation, and better utilization/assimilation of *in situ* and remotely sensed global oceanic, atmospheric, and terrestrial observations (e.g., better utilization of altimetric sea-level data and improved methodologies for assimilating soil moisture and sea surface salinity; see also Chapter 12).
- The most significant challenge in data assimilation for climate prediction is the bias in prediction models. To reduce model bias, CCSP will significantly improve physical formulations of climate models, through improved incorporation of existing and new observations as well as through results from new process studies.
- Observations of several new variables have been demonstrated to be critical for improving seasonal forecasts, including sea surface salinity (SSS), particularly in tropical oceans, and soil moisture over the land. Satellite missions for measuring SSS and soil moisture have been planned for the next decade (see Chapter 12). In conjunction with efforts to make these observations available globally, the program will invest in research and development on the use of salinity and soil moisture observations in data assimilation for initialization of climate forecast models.

Objective 1.5: Provide comprehensive observationally based model-assimilated climate data sets for climate process research and testing of climate model simulations and retrospective projections

There is a critical need for an ongoing effort to provide complete descriptions of the present and past state of the atmospheric and oceanic components of the global climate system, together with continually updated data sets compiled in consistent ways to enable comparison of models with observations (see Chapter 12). As new climate observations are obtained, it is essential to place them in a historical context to enable the accurate evaluation of departures from normal, and trends and change in variability. New observations can provide additional information about climate when they are put in the context of past observations at uniformly spaced points in space. This is accomplished through the systematic processing and integration of climate observations using the state-of-art climate models and data assimilation methods.

Near-Term Priority

- The program will support research and development of advanced data assimilation methods and the production of global climate time series to establish reliable climatologies, identify real versus fictitious trends, and develop techniques to minimize the effects of changing observing systems and model biases.

Objective 1.6: Accelerate the development of scientifically based predictive models to provide regional- and fine-scale climate and climate impacts information relevant for scientific research and decision support applications

Regional climate models (RCMs, also called mesoscale models) are used in conjunction with GCMs to provide "downscaling" of climate variables for regional-scale predictions or projections. RCMs operate on scales that could not be accessed directly in GCMs due to computational capability limitations. Thus, while global climate models will be approaching 100-km resolution in the foreseeable future, it is unlikely that in the near future they will reach the 10-km scale that regional models can simulate. RCMs represent the most prevalent current approach to dynamic downscaling. An alternative approach is statistical downscaling, for which the applicability to climate change is uncertain. Other dynamic approaches include stretched-grid GCMs, with uneven horizontal resolution, which can provide regional-scale resolution over certain domains.

RCMs can also be used for "upscaling" of information to test GCMs. To the extent that many small-scale processes, parameterized in general circulation models as "sub-grid scale," can be better simulated in regional models, they can provide feedback on the adequacy and limitations of the coarse-grid parameterization schemes.

While the finer resolution provided by regional-scale models is desirable, the quality of the resulting solutions is limited, except perhaps in better delineating orographic (mountain region) precipitation. There are several broad issues involved in the use of regional models. While such models may allow many physical processes to be incorporated on the scale at which they occur (e.g., rainfall in the vicinity of mountains), the physics on regional scales is often uncertain. The uncertainties have led to the existence of multiple versions of most RCMs. When utilizing them for climate change simulations, the different versions can produce results that differ from each other by as much as the observed climate changes they are attempting to simulate, which questions their use by policymakers and decisionmakers.

A second issue is consistency. Regional-scale models of necessity generally utilize physical parameterizations that differ from those of the GCM, if for no other reason than because parameterizations are often scale-dependent. The GCM supplies boundary conditions generated with one physics package, and the RCM utilizes these boundary conditions in conjunction with different physics. This effect is made worse in "two-way coupling," in which the mesoscale result is fed back to the GCM. Using a mesoscale model over the Rocky Mountains, for example, and not at the same time over the Himalayas, provides inconsistent forcing for the GCM planetary wave structure, which is affected by both mountain chains.

Near-Term Priorities

- Many challenging downscaling and upscaling research issues remain to be addressed in order to provide the most useful information possible for decision support. Although regional modeling is highly relevant to most of CCSP's participating agencies, research and applications are often centered on the missions and interests of individual agencies. This has resulted in a need for more focused leadership and coordination following the guidelines given in Chapter 16. Toward this end, CCSP will establish a process for coordination of regional modeling activities, including the development of methods to transfer regional decision support products into operations. Regional and sectoral climate and climate impacts research and modeling is a high program priority and its support will be accelerated.
- Reducing the uncertainty associated with such issues will require diagnostics and intercomparison of regional-scale models and application techniques. With CCSP support. regional models will be tested systematically when forced by real-world and GCM-produced boundary conditions, and the results quantified against regional observations for different locations. The different physical parameterizations being used on regional scales will be compared with observations when available, and assessments made of their realism. Regional reanalysis and observational data sets will be used for verification purposes when evaluating RCM output.
- To provide greater consistency and to help improve GCMs, CCSP will support upscaling of well-validated physics at the regional level to provide insight into parameterizations for the coarser grid GCM. For example, the mesoscale model resolution will be expanded systematically to learn how the results change, and help determine what is appropriate for GCMs. This approach is promising (e.g., for physically based cloud-resolving models).

Goal 2: Provide the infrastructure and capacity necessary to support a scientifically rigorous and responsive U.S. climate modeling activity.

The principal U.S. agencies that support climate model development and application commissioned NRC to analyze U.S. modeling

efforts as well as to suggest ways that the agencies could further develop the U.S. program so that the need for state-of-the-art model products can be satisfied. The NRC reports, *The Capacity of U.S. Climate Modeling to Support Climate Change Assessment Activities* (NRC, 1999b) and *Improving the Effectiveness of U.S. Climate Modeling* (NRC, 2001d), provided valuable guidance on how to improve U.S. climate modeling efforts. Also, the U.S. Global Change Research Program issued *High-End Climate Science: Development of Modeling and Related Computing Capabilities* (USGCRP, 2000), a report commissioned by the Office of Science and Technology Policy to make recommendations on climate modeling activities.

These documents emphasized four key points: (1) the acknowledged U.S. leadership in basic climate research that generates the knowledge base, which underpins both domestic and international modeling programs; (2) the limited ability of the United States to rapidly integrate the basic climate research into a comprehensive climate modeling capability; (3) the challenges, including software, hardware, human resource, and management issues, to routinely produce comprehensive climate modeling products; and most important, (4) the need to establish a dedicated capability for comprehensive climate modeling activities, including the global climate observations and data that support modeling.

Objective 2.1: Provide the computing, data storage and retrieval, and software engineering resources required to support a world-class U.S. climate modeling activity

The production of global model predictions of climate variability and change with sufficient resolution and veracity to provide useful regional information to decisionmakers requires a comprehensive computational infrastructure of computing resources, data centers, networks, and people. The success of this enterprise is predicated on a long-term commitment of the program to sustain the institutional support and investment required to maintain a resilient and state-of-the-art computational and information technology infrastructure.

Near-Term Priority

- CCSP will support researchers in developing more comprehensive coupled models that need to be evaluated, then exercised to produce ensemble projections of multi-century climate change scenarios. The results from these runs will be analyzed and employed by hundreds of researchers engaged in climate studies, impacts research, and assessment.

High-end computing needs are often divided into the two broad and overlapping categories of *capability* needs and *capacity* needs. The former refers to applications that require the *capability* to do sophisticated, cutting-edge simulations that were impossible just a few years ago because the computing platforms to execute them did not exist. Typically, these simulations require the dedicated use of the most powerful computer available for several weeks at a time. Ensemble simulations with the current and next generations of coupled models fall into this category. In addition to capability resources, CCSP requires a large amount of *capacity* resources to carry out the bulk (in terms of total computing cycles) of its modeling and analysis needs.

Near-Term Priority

- To meet its mandate, CCSP will provide researchers at the major modeling centers with access to steadily growing computational resources that increase by a factor of four each year. This will result in a 256-fold increase in available computing power over 4 years and a roughly 1,000-fold increase over 5 years. The factor of 1,000 will provide a three-fold increase in resolution which corresponds to a factor of 20 increase in computing requirements, a factor of two for improved process representation, a factor of four for increased comprehensiveness, a factor of three for increased ensemble size, and a factor of two for an increased number of scenarios. This level of enhancement will meet the computational requirements for the next-generation climate system models (described in Objective 1.2). This will be accompanied by appropriately scaling investments in software engineering, input/output systems, and local storage systems together with increasing investments in high-end analysis and visualization software development. A part of the growth will be met as computing equipment is replaced periodically every 3 to 5 years, with better technology at lower prices (Moore's "law").

Currently, capability resources needs are met through a combination of dedicated and shared resources. The dedicated resources—particularly the computers at the NOAA, NASA, DOE, and NCAR laboratories—do not fully meet the needs of the modeling community. The shortfall is made up by additional resources acquired at several shared-access supercomputer centers by the researchers themselves, through individual proposals. These centers are not supported by CCSP and provide computing through a competitive review process among many researchers from many fields of science and engineering. CCSP objectives cannot be met without these additional computing resources.

One example of such computing resources is DOE's National Energy Research Scientific Computing (NERSC) facility at the Lawrence Berkeley National Laboratory. To help meet computing needs for U.S. climate modeling, 10% of the computing cycles at NERSC will be made available to the broader scientific community in an open competition, with a special emphasis on climate modeling. Those cycles that might be allocated for climate modeling in that open competition would supplement the cycles at NERSC already used for climate modeling.

Possible collaborations with the Earth Simulator center in Japan, which has a computer with 20-50 times the capability of any existing U.S. machine for climate model applications, may help meet some of the near future needs. Unfortunately, this resource is now not directly accessible to U.S.-based researchers, because its lack of appropriate communication and mass storage and retrieval capacity requires users to be physically located at the center. Data generated at the center must be transferred to storage media that are then physically transported to data archives in the United States. This problem may be resolved in the near future.

Near-Term Priority

- In April 2003, the Office of Science and Technology Policy initiated a High-End Computing Revitalization Task Force

(HECRTF) to assess the current high-end computing capability and capacity within the federal agencies, and to develop a plan to revitalize high-end computing research and enable leading-edge scientific research using high-end computing. CCSP has been coordinating with HECRTF to ensure that the CCSP computing capability and capacity needs are considered.

The divergence in high-end computing architectures over the last decade has made developing models for high-end capability machines more labor-intensive, requiring the addition of a software engineering component to model development.

Near-Term Priority

- Recent CCSP projects, including the NASA Earth System Modeling Framework (ESMF) program (see Objective 3.2) and the DOE Community Climate System Model (CCSM) Software Engineering Consortium program address this requirement. CCSP will support their continuation.

In addition to these primary capability computational resources, CCSP requires a network of available capacity computing engines, data archives, and associated information technology infrastructure to make the model products readily available and accessible, so that further analysis and the development of secondary products, such as downscaled model information, can be used for research and assessment. While the current archive of model results totals several tens of terabytes (trillions of units of information), future model data archives, and associated observational data to evaluate the models, will consume tens of petabytes (thousand terabytes).

Near-Term Priority

- The information technology infrastructure will be tailored to meet the specific needs of the CCSP modeling community, which will require the development and maintenance of both the software and hardware components that form the backbone of the infrastructure. To accommodate the rapid rate of turnover in information technology, the infrastructure will be flexible and dynamic so that it can evolve over time to meet increasing demands and utilize the best available technology. Projects such as the DOE Earth System Grid (ESG) provide a start in this direction by cataloging and making a subset of the existing model archives available over the Internet, but far more is needed. Contingent on continued progress and merit of the ESG project, the DOE Office of Science will continue to directly support ESG through at least 2005.

Objective 2.2: Establish graduate, post-doctoral, and visiting scientist programs to cross-train new environmental scientists for multidisciplinary climate and climate impacts modeling research and applications

The development, testing, and application of climate system models requires environmental and computer technicians and scientists with expertise in a broad range of disciplines. The scientific disciplines include atmospheric physics and chemistry; ocean physics, chemistry, and biology; ice physics; biological ecology; geology; applied mathematics; and the interactions among them. The activity also requires computer software engineering to develop, test, and manage model code (which, for a state-of-the-art climate system model, is currently about 500,000 lines and is projected to grow to about one million lines over the next 5 years).

Near-Term Priority

• As climate modeling becomes ever more complex, a shortage of appropriately trained scientists and technicians has become one of the limiting factors to progress. To meet this need, CCSP will establish a graduate student, post-doctoral, and visiting scientist fellowship program for climate modeling research and applications. The program will offer cross-training opportunities in climate modeling and computer sciences/software engineering. A modest post-doctoral and visiting scientist program has been established in FY2003 and will be expanded in future years.

> **Goal 3**: Coordinate and accelerate climate modeling activities and provide relevant decision support information on a timely basis.

The NRC review of U.S. climate modeling (NRC, 2001d) recommended the following as high priorities for the nation:
• Centralized operations and institutional arrangements for delivery of climate services
• A common modeling infrastructure
• Human resources.

The dispersed and diverse nature of climate research, including climate modeling research, in the United States requires an integrating strategy to rapidly infuse new knowledge into the most complete models used to simulate and predict future climate states. At the same time, the ability to provide decision support requires a robust and ready modeling capability to perform specialized projections and simulations to inform policymaking. A multi-tiered CCSP strategy has evolved over the last several years to address shortcomings identified in NRC reports on U.S. integrated climate modeling efforts (NRC 1999b, 2001d), as well as to meet anticipated future demands. The strategy combines a "bottom-up" approach required to solve difficult basic research problems with a "top-down" approach to focus a component of modeling research on the needs for decision support.

At the most fundamental level of the strategy is the large number of basic research projects at universities, federal laboratories, and in the private sector, which, along with the larger centers, produce new knowledge required to further improve climate models.

The middle level includes modeling centers that conduct essential research and development for climate, weather, and data assimilation applications. These centers are of two types, the first of which are the large modeling centers that develop and maintain state-of-the-science global models but have primary missions other than century-scale global climate projection, although they may have some activities in that area. These centers include the NOAA National Center for Environmental Prediction, which conducts operational weather and climate prediction; the NASA Goddard Institute for Space Studies, which focuses on using satellite data to provide representations of

climate forcing fields and uses these to model and study climate sensitivity to natural and human forcings; the NASA Goddard Space Flight Center, which focuses on the use of satellite data to generate research-quality data sets, to improve climate models, and to improve weather and coupled model predictions; the International Research Institute for Climate Prediction, which prepares and internationally distributes seasonal-to-interannual climate prediction and impacts products; and the Center for Ocean-Land-Atmosphere Studies, which focuses on studies of the predictability of climate. The second type is a number of smaller centers that have focused research interests on specific questions in climate research.

The strategy includes, at its third level, two "high-end" climate modeling centers that will continue to develop, evaluate, maintain, and apply models capable of executing the most sophisticated simulations, such as those required for assessments by the Intergovernmental Panel on Climate Change (IPCC). These two centers—one based at the NOAA Geophysical Fluid Dynamics Laboratory (GFDL) and the other based on CCSM and coordinated by the National Center for Atmospheric Research (NCAR)—are complementary, cooperative, and collaborative. Both high-end modeling groups have a long legacy of successful climate change research that predates the IPCC process and have led U.S. participation in international modeling evaluations and assessments.

A healthy balance of resources (financial, human, and computer) among these three levels is essential to maintain a strong U.S. applied modeling program. Researchers collaborate extensively across all three tiers, ensuring the rapid flow of knowledge and understanding, as well as the definition of new problems. One example of such collaboration is support for a common modeling infrastructure (CMI) and the Earth System Modeling Framework to optimize modeling resources and enable meaningful knowledge transfer among modelers. By adopting common coding standards and system software, researchers will be able to test ideas at any of the several major modeling centers and the centers themselves will be better able to exchange model components.

The multi-tiered strategy provides a structure by which the diverse contributions of the basic research community can be quickly utilized and integrated in state-of-the-art models used for climate simulation and prediction. At the same time, the strategy supports the decision support requirements of CCSP to make model simulations available for policy and impacts studies.

Objective 3.1: Provide routine, on-demand state-of-the-science model-based global projections of future climate

A major CCSP objective is to develop scientifically based global, time-dependent, multi-century projections of future climate change for different scenarios of climate forcing caused by natural variations and human activities. These projections are a primary form of scientific information to support decisionmaking about options to address the potential consequences of climate change. Development of capabilities to produce world-class climate change projections on demand to meet the needs of international and national assessments and other decision support requirements will bring new cohesion and coherence to the efforts of the U.S. climate modeling community. Given current scientific uncertainty and gaps in knowledge, it is essential that the United States maintain more than one high-end modeling center focused on long-term climate change, so that differing approaches to unsolved problems can be explored.

Both CCSM and GFDL support ongoing development of comprehensive climate system models, involving chemistry, biogeochemistry, and ecological processes. The centers are engaged in the development and application of distinct models. Their methods, innovations, and working hypotheses differ regarding many of the outstanding unresolved theoretical and modeling issues. The different approaches are essential at this stage, given the highly complex nature of the climate system with its numerous feedback mechanisms across a broad range of temporal and spatial scales.

Further, a comprehensive U.S. climate modeling strategy benefits from a measure of differentiation between the roles of the two centers. Despite an apparent overlap in responsibilities, the missions and structures of the two centers are more complementary than duplicative. CCSM is an open and accessible modeling system that integrates basic knowledge from the broad, multidisciplinary basic research community for research and applications. The GFDL model development team participates in these community interactions and will focus on model product generation for research, assessments, and policy applications as its principal activity. GFDL models and products are integral to the development of the NOAA Climate Services program, which provides operational climate products and services to policymakers and resource managers. GFDL maintains dedicated computer resources that can be allocated flexibly to meet mission requirements.

Near-Term Priorities
*   Results from GFDL and CCSM models will comprise the primary U.S. contributions to IPCC assessments, as well as provide input to other assessments of the science and impacts of climate change. Independent century-long climate projections will be executed by each center on schedules to meet national and international assessment demands.
*   CCSM will maintain an open model development system with major changes developed through consensus from the broad scientific community engaged in climate research. Computer resources for CCSM research will continue to be provided mostly at shared-access supercomputer centers through allocations given to the many projects associated with CCSM. This arrangement works well for the IPCC and other major assessments that have long lead-times that allow for sufficient planning.
*   GFDL plans to procure additional supercomputing resources to enable the systematic generation of model products needed by the impacts, assessments, and policy communities to document and assess the regional and global impacts of long-term climate variability and change. In addition to products derived from GFDL model simulations, additional products will be generated using results from other modeling centers, including CCSM.
*   Although they will maintain separate model development paths, the two centers have developed, and will implement, a plan for extensive scientific interaction and collaboration. The first step is to understand why the two coupled models have significantly different climate sensitivity to increased

atmospheric carbon dioxide concentrations. Initial studies indicate that the prediction of changes in cloud amount in response to atmospheric warming is very different in the two models (see Figure 10-2). This suggests that an important factor leading to differences in climate sensitivity is the differences in representation of cloud processes.
*   The centers will work with the broader scientific community of university and other laboratory modeling groups to focus research, including climate process studies, to better understand and resolve the differences between the models.
*   Further enhancement of the collaborations will be enabled by the following actions, some of which are dependent upon sufficient resources: a program of focused model intercomparisons (see also Objective 1.3); a graduate student, post-doctoral, and visitors program to accelerate interactions (see also Objective 2.2); and development of common model diagnostics.
*   The centers will work cooperatively to develop methods for providing global model output for a variety of decision processes (see Chapter 11).

Objective 3.2: Develop mechanisms for effective collaborations and knowledge transfer
Climate Process and Modeling Teams. Climate scientists who conduct observational and empirical research into climate processes are often not well connected with modeling centers and model

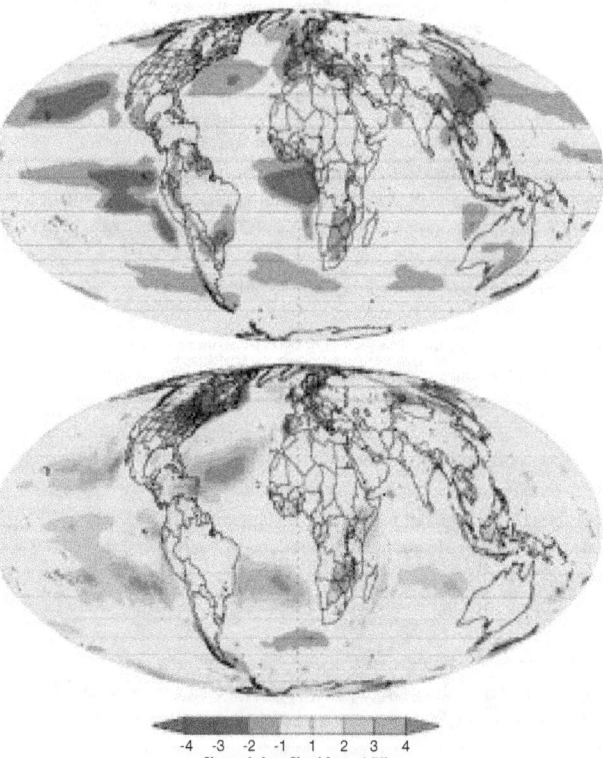

-4 -3 -2 -1   1   2   3   4
Change in Low Cloud Amount (%)

Figure 10-2: Changes in the amount of low clouds simulated by the GFDL model (top) and CCSM (bottom) resulting from a doubling of atmospheric carbon dioxide concentration. In many areas, the amount of low cloud is reduced in the GFDL model, where the amount of clouds increases over most areas in CCSM. This difference in behavior may be a major source of the models' differing climate sensitivities. Source: GFDL and NCAR.

developers. The U.S. Climate Variability and Predictability (CLIVAR) program has developed and promoted a research strategy that involves the formation of **C**limate **P**rocess and Modeling **T**eams in order to improve collaboration between researchers and modelers. CPTs consist of process-oriented observation specialists, researchers, and individual process and parameterization modelers, working collaboratively with climate model developers. The teams are organized around an issue, model deficiency, and/or parameterization(s) that are generic to all climate models. An important distinction between CPTs and other more conventional model development research is the emphasis on directed teamwork, demonstrated progress, and delivery of products that will be tested and possibly implemented in climate models.

Near-Term Priority

*   The goal of the CPT approach is to facilitate and accelerate progress in improving the fidelity of climate models and their predictions and projections. Specifically, CPTs will:
    *   Speed the transfer of theoretical and practical process-model understanding into improved treatment of processes in climate system models (e.g., coupled models and their component models, assimilation and prediction systems), and demonstrate, via testing and diagnostics, the impact of these improvements
    *   Identify process study activities necessary to further refine climate model fidelity
    *   Develop observational requirements for climate system models. Success of CPTs will be measured not only by advances in knowledge, but more importantly by the development of new modeling capabilities and products. Several pilot-scale CPTs are being funded by CCSP in FY2003 (see Chapter 4).

Common Modeling Infrastructure and Earth System Modeling Framework. One of the great strengths of atmospheric, oceanic, and climate modeling in the United States is the variety, availability, and wide use of models. But this diversity has also led to duplication of effort and a proliferation of models and codes that, due largely to technical reasons, cannot interoperate and have been unable to keep up with and exploit advances in computing technology.

Climate models are increasingly being used to support decisionmaking. The predictive requirements are becoming more stringent. The demand for interoperability of climate model components has intensified as various modeling groups are engaging in collaborative research. Without exchangeable model components, it is often difficult to point to a component as a clearly identifiable cause of divergent results when one model is compared against another or against observations. In order to optimize modeling resources and enable meaningful collaborations among modelers, it is necessary to build common and flexible modeling infrastructure at the major centers.

The common modeling infrastructure that will be implemented by the two centers will enable the exchange of model components between different modeling systems and facilitate analyses and intercomparisons of model results by adopting common coding standards for model components and common output formats, and developing common diagnostics packages. To achieve part of this commonality, the Earth System Modeling Framework project has been established. ESMF is a community-wide engineering effort to develop common software to facilitate interoperability of climate models on various hardware platforms, especially on massively

parallel architecture platforms (i.e., many individual computational units operating in parallel, connected by data communication links).

Near-Term Priority

*   CCSP will support further development of CMI/ESMF through multi-agency mechanisms that will ensure participation of the major U.S. climate modeling centers and groups. CMI/ESMF will: (1) facilitate the exchange of scientific codes (interoperability) so that researchers may more readily interface with smaller scale, process-modeling efforts and can share experience among diverse large-scale modeling efforts; (2) promote the implementation of standard, low-level software, the development of which now accounts for a substantial fraction of the software development budgets in many institutions; (3) focus community resources to deal with changes in computer architecture; (4) present the computer industry and computer scientists with a unified, well-defined, and well-documented task to address; (5) share the overhead costs of software development, such as platform-specific user libraries and documentation; and (6) provide greater institutional continuity to model development efforts by distributing support for modeling infrastructure throughout the community. Products will include more efficient and rapid transfer of research results into model applications, and human resources and dollar cost savings.

Objective 3.3: Provide for interagency coordination of CCSP modeling activities to improve implementation and external advisory processes to evaluate performance

Near-Term Priorities

*   CCSP modeling activities are carried out by a number of agencies. In order to improve the coordination of the implementation of these activities at the program level, CCSP will establish a process for coordination of CCSP modeling activities that lie beyond the boundaries of the missions or programs of single agencies, including coordination of the use of computer resources that can be shared between agencies.
*   CCSP will use advisory processes to facilitate its programs (see Chapter 16). In the case of its modeling strategy, CCSP will use a variety of advisory mechanisms to evaluate, guide, and provide feedback. Such mechanisms will include a continuing NRC relationship to examine strategic issues for future development, standing advisory committees, focused ad hoc working groups on technical issues (e.g., ESMF), and specialist workshops.

CHAPTER 10 AUTHORS
Lead Authors
David Bader, DOE
Jay Fein, NSF
Ming Ji, NOAA
Tsengdar Lee, NASA
Stephen Meacham, NSF
David Rind, NASA
Contributors
Randall Dole, NOAA
David Goodrich, NOAA
James Hack, NCAR
Isaac Held, NOAA
Jeff Kiehl, NCAR
Chet Koblinsky, NASA and CCSPO
Ants Leetmaa, NOAA
Rick Piltz, CCSPO

Decision Support Resources Development

The Role of Decision Support

In order to fulfill the scientific assessment requirements of the 1990 Global Change Research Act (P.L. 101-606)[3], and to enhance the utility of the extensive body of observations and research findings developed by the U.S. Global Change Research Program (USGCRP) since 1990, the Climate Change Science Program (CCSP) is adopting a structured approach to match, coordinate, and extend resources developed through the research activities to the support of policy and adaptive management decisionmaking. The USGCRP has made very large investments in research and observing programs since 1990. By comparison, the USGCRP investment in assessment activities and other decision support resources has been much smaller to date. The largest assessment program previously undertaken by the USGCRP was the National Assessment initiated in 1998, which produced overview reports in late 2000 and a series of specialty reports in the period 2001-2003.

The decision support approach for analyses and assessments adopted by the CCSP builds upon the "lessons learned" from earlier USGCRP assessment analyses, as well as other sector, regional, national, and international assessments. The Climate Change Research Initiative (CCRI) will place enhanced emphasis on the extraction of mature scientific

knowledge from the core research program for use in assessment and decision support. The principal guidelines for the CCSP decision support approach follow:

- *Analyses structured around specific questions.* This approach enhances consistent communications among all involved scientists and stakeholders addressing designated questions.
- *Early and continuing involvement of stakeholders.* Stakeholder feedback is essential in defining key science and observation questions, and in defining the key issues in each analysis.
- *Explicit treatment of uncertainties.* The CCSP and the general scientific community have the responsibility to define the applicability limits imposed on various projections and other analyses, as related to the uncertainties in the underlying data and analysis methods. The CCSP will consistently address the uncertainties and the limits of applicability (related to underlying uncertainties) associated with the decision support analyses it reports.

---

[3]"On a periodic basis (not less frequently than every 4 years) the Council, through the Committee, shall prepare and submit to the President and the Congress an assessment which:
1) Integrates, evaluates, and interprets the findings of the Program and discusses the scientific uncertainties associated with such findings
2) Analyzes the effects of global change on the natural environment, agriculture, energy production and use, land and water resources, transportation, human health and welfare, human social systems, and biological diversity
3) Analyzes current trends in global change, both human-induced and natural, and projects major trends for the subsequent 25 to 100 years." (from Section 106).

- *Transparent public review of analysis questions, methods, and draft results.* In the same manner that the CCSP published a *Discussion Draft Strategic Plan* for public review in November 2002, the CCSP will publish drafts of decision support analysis plans for open review. Draft results from the analyses will also be published for review before their completion.
- *Evaluate ongoing CCSP analyses and build on the lessons learned.* The CCSP plans to conduct, through a variety of mechanisms, a limited number of decision support case studies during the next 2 years, and to expand the scope of these analyses only after evaluating scientific and stakeholder community feedback from the initial experiences.

The CCSP activities in Decision Support Resources go beyond what has been accomplished in the past in the breadth of interagency activity and commitment to extend beyond traditional science assessments to new forms of stakeholder interactions that focus development and delivery of information in more effective and credible ways. The CCSP Decision Support Resources activities will build on the science foundation established by the USGCRP, the CCRI, and related international research programs, as well as the lessons learned from other assessments and stakeholder interaction projects conducted during the last decade. The planned decision support resource development will address key recommendations from the National Research Council (NRC), particularly those

discussed in *Global Environmental Change: Research Pathways for the Next Decade* (NRC, 1999a), *Climate Change Science: An Analysis of Some Key Questions* (NRC, 2001a), and *The Science of Regional and Global Change: Putting Knowledge to Work* (NRC, 2001e).

Priorities on decision support resources are guided by national and international priorities that have been established with stakeholder partnerships. National priorities include the management of carbon, energy, water, air quality, community growth, disaster, invasive species, and coasts, along with possible negative ancillary impacts associated with health and agricultural efficiency.

The analyses and development of other decision support resources are intended to support the decisionmaking process and to be capacity-building activities. By sponsoring these activities (conducted by government, academic, and other groups), the CCSP will enhance the capabilities of various interdisciplinary research groups to assist in the evaluation of the many different policy and adaptive management questions likely to arise in the coming years. Decision support resources are improved on an iterative basis, requiring a continuous process of incorporating new technologies, processes, and knowledge. The CCSP plan for evaluation is to systematically verify and validate the integration of each new generation of climate change research results into decision support resources and to determine the confidence in using the enhanced tools in a variety of applications.

BOX 11-1

WORKING DEFINITIONS

Decision Support Resources
Decision support resources refers to the set of analyses and assessments, interdisciplinary research, analytical methods (including scenarios and alternative analysis methodologies), model and data product development, communication, and operational services that provide timely and useful information to address questions confronting policymakers, resource managers and other stakeholders.

Policy Decisions
Policy decisions result in laws, regulations, or other public actions. These decisions are typically made in government settings (federal, state, local) by elected or appointed officials. These decisions, which usually involve balancing competing value issues, can be assisted by—but not specified by—scientific analyses.

Adaptive Management Decisions
Adaptive management decisions are operational decisions, principally for

managing entities that are influenced by climate variability and change. These decisions can apply to the management of infrastructure (e.g., a waste water treatment plant), the integrated management of a natural resource (e.g., a watershed), or the operation of societal response mechanisms (e.g., health alerts, water restrictions). Adaptive management operates within existing policy frameworks or uses existing infrastructure, and the decisions usually occur on time scales of a year or less.

Planning
Planning is a process inherently important for both policy decisions and adaptive management. It usually occurs in the framework of established or projected policy options.

Stakeholders
Stakeholders are individuals or groups whose interests (financial, cultural, value-based, or other) are affected by

climate variability, climate change, or options for adapting to or mitigating these phenomena. Stakeholders are important partners with the research community for development of decision support resources.

Assessments
Assessments are processes that involve analyzing and evaluating the state of scientific knowledge (and the associated degree of scientific certainty) and, in interaction with users, developing information applicable to a particular set of issues or decisions.

Scenario
A scenario is a coherent statement of a potential future situation that serves as input to more detailed analysis or modeling. Scenarios are tools to explore "If ..., then..." statements, and are not predictions of or prescriptions for the future.

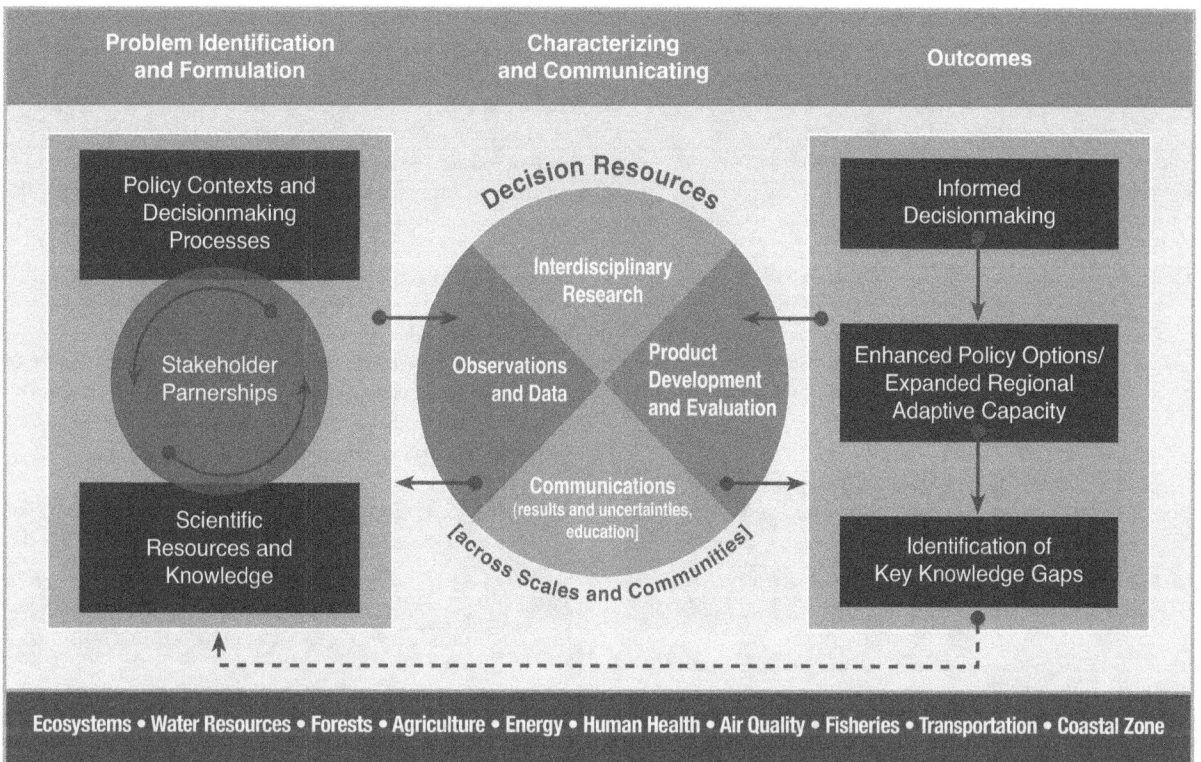

Figure 11-1: Schematic representation of decision support framework illustrating components of problem identification and formulation, development of decision resources, and final outcomes.

The planned CCSP Decision Support Resource activities respond to the following three goals:

1) *Scientific syntheses.* Prepare scientific syntheses and assessments to support informed discussion of climate variability and change issues by decisionmakers, stakeholders, the media, and the general public.

2) *Adaptive management for resources and infrastructure.* Develop information resources to support adaptive management and planning for responding to climate variability and climate change, and transfer these resources from research to operational application.

3) *Support for policymaking.* Develop and evaluate methods (scenario evaluations, integrated analyses, alternative analytical approaches) to support climate change policymaking and demonstrate these methods with case studies.

Management and advisory processes (involving both government and non-government reviewers) will be developed to ensure implementation of a coordinated CCSP decision support effort using review and feedbacks to identify and set priorities.

Three decisionmaking categories will be addressed by the CCSP: (1) public discussion and planning based on state-of-science syntheses and assessments; (2) operational adaptive management decisions undertaken by managers of natural resources and built infrastructure (i.e., "climate services applications"); and (3) support for policy formulation. Each of these decisionmaking categories has a unique set of stakeholders and requires different decision support tools. A common framework of activities will be used where appropriate for

all three categories as shown in Figure 11-1. The figure illustrates stakeholder partnerships with scientists to identify and formulate the problems to be addressed, the development of decision support resources, and expected outcomes. Key elements in this framework include:

• Involvement of stakeholders in question identification and formulation

• Science syntheses and assessments focused on the identified question(s)

• Formation of interdisciplinary research teams that interact with the decisionmaking communities, and that integrate the natural and social sciences

• Development, with users, of a decision support "toolbox" (a collection of validated and verified products and processes that can be used by decisionmakers)

• Quantification and communication of the level of confidence in reported findings

• Evaluation and review processes for the decision support analyses that engage the relevant decisionmaking communities.

The "decision support toolbox" refers to the collection of decision support products [including communication methods, integrated maps, geographic information system (GIS)-based analysis products, targeted forecasts for particular sectors, "decision calendars," scenarios, etc.] that have been validated, verified, and evaluated from the perspective of users. In its mature state, decisionmakers will be able to assess the extent to which analytical tools applied in a particular sector or circumstance could be applied or modified in their particular setting. The toolbox depends upon the physical, natural, social science, and, now, assessment foundations of the CCSP, including Earth observation networks and systems, Earth system models, the data and data-handling infrastructure of the CCSP research activities, and an evolving network of socioeconomic data.

The expected outcomes from the CCSP Decision Support Resources activities include:
- Improved science syntheses and assessments for informing public discussion of climate change issues
- Expanded adaptive management capacity to facilitate the responses of resource managers to climate variability and change
- Assessment information for evaluating options for mitigation of and adaptation to climate variability and change
- Identification of information needs to guide the evolution of the CCSP science agenda.

Goal 1: Prepare scientific syntheses and assessments to support informed discussion of climate variability and change issues by decisionmakers, stakeholders, the media, and the general public.

The Global Change Research Act of 1990 (P.L. 101-606, Section 106) directs the USGCRP to support research to "produce information readily usable by policymakers attempting to formulate effective strategies for preventing, mitigating, and adapting to the effects of global change" and to undertake periodic science assessments. Assessments are an effective means for integrating and analyzing CCSP research results with other knowledge, and communicating useful insights in support of a variety of applications for decision support. Assessments also help identify knowledge gaps and thus provide valuable input to the process of focusing research.

During the next decade, CCSP will continue to support assessment analyses. Given the broad set of policy, planning, and operational decisions that would benefit from climate and global change information, there are a wide variety of candidates for CCSP assessment analyses. A focused, systematic approach for selecting and producing a practical number of assessments—and for continuously addressing the "lessons learned" from each assessment analysis—will be developed and published by the CCSP.

Objective 1.1: Produce scientific synthesis reports
The CCSP participating agencies will coordinate their work to produce a number of synthesis reports that integrate research results focused on identified science and decision issues. These reports will provide current evaluations of the science foundation that can be used for informing public debate, policy, and operational decisions, and for defining and setting the future direction and priorities of the program.

The CCSP agencies and scientists funded by these agencies will also continue to participate in the principal international science assessments including the Intergovernmental Panel on Climate Change (IPCC) fourth assessment scheduled for completion in 2007, and the World Meteorological Organization (WMO)/United Nations Environment Programme (UNEP) assessments of stratospheric ozone depletion and associated environmental impacts.

Objective 1.2: Plan and implement designated assessment analyses in collaboration with the stakeholder and research communities
The CCSP will produce a set of assessments that focus on a variety of science and policy issues important for public discussion and decisionmaking. The assessments will be composed of syntheses, reports, and integrated analyses that the CCSP will complete by the third quarter of 2006. CCSP cooperating agencies will sponsor or carry out the analyses with interagency oversight to ensure that resources from the entire program are best utilized. This approach will cover the full range of CCSP goals and will provide a "snapshot" of knowledge concerning the environmental and socioeconomic aspects of climate variability and change. A list of the planned CCSP scientific synthesis and assessment reports is provided in Box 11-2. This list reclassifies the product summary in Table 2-1 (Chapter 2) by primary decision support purpose.

Goal 2: Develop resources to support adaptive management and planning for responding to climate variability and climate change, and transition these resources from research to operational application.

Adapting to climate variability and potential change poses challenges to management of resources, infrastructure, and the economy. The pressures of increased population densities and intensified land use, common throughout much of the United States and other nations, increase the demand for effective management of resources sensitive to climate in many regions. For example, information on short-term climate variability (i.e., weekly, monthly and seasonal projections) is relevant for the development of state and regional drought action plans, agricultural operations management, water resource system management, and fishery management. Much of the information from CCSP research is relevant to these decisions, but often is insufficiently focused on management applications to be directly useful. Thus, the CCSP decision support resource activities will play an important role in the "transition from research to operations" for major elements of the underlying research. In the transition process, particular attention will be placed on the establishment of validation and verification guidelines for the extension of the research, analyses and assessments, model and data products, and other resources into operational decision support.

CCSP research results, data products, forecasts, and model results are already being applied to adaptive management decision support in a limited number of regional and sectoral case studies. Elements of climate and associated ecosystem observations from satellite, ground-based, and *in situ* platforms are also being synthesized into useful data products for decisionmakers. Examples include a variety of maps for crop management, water quality management, and urban planning, and integrated products illustrating snowpack, precipitation, streamflow, and potential for drought conditions. Climate projections, especially those from El Niño-Southern Oscillation (ENSO) analyses (which have demonstrated elements of seasonal- to biennial-scale forecast skill), have provided information for state and local emergency preparedness organizations; water resource management plans for the western regions; agricultural planning for the southeast; and fire management for drought-stricken

BOX 11-2

CCSP TOPICS FOR INTEGRATED SYNTHESIS AND ASSESSMENT PRODUCTS CATEGORIZED BY PRIMARY END USE

Science Reports to Inform Evolution of the Science Research Agenda
- Temperature trends in the lower atmosphere—steps for understanding and reconciling differences
- Past climate variability and change in the Arctic and at high latitudes
- Updating scenarios of greenhouse gas emissions and concentrations, in collaboration with the Climate Change Technology Program (CCTP); review of integrated scenario development and application
- North American carbon budget and implications for the global carbon cycle
- Climate models and their uses and limitations, including sensitivity, feedbacks, and uncertainty analysis
- Climate projections for research and assessment based on emissions scenarios developed through the CCTP
- Climate extremes including documentation of current extremes; prospects for improving projections
- Relationship between observed ecosystem changes and climate change

- State of the science of socioeconomic and environmental impacts of climate variability

Synthesis and Assessment Products to Inform Adaptive Management Decisions
- Risks of abrupt changes in global climate
- Coastal elevation and sensitivity to sea-level rise
- Within the transportation sector, a summary of climate change and variability sensitivities, potential impacts, and response options
- Preliminary review of adaptation options for climate-sensitive ecosystems and resources
- Uses and limitations of observations, data, forecasts, and other projections in decision support for selected sectors and regions
- Best practice approaches for characterizing, communicating, and incorporating scientific uncertainty in decisionmaking

- Decision support experiments and evaluations using seasonal-to-interannual forecasts and observational data

Synthesis and Assessment Products to Inform Policy Decisions
- Re-analyses of historical climate data for key atmospheric features; implications for attribution of causes of observed change
- Aerosol properties and their impacts on climate
- Trends in emissions of ozone-depleting substances, ozone layer recovery, and implications for ultraviolet radiation exposure and climate change
- State-of-knowledge of thresholds of change that could lead to discontinuities (sudden changes) in some ecosystems and climate-sensitive resources
- Scenario-based analysis of the climatological, environmental, resource, technological, and economic implications of different atmospheric concentrations of greenhouse gases

---

regions. Decision support tools are also employed by federal agencies to serve the public in local and regional decisionmaking and include applications in the management of carbon, water, disasters, invasive species, and coastal ecosystems along with information on public health, agriculture efficiency, and energy use. All of these products have been co-developed by scientists and users after extensive dialogue and are potential resources for a "decision support toolkit."

Making use of information on variability and potential future changes in climate requires that decisionmakers be directly involved in shaping their key questions, and not passive consumers of general scientific information. User partnerships that actively engage scientists provide the opportunity for understanding where scientific resources and knowledge can best be used and what new research may be needed. Outputs from such interactions include decision calendars and assessments that frame the context in which the science will be used, determination of what products need to be developed using the science information base and experiential knowledge of stakeholders, and determination of the limits of existing knowledge to be applied to the problem.

Decision support for adaptive management requires advances in basic knowledge and progress in applying scientific information within adaptive management settings. Conducting research within a decision support framework can provide multiple benefits for both practitioners and scientists. Ideally, users of research information are served so that new options exist for minimizing negative impacts or pursuing opportunities, and researchers benefit from refinement and prioritization of research agendas through the identification of the uncertainties most relevant to decisionmaking.

CCSP will play an important role in generating processes and products relevant to adaptive management decision options. Examples of pilot products include historical data analyses and products; forecasts for particular sectors at key time periods; probabilistic climate variability and change information integrated with decision models; "decision calendars;" geo-referenced maps of critical climate and associated environmental parameters; and specific model runs or data sets. The CCSP will also develop mechanisms to sustain interactions between users and researchers in order to better understand how to optimize the delivery of research results, data products, and forecasts.

Objective 2.1: Conduct research to extend the uses and identify the limits of existing decision support resource capabilities for adaptive management

The CCSP's approach for accelerating and enhancing decision support for adaptive management will be based on the following:

- Enhancement of existing case studies of adaptive management decision support using a variety of approaches sponsored by CCSP agencies
- Implementation of evaluation processes to address (1) the role of scientific uncertainties in the analyses and (2) feedback information to help frame future research agendas
- Development and demonstration of elements of a decision support toolkit.

CCSP research will target adaptive management issues and information use, including the potential entry points and barriers to using climate information as well as the types of new information that would provide the greatest benefit to decision processes. This research will integrate natural and social systems within an application context of managed resources or infrastructure, utilizing climate and environmental observations, model outputs, socioeconomic data, and decision models. It will incorporate elements of regional/ sub-regional climate science and associated environmental processes, socioeconomic impacts, technological capabilities, management institutions and policies, and decision processes including evaluation.

CCSP will integrate lessons learned from current adaptive decision support case studies sponsored by CCSP participating agencies. These lessons will provide a mechanism for evaluating how scientific information is currently used by decisionmakers (to help frame problems) and for evaluating the quality of the scientific resources available to be applied to the problems.

Within a case studies framework, CCSP will support development of resources for decision support, and will develop methods to quantify uncertainty and its effect on the adaptive management process in a range of example cases. Illustrative resources to be developed are listed in Box 11-3.

CCSP will periodically organize workshops and forums to gather information on lessons learned from adaptive management decision support activities, and will prepare summary reports that help transition knowledge and resources across regions and sectors. The resources (processes and tools) that emerge from this research in decision support are the foundation for a "tool kit," a term that describes a range of products useful to individuals and institutions responding to the effects of climate variability and change. The CCSP will support the mechanisms to help users identify and use the capabilities in the tool kit, including web-based tutorials, workbooks, and interactive forums.

Objective 2.2: Promote the transition of resources from research to operations for sustained use

Once decisionmakers begin using new products, there is a need to ensure the continuity of that product through services entities. While the CCSP itself does not have a service mission, many of the CCSP collaborating agencies do. The CCSP will work to facilitate the successful transitioning, verification and validation, and maintenance of newly developed decision support products within its collaborating agencies or other non-federal service entities. CCSP will work to support collection of data, information, and other resources utilized by the decision support products and will aid in the transition of this collection to operational entities when appropriate. In the transition process, it is important to benchmark

---

BOX 11-3

ILLUSTRATIVE RESOURCES TO BE DEVELOPED FOR ADAPTIVE MANAGEMENT DECISION SUPPORT

New experimental long-lead (12-month) streamflow forecasts for major watersheds of the United States, coupled with improved decision-support for water managers and users [2-4 years].

Experimental to operational decision support systems for agriculture and ranching in selected regions (Southwest and Southeast) of the United States [2-4 years].

Prototype regional (Western and Southeastern) integrated "multi-stress" and multi-jurisdiction decision support systems for forest and wildfire management [2-4 years].

Development of a blueprint for the improved regional climate, hydrologic, and ecological observing systems needed for enhanced decision support, particularly in mountainous regions [2-4 years].

Tests of existing regional modeling capabilities, and definition of the improved regional modeling capabilities needed for enhanced decision support [2-4 years and beyond].

Improved public health decision support for major climate-modulated infectious disease threats in the United States, including mosquito-borne viral disease, Hantavirus, and Valley Fever [2-4 years].

Analysis of historical records in target areas to gain a better understanding of past and current climate variability across all time-scales for use in sensitivity analyses of existing and planned physical infrastructure [2-4 years and beyond].

Assessments of potential effects of climate change and land-use change on water and vector-borne diseases [2-4 years].

Assessment of the potential effects of climate change, land-use change, and UV radiation on aquatic ecosystems [2-4 years].

the improvement in performance of solutions that result from integrating research-quality observations with research-quality predictions and outlooks into operational decision support tools.

Two case studies of adaptive management decision support, summarized in Boxes 11-4 and 11-5, illustrate the transition of decision support analyses into the type of operational management resources anticipated by the CCSP Decision Support Resource development goals.

> **Goal 3**: Develop and evaluate methods (scenario evaluations, integrated analyses, alternative analytical approaches) to support climate change policymaking and demonstrate these methods with case studies.

Policy-related questions regarding climate change typically arise from numerous sources, for example from:
• Consideration of climate change policy within federal government

BOX 11-4

HANTAVIRUS PULMONARY SYNDROME IN THE SOUTHWESTERN UNITED STATES

This case study describes research and assessment activities undertaken to better understand the cause of outbreaks of hantavirus pulmonary syndrome (HPS) in the southwestern United States in the 1990s. The research and assessment efforts led to pilot production and evaluation of risk maps, which were then used by public health officials for on-the-ground interventions to prevent disease outbreaks and protect public health. This study illustrates how multidisciplinary, place-based research and assessment, conducted in response to questions raised by a particular user group (public health officials), can lead to the development of products (risk maps) that successfully increase regional adaptive capacity (enhanced public health care).

Problem Formulation
In 1993, a disease characterized by acute respiratory distress with a high death rate (greater than 50%) among previously healthy persons was identified in the southwestern United States. This disease, HPS, was traced to a virus maintained and transmitted primarily within populations of a common native rodent, the deer mouse (*Peromyscus maniculatus*). Public health officials wanted to understand the cause of the outbreak so they could develop effective techniques for intervening and preventing the disease.

Researchers hypothesized that the outbreak was due in part to the unusual weather in 1991-1992 associated with the El Niño-Southern Oscillation. Unseasonable rains in 1991 and 1992

during the usually dry spring and summer, and the mild winter of 1992, were thought to have created favorable conditions for an increase in local rodent populations. It was suggested that a cascading series of events from weather—through changes in vegetation, to virus maintenance and transmission within rodent populations—culminated in changes in human disease risk from HPS.

The Assessment
A study explored this hypothesis by comparing the environmental characteristics of sites where people were infected with those sites where people were not infected. The study used a retrospective epidemiologic approach to risk assessment. Satellite imagery (Landsat Thematic Mapper images), combined with epidemiologic surveillance, retrospectively identified areas at high risk for HPS associated with *Peromyscus* populations over broad geographic regions during the 1993 outbreak. Thematic Mapper data identified environmental conditions approximately 1 year before the outbreak that were measurably different near HPS sites than in rural, populated sites where the disease did not occur.

Pilot Production and Evaluation of Risk Maps as a Decision Support Tool
The assessment revealed that environmental conditions near HPS sites varied with the presence or absence of ENSO. The geographic extent and level

of predicted HPS risk were higher during ENSO, supporting the view that El Niño may increase the likelihood of HPS outbreaks.

It was then determined that high-risk areas for HPS can be predicted more than 6 months in advance based on satellite-generated risk maps of climate-dependent land cover. Predicted risk paralleled vegetative growth, supporting the hypothesis that heavy 1992 rainfall due to El Niño was associated with higher rodent populations that triggered the Hantavirus outbreak in 1993. Landsat satellite remote-sensing images from 1995, a non-El Niño "control" year, showed low risk in the region, whereas the images from the 1998 strong El Niño again showed high risk areas as in 1992-1993. Trapping mice in the field validated the satellite-generated risk maps with mouse populations directly related to risk level, with a correlation factor of over 0.90. Risk classification also was consistent with the numbers of HPS cases in 1994, 1996, 1998, and 1999.

Next-Generation Integrated Knowledge
This information was used to develop an early warning system, with intervention strategies designed to avoid exposure. These strategies, developed in partnership with the Centers for Disease Control and Prevention (CDC) and the Indian Health Service, are already being implemented by the U.S. Department of Health and Human Services for disease prevention in the southwest.

117

BOX 11-5

CLIMATE-ECOSYSTEM-FIRE MANAGEMENT

Wildland fires burn millions of acres each year and major resources are committed to fuel (live and dead vegetation) treatment, fire prevention, and fire suppression. Effective decision support products and tools can improve resource allocation decisions and maintain a high standard of safety for firefighters and the public. The fire-climate assessment tool, which is in essence a structured process, allows fire and fuels specialists and fire weather meteorologists in each of the National Interagency Fire Center's eleven Geographic Area Coordination Centers (GACCs) to work with climatologists to develop GACC-level assessments of fire risk at seasonal to shorter time scales. The tool also allows Predictive Services staff to develop and update a national map and discussion of fire potential for the fire season each year.

Problem Formulation

- La Niña conditions prompted the first climate-fire-society stakeholder workshops in 2000. These annual workshops have established a dialogue between stakeholders (fire managers and decisionmakers) and climatologists.
- The workshops evolved over 3 years, refining the contribution of climate information to seasonal fire outlooks at the regional level and focusing on "fire science," including the nature of the fire regime (frequency, size, intensity); conditions in the natural system (adaptive ecosystems, vegetation, fuels, watershed, soil, wildlife); and characteristics of the human systems (property, economic sectors affected, policy and land-use planning, multi-agency jurisdictions).
- The workshop process is supported by interdisciplinary teams of research scientists interested in focusing climate impacts research on information and insights essential to decision challenges influenced by climate variability and change.

Research Modules

- Design and communication of climate information and forecasts useful for fire

management involves collaboration between decisionmakers and scientists in the conduct of research to improve institutional capacity to integrate scientific information and predictions into planning and operations; analysis of individual stakeholder perceptions of fire risk and of their capabilities and willingness to use scientific information in their decision processes; and identification of the factors (environmental, economic, and public health) that are most relevant in the context of overall public good, as viewed across geographic and agency boundaries.

- Climate and ecological processes including improving understanding of the spatial variability of fire. Recent investigations have found strong associations between the Palmer Drought Severity Index (PDSI) several months to 2 years earlier and fire season severity. Correlating anomalous wildfire frequency and extent with the PDSI illustrates the importance of prior and accumulated precipitation anomalies to future wildfire season severity.
- Risk assessment and mitigation including strategic planning for fire use (i.e., prescribed burns and allowing selected fires to burn) and fire suppression; improving predictive capabilities based on improved understanding of relationships between wildland fire and climate before, during, and after events; and determination of how science can inform development of wildland fire objectives important to interagency preparedness planning.
- Development of an integrated model called Fire-Climate-Society (FCS-1) to provide a planning tool, accessible to fire managers and community members, that would integrate the climate, fuels, fire history, and human dimensions of wildfire behavior for strategic management.

Next-Generation Integrated Knowledge

The objective is to develop an understanding of the interactions among climate, ecology (e.g., fuel load), and human factors (e.g., real estate, land use, recreation, conservation, jurisdiction, law) such that decision support insights reflect the true "multiple-stressor" realities of wildland fire risk and management. Particularly important are questions such as:

- What federal/state policies and programs increase fire risks and severity; what policies pose barriers to adoption of innovations such as new decision support tools?
- What are the scales of impacts (temporal and spatial) that influence fire regimes?
- What are the connections between multi-year drought and fire risk?

Pilot Product Development and Evaluation

- A new Predictive Services program has been launched to anticipate where fires are most likely to occur in order to allocate the appropriate firefighting resources to these areas. Geographic area predictive services units, established by the 2000 National Fire Plan, are tasked with integrating information about climate, weather, fire danger, and firefighting resources to provide decision support to fire managers on the location, timing, and severity of fire potential.
- The National Wildland Fire Outlook is the compilation of the 11 geographic area outlooks generated at the National Seasonal Assessment Workshop, and provides the first national-level, interagency, climate-based seasonal fire outlook.

In the spirit of the adaptive learning approach built into many of the regional projects, pilot products are inspiring a new round of research into understanding decision structures and constraints in order to transfer knowledge gained in this particular decision support experience.

- Proposals advanced by private and non-governmental organizations
- Preparation for international negotiations
- Consideration of legislative proposals
- Priority-setting processes for science and technology programs.

The CCSP will work in close collaboration with the Climate Change Technology Program (CCTP) to develop evaluations of relevant policy questions that incorporate up-to-date knowledge of both scientific and technology issues. The CCSP will focus on two objectives in this area: (1) developing scientific syntheses and analytical frameworks ("resources") to support integrated evaluations, including explicit characterization of uncertainties to guide appropriate interpretation, and (2) initially conducting a limited number of case studies with evaluation of the lessons learned, to guide future analyses.

Objective 3.1: Develop scientific syntheses and analytic frameworks to support integrated evaluations, including explicit evaluation and characterization of uncertainties

One of the challenges of developing scientific syntheses is providing a systematic way of integrating knowledge across disciplines, each having their own methodologies, resolutions, and degrees of certainty of scientific information. Meeting this challenge requires defining and meeting information needs across these borders, and developing methods and approaches to put information from different disciplines in compatible formats. Integrated models are an important tool for synthesis and comparative evaluation because they impose stringent standards of cross-disciplinary consistency and intelligibility. The CCSP supports the development of a number of integrated modeling frameworks that are useful for exploring many dimensions of climate and global change. The CCSP will also adopt other approaches for synthesis, including integration of expert knowledge across the relevant fields.

The CCSP will structure its syntheses and integrated analyses of policy questions related to climate variability and change using four types of approaches and drawing on research results produced throughout all areas of the program. These four approaches are:
1) Evaluations of net greenhouse gas flux and uptake using a variety of methods
2) Climate system analyses to study sensitivities and quantify ranges of climate variability and change
3) Analyses of the effects of climate variability and change
4) Integrated analytic frameworks.

Evaluation of net greenhouse gas flux and uptake in the Earth system (including human activities, the land surface, ecosystems, the atmosphere, and the oceans). The CCSP will use several methods to evaluate historical, current, and projected future patterns of greenhouse gas flux uptake, and consequent concentrations. These methods include state-of-science syntheses for emissions and carbon cycle information, evaluation of the effects of future technology adoption in the United States and globally (in collaboration with the CCTP), use of expert working groups (including both government and non-government specialists) to evaluate historical and projected greenhouse gas emission information and uncertainties (including uncertainties arising from different assumptions about human driving forces), and various inverse-calculation methods to verify greenhouse gas flux rates compared to recent and current observations of

greenhouse gas levels. Consistent with overall CCSP guidelines, these analyses will be developed in response to specific questions, and will be released for public review prior to publication in final form.

Climate system analyses. The CCSP will examine the range of natural variability (short- and long-term), responsiveness of the climate system to changes in net greenhouse gas fluxes and concentrations, and the potential for abrupt climate changes. CCSP analyses will use analytic approaches to improve the evaluation of uncertainty in important variables. It is well-recognized that there are significant questions about climate model sensitivity, as well as questions about verification of climate model projections when compared to long-term observation records. The CCSP will prepare an updated analysis on the uses and limitations of climate models for various policy support applications, and this CCSP analysis will guide the use of climate models in other CCSP analyses. In addition to computer-based climate models, several other analytical techniques will be used by the CCSP in developing policy-support analyses. These include atmospheric and oceanographic process research, historical and analog evaluations, and various data analysis and projection techniques. The CCSP supports a major program of climate model development and verification (see Chapter 10), and results from this program will be used in support of syntheses as appropriate.

Analyses of the effects of climate variability and change. Evaluation of the potential impacts associated with different atmospheric concentrations of greenhouse gases and aerosols is an important input to weigh the costs and benefits associated with different climate policies. Further research is required to integrate our understanding of the range of effects of different concentration levels and to develop methods for aggregating and comparing those impacts across different sectors and settings. Working with external advisory groups and the broad range of CCSP scientists, the CCSP-supported research will analyze a range of possible climate change impacts determined from climate system modeling and arising from different assumptions about natural and human influences, including (among many others) implications for agriculture, forestry, drought, fire, water resources, fisheries, coastal zones, and built environments such as ports. It will also address (to the extent possible given uncertainties) the potential implications of various response options for both the climate system and the economy.

Integrated analytic frameworks. Integrated analysis of climate change is essential for bringing together research from many contributing disciplines and applying it to gain comparative insight into policy-related questions. Full integration of information including research on human activities, greenhouse gas and aerosol emissions, land-use and land-cover change, cycling of carbon and other nutrients, climatic responses, and impacts on people, the economy, and resources is necessary for analysis of many important questions about the potential implications (both economic and environmental) of different greenhouse gas concentrations and various technology portfolios. Development and use of techniques for scenario and comparative analysis is useful for exploring the implications of different hypothetical policies for curbing emissions growth or encouraging adaptation. Answers from integrated analysis can only reflect the existing state of knowledge in component studies, but it is important to develop frameworks and resources for integration, exercise them, and learn from analysis of the results. CCSP will encourage innovation and development of approaches to integrated analysis, and test these approaches in case study evaluations.

Evaluation of uncertainty. For all four of the analytical approaches described above, the issue of evaluation and communication of uncertainty and levels of confidence is fundamental. Uncertainties can arise from lack of knowledge; from problems with data, models, terminology, or assumptions; and from other sources. Integrated models are strong tools for examining uncertainty through repeated model runs with variation of key parameters. Use of scenarios, sensitivity analysis, and the specification of probability distributions for many inputs coupled with model runs are among the ways in which integrated models can be used to explore uncertainty. CCSP will use these and other techniques to evaluate uncertainty, and couple this analysis with its commitment to reporting levels of confidence and uncertainty clearly and transparently. The approaches to uncertainty evaluation and communication will enable users of CCSP analyses to understand the uses and limits of the information.

As indicated elsewhere in this plan, broad guidelines for consideration of scientific uncertainty by the CCSP include the following:

- Uncertainty by itself (i.e., without regard to its magnitude in each specific case) should not prevent the development of analyses that may provide useful information for policy considerations.
- The magnitude (or importance) of uncertainty should be directly reported as a key element of analytical results in all cases where policy-supporting analyses are conducted.

Analytic approaches. Within the four analytical domains described above, a variety of approaches will be used and tailored to the study of a particular set of policy questions. The approaches are a means for examining proposed courses of action that incorporate knowledge of important key factors and uncertainty. In addition to development of decision support-related resources, the CCSP will also support research that furthers development of resources for integrated and comparative analysis, including:

- Use of historical climates (climate analogs)
- Techniques for posing and analyzing "If …, then…" questions, and reporting analytical results in the form of probability distributions of possible outcomes, thereby incorporating uncertainty information directly in the projected outcomes
- Models and qualitative frameworks for integrating scientific and technical information to compare environmental and socioeconomic effects of alternative response options
- Techniques for developing and analyzing scenarios using only a few of the most important variables (parametric analysis techniques)
- Techniques for "inverse analysis" that start from the study of temperature, precipitation, and other climate inputs required to maintain a system or activity, then identify rates or levels of climate change that would lead to discontinuities or thresholds.

An emphasis will be placed on the development of the greenhouse gas net flux scenarios through collaboration with CCTP. The distinctive feature of scenarios is that they integrate knowledge from the full range of relevant sources into a consistent description of potential future events. The specific variables incorporated in scenario development depend on the question being addressed. Scenarios for atmospheric chemistry, climate, impacts, adaptation, and mitigation

models all require different techniques and variables. Scenarios will be constructed using up-to-date information on projections for key variables (e.g., demography, technology characteristics and costs, and economic growth and characteristics) and the relationship of key driving forces to environmental change (e.g., land use and land cover). CCSP will coordinate its scenario development plans with CCTP plans for analysis of different plausible technology portfolios and with the scenario efforts of the IPCC.

Objective 3.2: Conduct a limited number of case study analyses and evaluate the lessons learned in order to guide subsequent analyses

During the next 2 years, CCSP will conduct a limited number of case study integrated analyses using the approaches described above. For each case study analysis, a project team will be established that is responsible for design and implementation of the decision support methodology and products. Diverse input will be solicited to frame the questions, determine analytic methodologies to be used, identify needed observations to address the problem, and determine what resulting analyses and products are to be prepared. Stakeholder involvement will be sought throughout the case studies to aid in question development, selection of analytical methods, review of products, and guidance for communication of results. CCSP will provide support and coordination of the scientific community, stakeholder, and public interactions. End products of each case studied will include appropriate assessment reports, as well as the related "lessons learned" documents.

The lessons learned during the decision resource case studies conducted in the next 2 years will be used to guide the definition of a wider set of analyses to be completed in a 2- to 4-year time period. Both the initial and subsequent case studies will be specified (e.g., by choice of technical issues addressed, and by parameters selected for analysis) to reflect relevant climate variability and change issues. An illustrative case study demonstrating the application of the policy decision support process to an examination of technology mitigation is summarized in Box 11-6. Note that this illustration does not constitute a specific plan for analysis. As mentioned previously, CCSP will present specific plans for decision support analyses in separate announcements when ready, and will solicit public comment on each proposed plan.

Other case studies being considered for short-term (within 2 years) or longer term (2-4 years) analysis include:

- Possible climate and ecosystem responses to long-term greenhouse gas stabilization at various specified levels
- Projected climate, socioeconomic, and environmental outcomes of various carbon sequestration initiatives adopted both in the United States and globally
- Application of scientific information on the carbon cycle, ecosystems, and other research elements for the development of guidelines for crediting carbon sequestration activities—for example, as related to land use and soil type, cover species type, management practices, and so on
- Possible outcomes (in terms of projected greenhouse gas concentrations and climate system response) projected for various scenarios of new energy technology penetration both in the United States and globally
- Risk analyses related to various examples of abrupt climate change

BOX 11-6

TECHNOLOGY SCENARIO CASE STUDY: AN ILLUSTRATIVE CASE STUDY (IN CONJUNCTION WITH THE CCTP)

Evaluate Two Categories of Scenarios

1) What combinations of technologies can be expected to provide energy consistent with different emission levels between now and 2050?

2) What are the range of plausible consequences (on the climate system and socioeconomic parameters) of different emission scenarios reflecting the technology options responsive to the previous question?

Analysis Planning

A project workshop—with expert stakeholders including climate scientists, energy technology engineers and scientists, energy economists, and industry, government, and non-government organization representatives would be held (late 2003/early 2004) to gain alignment and organize cooperative research on deliverables. Additional project working groups would be formed for continued dialogue throughout the project time frame. A draft study plan, including competitive proposal processes and federal in-house research assignments, and a timeline for completion of tasks would be prepared by early 2004. CCSP would post the draft plan for public comment on its website, <www.climatescience.gov>.

Scenarios for Technology Performance with Alternative Profiles

Each profile would include technologies that are presently part of the global energy system and that are expected to have their performance improved over time, as well as new technology options presently under development (e.g., carbon capture and disposal, hydrogen systems, fusion energy, and biotechnology). Performance improvements with alternative technology options would also be explored. Estimates of potential benefits in terms of greenhouse gas emissions, energy security, and oil dependence would be assessed. The

scenarios would also explore the potential for unintended consequences—both positive and negative. For example, if wind energy were extracted over large areas, it could reduce conventional air pollution as well as reduce greenhouse gas emissions, but it could also conceivably affect regional atmospheric circulation. Interaction with CCTP would be essential in developing these baseline scenario profiles. Incorporation of current scientific information about socioeconomic, technological, climatic, and environmental factors would be undertaken, with a characterization of the uncertainty in the scenarios. The scenarios would be completed by the end of 2004.

Modeled Climate Change for Each Greenhouse Gas Scenario, including Dynamic Carbon Cycle and Reasonable Projections of Land Use

This climate modeling task would be the responsibility of the large U.S. modeling groups. The model runs would also estimate the uptake of greenhouse gases by natural systems. This modeling must be informed by the results of carbon cycle research. The veracity of the model simulations for carbon sources and sinks needs to be tested ahead of time by comparisons to historical observations of the carbon cycle and the contribution of land-use changes. The model results would likely be presented in probabilistic form, and would be required to achieve agreed confidence limits to be considered usable for decision support. The large baseline computer modeling would be completed by mid-2005.

Modeled Environmental Impacts on Soil Moisture, Streamflow, and Vegetation

It is likely that one or both of the high-end U.S. climate models used for these studies would include fully interactive hydrologic-carbon-biogeochemical cycles (i.e., the major natural systems being impacted by climate). Workshops

would be held to engage the broader community of researchers studying climate impacts in the analysis of these experiments, and provide guidance as to how process-driven impacts models can be interfaced with these simulations. The baseline global impacts modeling would be completed with probability distributions of projected outcomes by the end of 2005. Higher resolution (50-km or greater) simulations may be available for portions of the baseline scenarios starting in FY06, but only if such higher resolution models are shown to have useful confidence limits by that time.

Analyses, Assessments, and Reports

Throughout the process, several interim analyses, assessments, and reports would be generated. Example products for this case could include the following:

- Synthesis report on the scenario development and the characterization and evaluation of the uncertainties in the scenarios (2005).

- Evaluation of the impacts on the climate system for the baseline and alternative technology pathways. Report on the analysis of the scenarios of response of the climate system to various emission scenarios with emphasis on intercomparison of the different pathways (2006).

- Analysis of the scenarios of environmental responses to those response scenarios of the different climate states. Reports on the inter-comparison of those responses to the different climate states. One example may be an evaluation of the potential for carbon sequestration in different ecosystems and agricultural systems, including initial greenhouse gas accounting analyses and guidelines for agriculture and forestry (2006).

- Final synthesis report on the inter-comparison of different technology pathways with the baseline pathways including an analysis of potential environmental benefits and associated costs of each scenario (2007).

- Projected climate and ecosystem outcomes related to significant changes in land use and land cover, both in the United States and globally
- Analysis of adaptation strategies related to specified changes in climate or ecosystem parameters.

Decision Support Management Strategy

CCSP management and advisory processes will ensure implementation of an open and credible process for development of decision support resources.

Management Structure

Leadership and direction will be provided by the CCSP interagency governing body, working with representatives of the Interagency Working Group on Climate Change Science and Technology (IWGCCST) and CCTP to:
- Select syntheses or assessments to be conducted by the program, and provide oversight for these activities
- Review needs for decision support focused on adaptive management and promote development of CCSP research and resources to respond to these needs
- Periodically identify and define topics of importance to national decisionmaking to be addressed by the CCSP case studies of integrated analyses and scenario evaluations
- Establish external advisory mechanisms.

CCSP decision support activities will be implemented through an interagency working group with management and coordination support from the Climate Change Science Program Office (CCSPO). Specific responsibilities of the working group include:
- Carry out approved activities (e.g., identify resource requirements, develop implementation plans) under supervision of the CCSP interagency governing body
- Evaluate program needs and develop initiatives that respond to these needs
- Maintain an inventory of ongoing decision support and assessment activities within the CCSP agencies
- Coordinate agency activities
- Support advisory mechanisms as directed by the CCSP, including workshops, committees, or NRC activities.

CCSPO will support development of decision support activities, under the supervision of the CCSP interagency governing committee. CCSPO will be responsible for helping to coordinate the preparation of assessment and synthesis products; connecting the assessment activities, lessons learned, and decision tools development to broader

interests and communities; and evaluating, reporting, and communicating results from the decision support activities.

Stakeholder Input and Evaluation

CCSP decision support activities will be conducted openly with input from external technical experts and other stakeholders. Advisory processes will be structured to meet specific requirements for each activity. Past experience has indicated that open framing of the questions to be addressed and methods to be adopted are crucial to establishing an open process and credible product. Ongoing advisory processes will provide independent review and oversight and ensure that products developed bring in all relevant perspectives. Independent evaluation of both products and processes will be included as a component of decision support efforts to ensure that future activities are improved by consideration and application of experience garnered through initial case study activities.

A first step that will be taken in developing stakeholder input is to hold a focused workshop to provide comments on (1) initial selection of topics/questions for policy decision support activities; (2) possible structures of "If ..., then..." scenario analyses and other approaches for providing insight into the identified topics; (3) suggestions regarding the appropriate role of CCSP-supported research in fostering problem-oriented/solution-based adaptive management decision support; and (4) suggestions for ongoing external advisory and review mechanisms.

CHAPTER 11 AUTHORS

Lead Authors
Susan Avery, NOAA and CCSPO
James R. Mahoney, DOC
Richard Moss, CCSPO
Jae Edmonds, PNNL
Shahid Habib, NASA
John Houghton, DOE
Claudia Nierenberg, NOAA
Joel Scheraga, USEPA

Contributors
James Andrews, ONR
Ron Birk, NASA
David Goodrich, NOAA
John Haynes, NASA
Ants Leetmaa, NOAA
Robert Marlay, DOE
Robert O'Connor, NSF
Rick Piltz, CCSPO

Observing and Monitoring the Climate System

CHAPTER CONTENTS

For each of six overarching goals, this chapter introduces objectives that provide products and milestones to be addressed in the coming decade based upon current knowledge.

Goal 1: Design, develop, deploy, integrate, and sustain observation components into a comprehensive system.

Goal 2: Accelerate the development and deployment of observing and monitoring elements needed for decision support.

Goal 3: Provide stewardship of the observing system.

Goal 4: Integrate modeling activities with the observing system.

Goal 5: Foster international cooperation to develop a complete global observing system.

Goal 6: Manage the observing system with an effective interagency structure.

The Global Change Research Act of 1990 specifically calls for "global measurements, establishing worldwide observations necessary to understand the physical, chemical, and biological processes responsible for changes in the Earth system on all relevant spatial and time scales," as well as "documentation of global change, including the development of mechanisms for recording changes that will actually occur in the Earth system over the coming decades." The program continues to respond to this call by following a strategy for the development and deployment of a global, integrated, and sustained observing system to address science requirements and decision support needs at appropriate accuracies and space and time resolutions.

The Climate Change Science Program (CCSP) strategy for observations and monitoring includes guiding principles, identification of priorities, and effective management of available resources. The purpose of this chapter is to describe how all the disparate observations described by the CCSP plan will be systematically organized and managed to improve our understanding of the climate system. The data management discussion that follows (see Chapter 13) describes the plan for archival and distribution of these data.

The CCSP observations and monitoring component seeks to address the following overarching question:

How can we provide active stewardship for an observation system that will document the evolving state of the climate system, allow for improved understanding of its changes, and contribute to improved predictive capability for society?

The development of space-based and *in situ* global observing capabilities was a primary focus of the program's first decade. Several new Earth-observing satellites, *in situ* networks, reference sites, and process studies are now producing unprecedented high-quality data that have led to major new insights about the climate system. The observing system for the future will build upon this success. For the purposes of this document, surface-based remote-sensing observations, as well as aircraft or suborbital measurements—in addition to direct observations within the atmosphere, ocean, ice, or land—will be considered as *in situ* measurements, and any references to *in situ* measurement networks should be considered as applying to networks using these techniques wherever appropriate.

The challenge for the coming decade is to maintain current capabilities, implement new elements, make operational the elements that need to be sustained, and integrate these observations into a comprehensive

"*Knowledge of the climate system and projections about the future climate are derived from fundamental physics and chemistry through models and observations of the atmosphere and the climate system. Climate models are built using the best scientific knowledge of the processes that operate within the climate system, which in turn are based on observations of these systems. A major limitation of these model forecasts for use around the world is the paucity of data available to evaluate the ability of coupled models to simulate important aspects of past climate. In addition, the observing system available today is a composite of observations that neither provide the information nor the continuity in the data needed to support measurements of climate variables. Therefore, above all, it is essential to ensure the existence of a long-term observing system that provides a more definitive observational foundation to evaluate decadal- to century-scale variability and change. This observing system must include observations of key state variables such as temperature, precipitation, humidity, pressure, clouds, sea ice and snow cover, sea level, sea surface temperature, carbon fluxes, and soil moisture. Additionally, more comprehensive regional measurements of greenhouse gases would provide critical information about their local and regional source strengths.*"

Climate Change Science:
An Analysis of Some Key Questions (NRC, 2001a)

global system to address the objectives of the CCSP research elements and decision support activities. Fundamental questions about the climate system and societal benefits that can be addressed are described in Chapters 3 through 11; these provide the basis for the observing system design and implementation. Illustrative examples of CCSP observation needs and milestones that address the research questions are described in Appendix 12.2. In addition, the overall climate observing system must address the five basic integrating goals for CCSP outlined in the introductory chapter:

- Improve knowledge of the Earth's past and present climate and environment, including its natural variability, and improve understanding of the causes of observed variability and change
- Improve quantification of the forces bringing about changes in the Earth's climate and related systems
- Reduce uncertainty in projections of how the Earth's climate and related systems may change in the future
- Understand the sensitivity and adaptability of different natural and managed ecosystems and human systems to climate and related global changes
- Explore the uses and identify the limits of evolving knowledge to manage risks and opportunities related to climate variability and change.

A system that integrates atmospheric, oceanographic, terrestrial, cryospheric, and cross-cutting observations does not currently exist *per se*. However, many components are available. For example, the Global Climate Observing System (GCOS) is a fairly well documented but not completely implemented international approach that has components in place to satisfy some climate requirements (see GCOS, 2003). GCOS is intended to provide a focused set of

observations from a subset of established measurement sites that are considered to have a sufficient climate history and spatial distribution. The system discussed here, however, goes beyond GCOS. CCSP has expanded the initial inventory of important climate observations (see Appendix 12.1) to encompass the needs of research on and applications to the global cycles of carbon, water, energy, and biogeochemical constituents; atmospheric composition; and changes in land use.

Building on the CCSP mission, the United States has also taken a leading role in fostering the development of a more broadly defined and integrated global observing system for all Earth parameters— for example, including geological as well as climate information. The United States hosted a ministerial-level Earth Observation Summit in July 2003, with participation by many developed and developing nations as well as many intergovernmental and international nongovernmental groups. This summit initiated a 10-year commitment to design, implement, and operate an expanded global observing system that builds on the major observational programs currently operated by the U.S. and many other governments and international organizations. CCSP agencies have provided the leadership, definition, and support for the Earth Observation Summit, and CCSP will closely integrate the U.S. observation and data management programs with the international programs launched at the summit.

The global observation system needed to fully implement CCSP includes the evolution of the observing capability provided through the U.S. Global Change Research Program (USGCRP) and the operationally oriented monitoring systems routinely provided by several federal agencies. The latter were never included as part of USGCRP, and may or may not be included as part of CCSP in budget inventories. In fact, as noted below, a critical issue associated with the implementation of the observing system for CCSP is the transitioning of research observations, typically made through USGCRP, into operations, not currently included as part of USGCRP (see Objective 1.3). Any consideration of budgets associated with global observations should be based on those for all the component programs, and not the historic USGCRP budget.

The basic elements of a global observing system must consist of:
- Routine and continuing measurements of selected variables, collected using established principles
- Shorter term exploratory observations carried out with satellites, process-oriented field campaigns with *in situ* techniques, and other, finite duration, research observations, collected using established principles
- A comprehensive and reliable distribution network and long-term archive
- Analysis and integration activities, including the use of four-dimensional (space, time) data assimilation and scientifically validated models.

These elements must be managed by an effective national entity, and coordinated at the international level. The management must have the capability to establish observing protocols, provide oversight, address deficiencies, and mobilize resources as required to maintain the integrity of the entire end-to-end system. A major challenge for CCSP in this decade will be the transition of many observation elements developed in a research mode to a sustained and operational environment [e.g., NASA research satellites to National Polar-orbiting Operational Environmental Satellite System (NPOESS) and some *in situ* networks]. The resulting global system must obviously engage many countries in a cooperative enterprise. Developing such a system presents a daunting challenge that must be met to provide essential information for decisionmakers.

In order to move toward a global observing system, CCSP will focus on six goals:

1) Design, develop, deploy, integrate, and sustain observation components into a comprehensive system
2) Accelerate the development and deployment of observing and monitoring elements needed for decision support
3) Provide stewardship of the observing system
4) Integrate modeling activities with the observing system
5) Foster international cooperation to develop a complete global observing system
6) Manage the observing system with an effective interagency structure.

Goal 1: Design, develop, deploy, integrate, and sustain observation components into a comprehensive system.

A system is the integration of interrelated, interacting, or interdependent components into a complex whole. The path to an overall observing system must address a number of key issues in addition to the observation components themselves, including the priorities for development, the implementation strategy, assessment, utilization, cooperation with other providers of environmental measurements, data management, international development, and system management. Some of these activities are considered in this goal, but others are of sufficient complexity that they are a goal unto themselves and are discussed later in this chapter.

Prioritization. Observing elements are in place within existing research and operational programs that partially fulfill the requirements for meeting these objectives. Other key sensors and observing networks still need to be developed and implemented. Priorities for

these augmentations are required because resources are limited. The CCSP research element questions and decision support goals provide the basis for determining priorities. In CCSP implementation plans, the research and decision support elements will provide a link between research question or decision support goal, measurement requirements, and the observation elements that meet the requirements. The prioritization criteria include: benefit to society, scientific return, partnership opportunities, technology readiness, program balance, and implementation of the climate monitoring principles (see Appendix 12.4). The management of the program will recommend priorities in consultation with the scientific community using the management mechanisms outlined in Chapter 16.

Evaluation. A key lesson learned over the past decade is that observing systems and networks must be implemented in a way that allows

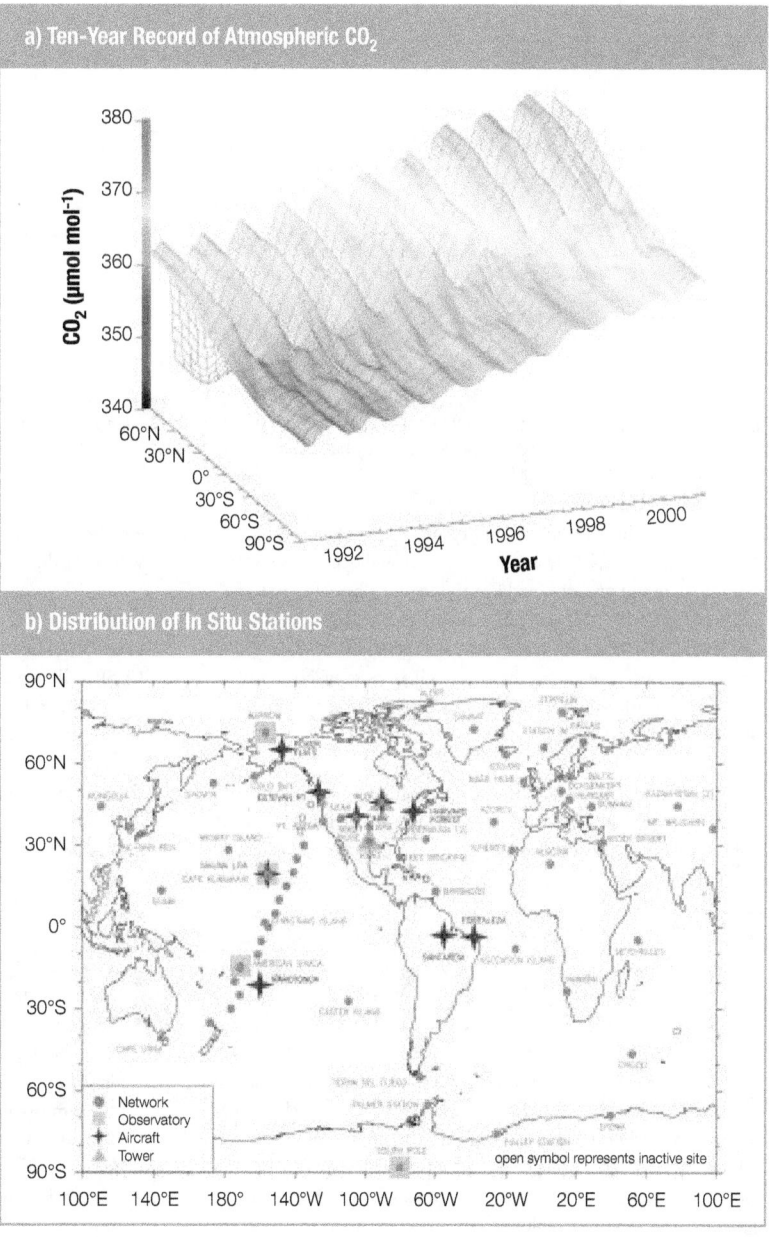

Figure 12-1: (a) Ten-year record of atmospheric carbon dioxide (CO_2) at a variety of stations as a function of latitude, and (b) distribution of *in situ* stations collecting data on CO_2 and other greenhouse gases. Source: Pieter Tans, NOAA CMDL. For more information, see Annex C.

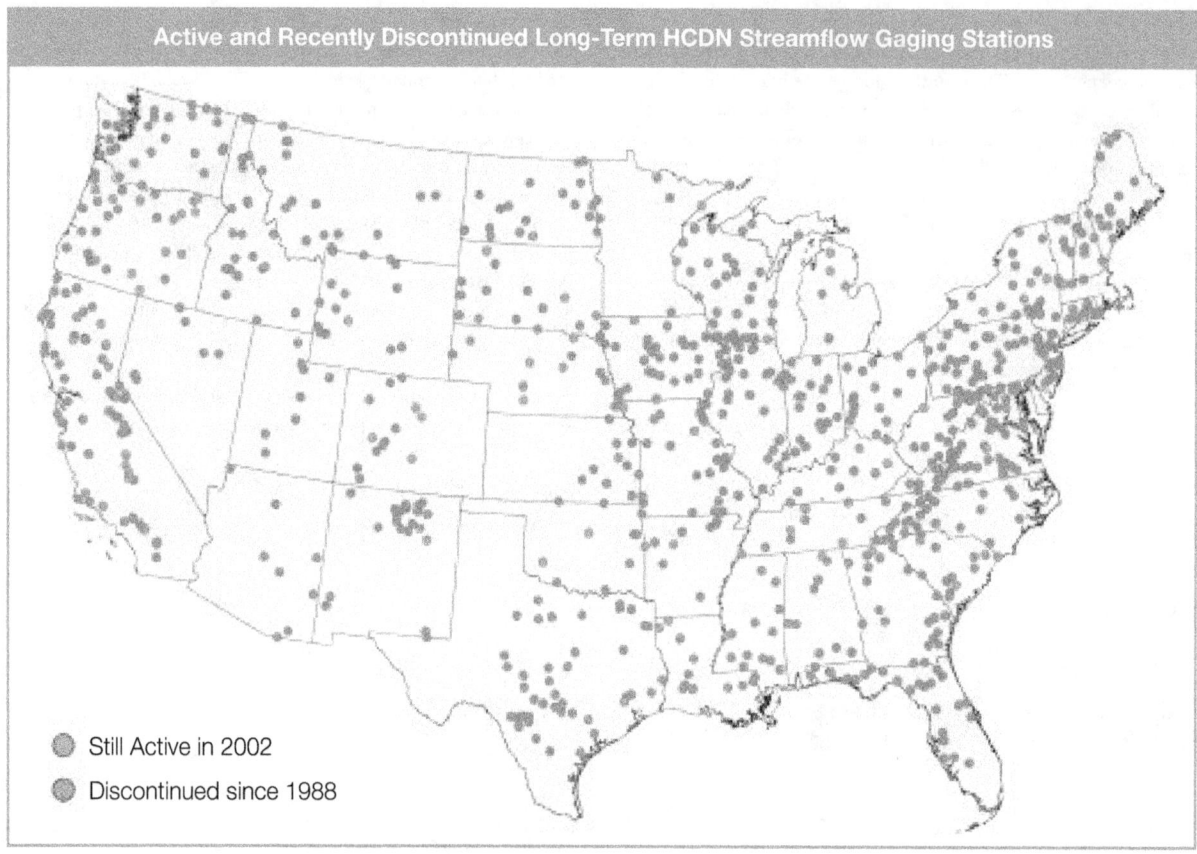

Active and Recently Discontinued Long-Term HCDN Streamflow Gaging Stations

- Still Active in 2002
- Discontinued since 1988

Figure 12-2: The U.S. Geological Survey (USGS) Hydro-Climatic Data Network (HCDN) was compiled in 1988 and included 854 gauges that were active and had at least 36 years of record. Stations that are still active are shown in green. Stations discontinued since 1988 are shown in red. These stations would have had at least 50 years of record as of 2002, if they had been kept in operation. Source: William Kirby, Global Hydroclimatology Program, USGS. For more information, see Annex C.

flexibility as both requirements and technology evolve. Therefore, the program will regularly assess the science and decision support requirements and propose modifications to the observing systems required for CCSP to execute its research plans. This process will involve the scientific community and program managers working on each research element, as well as those involved in modeling, scientific assessment, and other integrative activities (see Goal 6).

Cooperation. Many important observing systems are developed and operated by organizations that are not formal participants in CCSP, making the development of strong cooperative relationships that extend beyond the current CCSP a necessity. CCSP will work with observing system partners and the scientific community to identify climate requirements for these observing systems and to set priorities in light of available resources and competing needs.

Objective 1.1: Develop a requirements-based design for the climate observing system

A requirements-based design will be developed to identify baseline and minimum requirements that will address CCSP research element questions and decision support goals. U.S. needs and contributions will be weighed in the global context of the requirements and contributions of other nations and coordinated with them in a manner that will lead to a global system. The most efficient and

effective observation elements and networks will be selected that will meet the requirements.

Objective 1.2: Stabilize and extend existing observational capabilities

CCSP will maintain and improve basic research observing facilities, networks, and systems (both space-based and *in situ*). Climate-quality data requires long-term continuous records (see Chapter 4, Figure 4-1). It is critical to maintain records required to answer research questions before they are lost for other reasons (see Figure 12-2). Long-term observations require a focus on maintenance and replacement to sustain the capability at a sufficient level of accuracy to detect climate change over decades. To meet this objective, CCSP will:

- Continue satellite missions (see Figure 12-3) that are critical to answering the research element questions and upgrade the quality of their data to climate standards
- Extend and stabilize *in situ* networks for global coverage with consistent data quality, including moored, drifting, and ship-based networks in the ocean; surface and upper air networks in the atmosphere; and the major terrestrial networks
- Provide a uniform global set of surface reference sites of key ocean, land, atmosphere, and hydrology variables [see Figure 12-1 as an example of a global system for carbon dioxide (CO_2)]

- Provide careful calibration and overlapping operation of new and old technology during transitions to maintain quality control of data records
- Extend the capacity in terrestrial inventory programs to provide comprehensive information for key ecosystems within the United States.

Objective 1.3: Develop and implement a strategy for the transition of proven capabilities to an operational mode

The transition of proven research observation elements to sustained or operational status will make these components more cost-effective and sustainable for long-term benefit. CCSP will work toward an effective integration for the planning and development of research and operational systems. CCSP will develop a strategy to transition from research to operations that will adhere to the climate monitoring principles (see Appendix 12.4), including continuity, in order to have the most benefit to its objectives. Management structures will be developed to provide for a handoff from research management to operational systems management. The operation management will be responsible for continuing the research mission and producing and delivering day-to-day operational results and products.

Objective 1.4: Incorporate climate and global change observing requirements in operational programs at the appropriate level

Operational observation networks continue to be the backbone of climate system measurements. These networks, with only modest incremental costs, will satisfy significant parts of the climate observing requirements as described in the climate monitoring principles (see Appendix 12.4). Scientific, as well as decision support, oversight of operational systems for climate must be implemented using the concept of Climate Data Science Teams described in Goal 6 below.

Objective 1.5: Identify, develop, and implement measurement improvements

CCSP must maintain a sustained research and technology development program to address major deficiencies in observing systems and the paleoclimate record identified in Chapters 3 to 9 (e.g., closing the budgets for carbon, energy, and water cycles; integrating the coastal ocean monitoring systems; and completing records about decadal- and century-scale climate variability prior to the instrumental record). To the extent possible, new or improved measurements will be integrated into existing networks so as to minimize redundant operations and costs. Future global measurements from satellites and regional *in situ* networks will be developed that dramatically improve quality and vertical, spatial, and/or temporal resolution, especially to enhance regional coverage for decision support applications.

Objective 1.6: Continue intensive field missions

The science chapters have described and justified numerous field studies that integrate airborne (*in situ*), surface, and satellite observations over regional scales and durations from days to several weeks. These intensive observation periods provide valuable data for

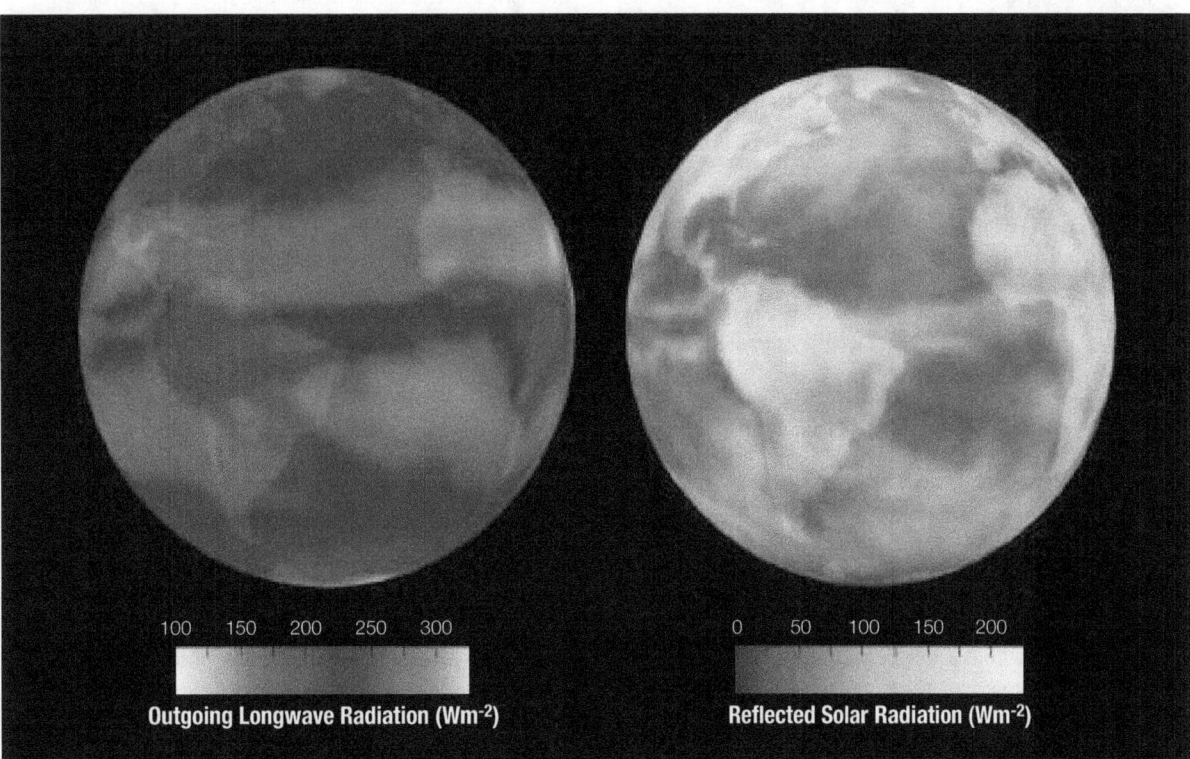

Outgoing Longwave Radiation (Wm⁻²)

Reflected Solar Radiation (Wm⁻²)

Figure 12-3: For scientists to understand climate, they must also determine what drives the changes within the Earth's radiation balance. From March 2000 to May 2001, the Clouds and Earth's Radiant Energy System (CERES) satellite measured some of these changes and produced new images that dynamically show heat (thermal radiation) emitted to space from Earth's surface and atmosphere (left sphere) and sunlight reflected back to space by the ocean, land, aerosols, and clouds (right sphere). Source: CERES Project, NASA (<http://asd-www.larc.nasa.gov/ceres/ASDceres.html>).

testing and validating satellite retrieval algorithms, and for the fine-scale resolution necessary to test, validate, and constrain physical processes in climate models. These coordinated observation efforts will need to become even more sophisticated as satellites evolve towards formation flying, onboard processing, and smart sensor technology. Aircraft remain a valuable platform for CCSP and mechanisms to make efficient use of these facilities across the CCSP agencies will be investigated.

Objective 1.7: Assess observing system performance with uniform monitoring tools and evaluation standards

A global effort of this magnitude must be managed as a system in order to be effective. For agencies to assess the performance of the observing system, an institutional mechanism must be put in place to monitor the status of the globally distributed components and to evaluate their combined capabilities.

CCSP, working with international partners, will develop a system architecture for monitoring distributed global operations that establishes and maintains links among the *in situ* and satellite elements and the data and modeling activities that are essential components of climate observation. The purpose is to provide a framework so that the nation can manage its observing system more efficiently and effectively, and:

- Provide an integrated view of the observing system linked to the CCSP mission
- Develop a more cost-effective observation system
- Allow all observations to be accessible by all customers when needed
- Provide a framework for examining future requirements and costs
- Allow for evolutionary improvements
- Identify gaps and overlaps
- Identify opportunities to migrate observations from the research elements into a sustained operational status.

The guiding principles for development of this uniform interagency and international monitoring capability include:
- Provide a system that is requirements-based
- Provide elements that are standards-based and interoperable
- Provide for system evolution that is minimally intrusive to other national and international missions
- Provide ample margins to accommodate changing requirements
- Provide a basis for continual evaluation and evolution.

Objective 1.8: Generate climate information through analysis and assimilation

Many countries have gathered climate system data to document climate system variability using many different instrument types during the past 150 years. In order to document and understand change from a historical perspective, we need to develop global, comprehensive, integrated, quality-controlled databases of climate system variables based on historical or modern measurements, and provide the user community with open and easy access to these databases. We must integrate these records as far into the past as is practical to reduce uncertainties in the climate trend estimates of individual parameters.

Our understanding of some of the changes in the physical climate system over the past 50 or more years has improved because of sys-

tematic reprocessing and integration of climate observations using state-of-the-art climate models and data assimilation technology. This must be continued through a routine and iterative improvement process that incorporates a rigorous research and validation component.

Objective 1.9: Initiate or participate in end-to-end pilot demonstrations of atmosphere, ocean, and terrestrial hydrologic observing systems

The integration of satellite and *in situ* observing networks for synthesis of the ocean, atmosphere, or terrestrial spheres in a systematic manner will be addressed through focused pilot programs. The Global Ocean Data Assimilation Experiment, while not specifically focused on climate issues, is an example of a pilot approach to the synthesis of ocean observations.

Objective 1.10: Develop a requirements-based program for collecting, integrating, and analyzing social, economic, and health factors with environmental change

Across the range of research on human response and consequences to climate change there is a particularly strong need for the integration of social, economic, and health data with environmental data (see Chapter 9). Observations will be used to address gaps in understanding, modeling, and quantifying the sensitivity and vulnerability of human systems to global change and measuring the capacity of human systems. Using retrospective analyses of consequences of shifts in climate will also help model future ability of hazard and resource management institutions to respond.

> **Goal 2**: Accelerate the development and deployment of observing and monitoring elements needed for decision support.

CCSP provides resources to develop observation systems and processing and support systems that will lead to reliable and useful products. These products will provide critical policy-neutral information for decision support and policymakers in areas such as climate and weather forecasting, human health, energy, environmental monitoring, greenhouse gases, and natural resource management.

CCSP will enhance the existing long-term monitoring elements with accelerated focused initiatives to provide a more definitive observational foundation for determining the current state of the climate. Many shortcomings of the current climate observing system relate to understanding climate forcings. In addition, fundamental observations for characterizing and understanding the state of the climate are needed for the global ocean, atmosphere, land surface, and ice variables. For the atmosphere, only half of the GCOS Upper-Air Network (GUAN), established for climate purposes, has been reporting regularly, and the GCOS Surface Network (GSN) for climate has had similarly disappointing results. The ocean is poorly observed below the surface, which limits our understanding about the ocean's response to a warming planet and its ability to naturally sequester greenhouse gases. Over land, the spatial heterogeneity requires detailed measurements and presents a major challenge.

In the budget requests for the past 2 years (FY2003 and FY2004), progress has been made to address many of these deficiencies, as described in Appendix 12.3. The objectives in this goal outline specific CCSP elements that will expand on this progress and support a directed strategy to focus resources and accelerate observations for climate change and decision support.

Objective 2.1: Complete the required atmosphere and ocean observation elements needed for a physical climate observing system

Within the climate monitoring arena, atmosphere and ocean observation elements to measure the physical aspects of the climate system are the most complete, and are ready to be brought together as a system. For example, the detailed open ocean observing system has been designed at both national and international levels through numerous community workshops, and implementation is about 40% complete (see Figure 12-4). The atmospheric system, which is a mixture of both climate and non-climate elements, has been recently examined (GCOS, 2003) and needs support for fixing degradations, sustaining current capabilities, and adding improvements. An accelerated effort to complete these two subsystems will improve understanding of climate characterization, forcing, and prediction, as well as facilitate the implementation and testing of the interagency management of complete observing systems. Focused new data management practices are being developed and

will promote efficient acquisition, validation, delivery, and archival of the measurements and data products. In order to complete these systems the following steps need to be taken in conjunction with international partners:

- GSN measures surface temperature, pressure, and precipitation at about 1,000 sites within the World Weather Watch (WWW) surface climate network. The network is not providing useful data at many sites in underdeveloped countries because of a lack of resources and training. Furthermore, only one-third of the network has provided historical data for examining climate extremes. CCSP will improve GSN data reporting.

- GUAN collects upper atmosphere temperature, wind, and humidity at 150 stations around the world. Performance at about one-third of the stations is poor. CCSP will improve *in situ* atmospheric column observations in GUAN. In addition, CCSP will identify techniques and implementation strategies for *in situ* atmospheric column observations of temperature, wind, and water vapor over the ocean, where GUAN cannot be used.

- CCSP will complete the U.S. Climate Reference Network of 250 stations nationwide to provide long-term homogeneous observations of temperature and precipitation that can be coupled to past long-term observations for the detection and attribution of present and future climate change.

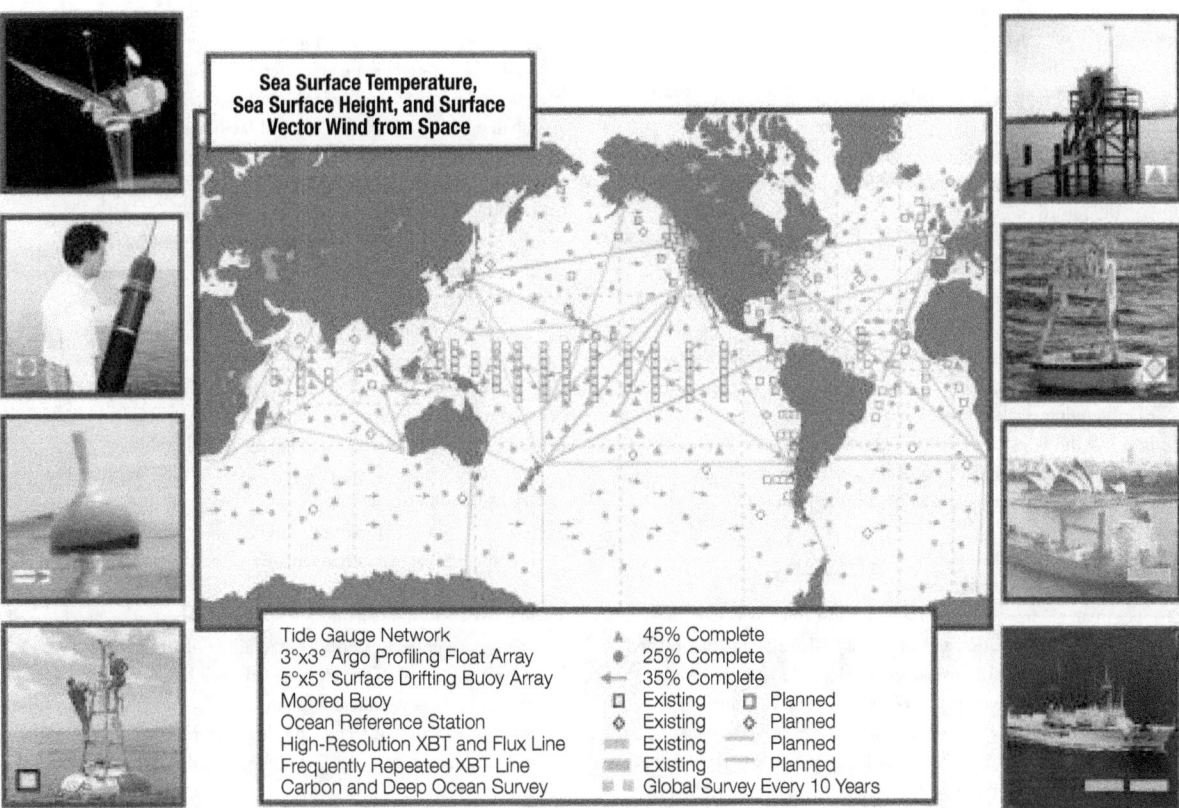

Figure 12-4: In 2003, the international global ocean observing system for measurements of the physical climate system was 40% complete. Source: Michael Johnson, NOAA Office of Global Programs.

- The United States will increase contributions to monitoring upper-ocean temperature and salinity structure that will improve understanding of the ocean's response to a warming planet. CCSP will support 50% of the international ocean profiling float program (Argo) and additional expendable bathythermographs from ships-of-opportunity to observe changes in heat and freshwater content.
- CCSP will improve estimates of global sea surface temperature for climate model initialization for better prediction capability, as well as regional barometric pressure and surface current velocity for improved model validation, by completing the global distribution of 1,250 surface drifting buoys.
- CCSP will reduce uncertainty in sea-level rise estimates by obtaining absolute positions for sea-level stations that are required to improve the calibration of satellite altimeters, used for the detection of long-term sea-level trends.
- The United States will continue to monitor the state of the tropical atmosphere and oceans in the Pacific and Atlantic Oceans with instrumented moored buoys and satellites, and contribute to the development of a similar network in the Indian Ocean for improved climate prediction and research.
- CCSP will improve model-based global air-sea flux estimates with surface flux reference moored buoy sites and Volunteer Observing Ships (also collect routine surface meteorological observations) with upgraded instruments for climate-quality observations.

Objective 2.2: Retrieve important paleoclimate records to provide a global long-term perspective on historical changes in climate

There are three aspects of the paleoclimatic record that need to be addressed by accelerated and focused programs within CCSP in order to improve characterization of long-term climate change and provide a valuable benchmark for testing models that are used to project the future (see Goal 4):

- Increase priority for retrieving rapidly disappearing paleoclimate information (e.g., melting glaciers, loss of corals and trees whose growth patterns are used for dating purposes, etc.) before these records are permanently lost
- Accelerate efforts to obtain interannual climate information in the Southern Hemisphere and tropics to develop a global record of millennium-scale climate variability
- Improve the integration of paleoclimatic observations and measurements with historical and modern climate data to form continuous time series of climatic information.

Objective 2.3: Develop new capabilities for ecosystem observations

Changes in an environmental variable—most often warming, but also changes in precipitation and air quality—have often been related to observed changes in biological and ecological systems. Several examples were mentioned in the Working Group II contribution to the Intergovernmental Panel on Climate Change's Third Assessment Report (IPCC, 2001b), including thawing of permafrost, lengthening of the period of active photosynthesis in mid- and high-latitude ecosystems, poleward shifts of plant and animal species ranges, movement of plant and animal species up elevation gradients, earlier spring flowering of trees, earlier spring emergence of insects, earlier

egg-laying in birds, and shifts in a forest-woodland ecotone (the boundary between the forest and the woodland).

These changes in ecosystems and organisms are consistent with warming and changes in precipitation, but the possibility remains that the observed biological and ecological changes were caused (in part) by other factors such as biological invasions or human land and marine resource management. Because of this, the attribution of the causes of biological and ecological changes to climatic change or variability is extremely difficult. Moreover, because many ecosystem-environment interactions play out over long periods—ultimately involving evolutionary changes and adaptations within ecosystems—long periods of study are needed in many cases to draw firm conclusions about relationships between environmental change, effects of that change on biological and ecological systems, and the significance of any observed biological or ecological changes for the functioning of ecosystems (see Chapter 8).

New research is needed to provide a significantly more complete picture of how biological and ecological systems may have responded to recent climatic change and variability, including possible biological or ecological responses to extreme events. New observational systems will also be needed to appropriately monitor potential future changes in the environment and accompanying biological or ecological changes (if any). A key challenge will be to provide organization, guidance, and synthesis for the emerging field of observed effects of climate change on biological and ecological systems.

CCSP will initiate studies of early effects and indicator systems across diverse ecosystems and geographic regions. A substantial amount of existing climate and effects data, a variety of monitoring efforts, and comparisons to scenario-based effects studies can be marshaled in this effort. CCSP will facilitate linked analyses of climatic trends and observed biological and ecological effects by supporting identification of appropriate past and ongoing monitoring efforts, design of needed new monitoring systems, and synthesis of results across ecosystems and regions. Research efforts will target those ecosystems that are subjected to the most rapid or extensive environmental changes and/or are most sensitive to possible environmental changes.

Long-term, spatially explicit, and quantitative observations of ecosystem state variables and concomitant environmental variables are needed. Initial activities and products will include:

- Define ecosystems sensitive to climate change and thresholds for measurable impacts
- Identify ecosystems and the interfaces between ecosystems (ecotones) that are either sensitive or resilient to environmental change
- Identify ecosystems experiencing the most rapid environmental changes (e.g., ecosystems located at high latitudes and high elevations or coastal ecosystems affected by ongoing sea-level rise and intensive human influences)
- Identify concurrent trends in other factors, such as population and land-use change and provide links to data sets that document these trends
- Identify links to biological and ecological data sets from monitoring programs, including those from remote-sensing platforms

- Validate impacts studies done with climate change scenarios over the near term or for small amounts of warming using observed climate and impact data
- Develop observational design criteria related to risk assessment and identification of causes of changes in distribution of pests and pathogens (e.g., climatic change interacting with weather)

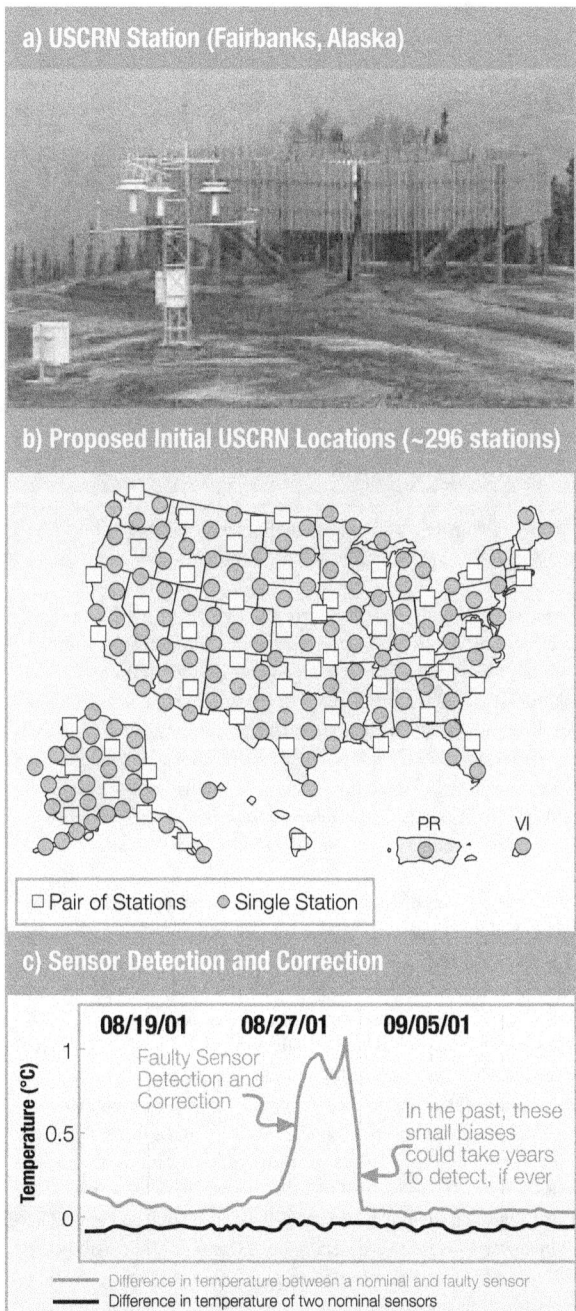

Figure 12-5: The U.S. Climate Reference Network is being established to reduce uncertainty and biases in long-term trends of key atmospheric variables at 296 stations (a&b). The monitoring system is designed such that multiple independent measurements are being made to capture sensor drift or biases (c). Source: NOAA National Climate Data Center.

- Develop design criteria for remote and *in situ* observations of biological and ecological systems that will help determine whether any observed ecological changes are attributable to global change
- Produce global, synoptic observational data products from satellite remote sensing documenting changes in biomass, albedo, leaf area and duration, and terrestrial and marine ecosystem composition for use in geographic information system (GIS)-based decision support systems
- Produce climate data at appropriate temporal and spatial scales for impacts studies.

Objective 2.4: Provide regular reports documenting the present state of the climate system components

CCSP will initiate a regular reporting program on the state of the climate system. Reports on various components of the system will highlight analysis products from the observing system to address the overarching question stated at the beginning of the chapter:

- Provide regular *State of the Climate* reports that include analyses of trends and variability in climate throughout the historical record using instrumental and paleoclimate records
- Evaluate the capabilities of existing and planned observing networks for providing data that will support the analysis of changes and trends in climate extremes and hazards
- Calculate the number of Climate Reference Network observing sites required to reduce the uncertainty in the observed climate signal for surface temperature to less than 0.1°C per century and precipitation to less than 1% per century on national and regional scales (see Figure 12-5)
- Regular reports evaluating the reduction in time-dependent biases in the space-based observing system obtained through implementation of the GCOS satellite monitoring principles and scientific data stewardship including the development of complementary independent observation and analysis techniques for critical climate variables
- For all operational monitoring networks, develop the tools necessary to identify time-dependent biases in the data as close to real-time as possible
- For climate monitoring networks, develop an operational system to identify non-climatic biases in the observing system for those climate system variables identified by the NRC (1999g) report on the adequacy of climate observing systems as relevant to detection, attribution, and direct societal impacts
- Identify and quantify the source of biases in climate system variables in existing climate reference data sets.

Goal 3: Provide stewardship of the observing system.

Observations of the climate require careful scientific oversight because climate signals are usually quite small relative to higher frequency phenomena in the data. Consequently, the climate science community has developed a set of principles that can be uniformly applied to all relevant measurements. In addition, special rigor must be applied to the algorithms, which are used to translate fundamental physical measurements into useful geophysical products, so that consistent results can be obtained and improved. Finally, scientific oversight of the algorithm, instrument calibration, and data processing and validation is critical to understanding the differences between instrument error and climate signal, and reduction of the error bars that define the uncertainty. This approach is captured in the four objectives of this goal.

Objective 3.1: Follow climate monitoring principles

Efforts in the last decade to use current research and operational data sets in global climate change research have provided a critical set of lessons learned. These lessons have been gathered into two sets of climate monitoring principles (see Appendix 12.4). These principles and experience will be used to guide the major improvements needed for the observing system. Instrument calibration, characterization, and stability become paramount considerations. Instruments must be tied to national and international standards such as those provided by the National Institute of Standards and Technology (NIST). When observations cannot achieve sufficient absolute accuracy or changes in spatial or temporal sampling occur, overlapping observations with high stability are required to ensure accurate monitoring of global change. Agreed-upon measurement protocols and procedures (e.g., continuous data validation and intercomparisons) are required to produce climate-quality data.

Objective 3.2: Provide independent measurement and analysis

To be policy-relevant, climate data must have reliable confidence intervals. Narrow intervals are most effective for decision support and policy development. Experience from previous scientific assessments, NIST, and other national standards laboratories have shown that actual accuracy is only known after comparison of independent measurements and analysis from multiple laboratories. Most climate observing systems are pushing the edge of capability in calibration. This suggests a final climate observing principle that climate observations should strive to address: *Each key climate variable will be measured using independent observations and examined with independent data analyses.*

Observation independence verifies instrument accuracy, while analysis independence verifies algorithms and computer code. For sea surface temperatures, examples of independent observations include satellite-derived measurements via infrared [Advanced Very High-Resolution Radiometer (AVHRR)] and passive microwave [Advanced Microwave Scanning Radiometer (AMSR)], and *in situ* instruments on ships, buoys, and floats. But many climate variables do not yet have independent sources. When surprises in climate data sets occur relative to current theory, independent confirmation is essential to ensure policy-relevant confidence in the results. A recent key example of this need is the air temperature record from radiosondes and the Microwave Sounding Unit (MSU) satellite data, as well as the different results from two analyses of the MSU satellite record. Using a single measurement of a key climate variable is a high-risk approach to reducing uncertainty for policymaking decisions.

Objective 3.3: Provide a sound foundation for climate-quality data products that will maintain integrity over time using well-characterized sensors and validated algorithms

An instrument radiometric mathematical model, termed the "measurement equation" by metrology experts, is developed to define, describe, and represent the relationship between the sensor output and desired physical observable (e.g., volts to temperature, counts to radiance, etc.). Additional models are used to reference the results in space and time (geo-location), and, finally, models are used to derive desired physical parameters (ozone profile, chlorophyll concentration, etc.). This process of characterization has been very successful in the past, especially for satellite programs. Its practice will continue to be encouraged, the results promulgated, and the instrument designs critically evaluated so that future measurements will produce the desired results.

It is important to understand that a sensor does not have to be ideal, but it must be well-characterized so that systematic effects can be evaluated and removed, even years after the mission, if new effects are identified. If these data are not available, the utility of the measurements is greatly compromised. The measurement equation approach defines the problem and places the entire procedure on a scientific basis. In the future all climate-quality observations need to be linked to their climate data records through an approved algorithm model that is described, for example, in an algorithm theoretical basis document (ATBD). These documents were pioneered by NASA in its implementation of the Earth Observing System (EOS) and provide a valuable lesson learned for future measurements.

Objective 3.4: Scientific stewardship for production of climate data records

Achieving climate accuracy with global and decadal sampling represents a unique scientific and organizational challenge. Calibration and stability requirements for climate data are often an order of magnitude more stringent than weather data. This accuracy is often achieved in research *in situ* and satellite data sets, with the length of this accuracy being dependent on the lifetime of the platforms and rate of instrument degradation; while in the longest satellite case (Earth Radiation Budget Satellite) this may approach 2 decades, in other cases it is much shorter. Global decadal sampling is often achieved in operational data sets, but without the accuracy and stability. Scientific data stewardship addresses this challenge through the production of Climate Data Records (CDRs).

Ultimately, the production of CDRs includes setting climate accuracy and stability requirements, instrument calibration and characterization, algorithm selection and development, end-to-end data validation and error analysis, quality control, and data production. Long-term archival and effective distribution are also essential and are addressed in Chapter 13. Experience in the last decade has shown that achieving CDR stewardship requires active and continual collaboration with the science community including both data producers and data users. Examples include Atmospheric Radiation Measurement (ARM) and Aeronet surface observations, the global surface temperature record, Tropical Atmosphere-Ocean buoys, and many forms of satellite observations for ocean, land, and atmosphere climate variables. Climate Data Science Teams (CDSTs) will be discussed in Goal 6 as a critical implementation strategy for the climate observing system.

Long-term CDRs will come about through the harmonious integration of highly accurate, research-quality observing systems with longer term operational and paleoclimate data. To reach the desired level of climate quality will also require additional efforts for current operational systems. We anticipate that CDSTs will be required to produce climate-quality data and represent the organizational link that provides operational level continuity with research-quality science direction.

The stewardship of scientific data is required to produce the climate-quality data sets that are a critical resource for policymakers and a legacy for future generations. As the length of CDRs increase, the power of the data to narrow uncertainties will increase and provide a basis for future decisions.

Goal 4: Integrate modeling activities with the observing system.

There are two primary roles for observations in the improvement of climate models: evaluation and initialization. At the same time, climate system models provide important insights for the improvement and interpretation of measurements through the use of simulations. These two-way interactions of observations and models affect not only the design but also the improvement of the observing system.

Objective 4.1: Develop protocols to evaluate models with climate-quality observations

Model testing uses observations to determine the uncertainties in current climate models and to prioritize needed improvements in future models. This is the most critical function of climate data records and depends primarily on two factors: (1) accuracy and stability of the climate data; and (2) the intrinsic background variability or noise in the Earth's climate system. Both factors vary with temporal and spatial scale. Climate system noise is typically largest at short time scales and small spatial scales (e.g., seasonal/regional scales), and smallest at long time scales and large spatial scales (e.g., century/global scales). Useful climate model tests require that both the data accuracy and the climate system noise be carefully assessed. As a result, climate accuracy requirements cannot be specified as a single number per variable, but must also be defined at several relevant temporal and spatial scales. Examples of relevant temporal/spatial scales include monthly/local, seasonal/regional, annual/zonal, and decadal/global. Some climate data records will have sufficient accuracy or stability to resolve regional climate change but inadequate coverage to resolve global changes and *vice versa*.

Consequently, observations and models of the climate system cannot make progress without continuous interaction between and understanding of the uncertainties in each research area (see also Chapter 10, Objective 1.3). Rapid progress requires regular interaction between the modeling and observation communities through CDSTs. One of the most critical collaborations between climate model and measurement scientists in the future will be the definition of prediction accuracy metrics capable of constraining key model uncertainties, such as cloud and water vapor feedback. Currently, such metrics are rather ad hoc because of the difficulty in unraveling the complex nonlinear climate system and in documenting climate data accuracy as a function of time and space. In the simplest sense, how would you recognize a perfect climate model if you had one? What tests would it pass? How will we know when the observing network is sufficient to characterize, attribute, and predict climate

changes accurately? The answers are a function of both climate models and the climate observing system design.

The definition of climate model evaluation metrics will affect observing system design. As models and data improve in quality, these metrics will evolve in an iterative fashion over the next decade.

Objective 4.2: Use observations to initialize climate variability models

Climate model initialization is primarily effective only for seasonal-to-interannual time scales, and for initialization of the state of the world's oceans, soil moisture, and vegetation. These latter conditions reflect climate processes with sufficiently long time scales that their initial state can affect seasonal-to-interannual climate states. The prime initialization example is prediction of El Niño events using initial ocean conditions. To improve short-term climate forecasts, it will require new and improved technologies in data assimilation and better utilization of *in situ* and remotely sensed global ocean, atmosphere, and terrestrial observations of key physical variables, such as ongoing observations of temperature and new measurements of salinity and soil moisture (see Chapter 10, Objective 1.4).

Objective 4.3: Utilize climate system models to assist in the design of observation systems

While climate models are not perfect, for some variables they can strongly support the estimation of accuracy requirements for CDRs. Consider the following simple paradigm of climate change: **forcing => feedback => response => climate change**. If the Earth were such a simple linear system, then designing observing systems and testing climate model predictions would be a relatively straightforward process. But the climate system is highly coupled and fundamentally nonlinear. Consequently, intrinsic internal variability is an inherent part of the real climate system. Climate change must be detected and understood. However, this signal is usually smaller than the background climate variability (e.g., year-to-year climate variability). CDRs will require accuracies at a temporal and spatial scale greatly below the level of natural variability.

Unfortunately, most quantitative observations of the climate system only exist for the last century and are a combination of background and climate variability, climate change, and observation error. Using the total variability of the system to define measurement requirements represents only an upper limit on required accuracy and is not likely to be sufficient. Climate model simulations, on the other hand, do not include the full complexity of the Earth's climate system. The models typically represent an underestimate of background variability. As a result, climate model ensemble simulations can be used to estimate the required accuracy for the observation system. The ensembles can also specify intrinsic variability of the system as a function of temporal and spatial scale. These simulations can use varying climate forcing and can separate background variability from the climate signal by running large numbers of simulations with slightly varying initial conditions, but the same boundary conditions.

Objective 4.4: Develop protocols for validating data assimilation and reanalysis products from the observing system

Increasingly, global and regional data sets derived from data assimilation models and retrospective reanalysis models are being used for climate research, monitoring (time series), diagnostics, applications, and impact assessments (see Objective 1.8 and also Chapter 10, Objective 1.5). Evaluation protocols must be developed to assess the accuracy and stability of these reanalysis products for climate applications. Acceptable error standards need to be established that will provide guidance on whether the data sets are usable or not, especially at regional scales and for model-derived/computed parameters.

> **Goal 5**: Foster international cooperation to develop a complete global observing system.

The range of global observations needed to understand and monitor climate processes, and to assess human impacts, exceeds the capability of any one country. Cooperation is therefore necessary to address priorities without duplication or omission (see Chapter 15). Satellite missions and *in situ* networks require many years of planning. Observations of the state of and trends in planetary processes cut across land, water, air, and oceans. National programs need to fit into larger international frameworks. At the international level, participation in both science and operational oversight committees must be continued, and strengthened or developed in disciplines that have not yet developed a global system.

In 1998, Parties to the United Nations Framework Convention on Climate Change (UNFCCC) noted with concern the mounting evidence of a decline in the global observing capability and urged Parties to undertake programs of systematic observations and to strengthen their capability in the collection, exchange, and utilization of environmental data and information. The United States supports the need to improve global observing systems for climate, and will join other Parties in submitting information on national plans and programs that contribute to the global observing capability.

Objective 5.1: Foster and support international partnerships in observations

The United States is an active and leading partner in the development and support of a global observing system that assembles key elements from a number of observing networks under the aegis of appropriate international organizations. With regard to climate, GCOS has fostered the integration of key elements including meteorological observations from the WWW Global Observing System, atmospheric constituents from the Global Atmosphere Watch, hydrological observations from the World Hydrological Observing System, critical oceanographic climate variables from the Global Ocean Observing System (GOOS), and several terrestrial variables from the Global Terrestrial Observing System (GTOS). Coordination of global satellite observations is carried out through the Committee on Earth Observation Satellites (CEOS). This is particularly important as multi-national, multi-spacecraft constellations and increased use of higher altitude orbits (e.g., geostationary) are used to improve the temporal resolution of global observations that may further enhance decision support.

Given the importance of validating satellite observations under a broad range of geophysical and biogeochemical conditions, participation of international partners—in providing coincident and correlative information that can be used to test and improve satellite

observations—is especially important. International partners are also important in the implementation of field campaigns that are best carried out with the full scientific involvement and logistical support of the host countries.

The full implementation of a global system for climate will require enhanced international coordination and commitment. Components for atmospheric, oceanic, terrestrial, and satellite observations are supported at varying levels depending on scientific priorities, availability of national contributions, and the sophistication of the relevant observing technologies. A specific focus on climate variables is essential to provide an adequate database to meet climate needs.

Recently, these observing systems, their sponsors, and the satellite community have developed the International Global Observing Strategy (IGOS). IGOS is a strategic planning process, uniting the major satellite and surface-based systems for global environmental observations of the atmosphere, oceans, and land in a framework for decisions and resource allocations by individual partners. The IGOS Partnership (IGOS-P) focuses specifically on the observing dimension in the process of providing environmental information for decisionmaking. This includes all forms of data collection concerning the physical, chemical, and biological environments of our planet, as well as data on the human environment, pressures on the natural environment, and environmental impacts on human well-being. IGOS-P is currently focusing on identifying and gaining commitments for essential requirements for observing oceanic processes, global carbon, atmospheric chemistry, the global water cycle, geo-hazards, and coral reefs as part of a future coastal initiative.

Bilateral agreements provide another strategy for achieving partnerships in observations and monitoring. Additionally, activities such as the U.S.-led Earth Observations Summit, the Subsidiary Body for Scientific and Technical Advice (SBSTA) meetings, and UNFCCC Conferences of the Parties provide opportunities to build international consensus—both scientific and political—for the implementation of a more integrated global climate observing system.

Objective 5.2: Provide support for key observations in developing countries

Developing countries provide key opportunities and challenges for observing systems. While many developing countries have the potential to make routine weather observations, and do so on a regular basis, many do not have the capability to collect and disseminate reliable observations of other variables that are critical for climate characterization and understanding. Further, many developing countries do not have adequate human resources to take full advantage of climate projections that would yield many benefits to their citizens.

It has been established that many developing countries lack either the capital or human resources to support high-quality observations or to sustain data and information systems. Countries often lack adequate capital for investment in equipment and supplies, trained technical staff, or maintenance capability.

The U.S. research community can point to numerous examples where it has contributed to measurements of key variables in developing countries. Many of these have provided valuable long-term data. The networks to measure atmospheric constituents through flask sampling and through vertical soundings have contributed a global database of important information on greenhouse gases. The oceanographic community has successfully engaged many coastal nations to participate in one or more of its observing systems (e.g., sea level, drifting floats and buoys, volunteer ships, etc.). International programs addressing terrestrial processes have demonstrated similar successes in observing ecosystems, obtaining hydrologic information, and initiating cryospheric measurement.

CCSP has specifically noted the key role that the United States can play in improving observational networks. In the President's June 2001 Rose Garden speech, he stated that "We'll also provide resources to build climate observation systems in developing countries and encourage other developed nations to match our American commitment."

The United States can contribute further by:
- Evaluating existing networks' capability to meet established climate requirements
- Improving existing networks through direct contributions to international programs [e.g., the World Meteorological Organization's (WMO) Voluntary Cooperation Programme]
- Forming partnerships with other developed countries to make directed investments to meet developing country inadequacies
- Providing direct assistance through U.S. programs of aid to specific developing country activities, such as the U.S. Agency for International Development (USAID)
- Continuing and expanding the collaborative international scientific programs that address critical climate variables.

Objective 5.3: International coordination through membership in key international groups

It is essential for the United States to maintain a leadership role in those international programs, both research and operational, that support climate observations of importance to U.S. programs. The principal international research programs (e.g., the World Climate Research Programme, the International Geosphere-Biosphere Programme, and the International Human Dimensions Programme) invite members from the U.S. scientific community and often from federal agencies to serve on relevant steering committees and working groups. These committees and working groups provide a forum for planning future research campaigns and for developing observational components that often lead to continuing measurement strategies. Individuals participating in these groups provide key links between U.S. and international research programs.

With regard to observations, U.S. scientists serve on the steering committees and working groups of the global observing systems and provide continuing advice to them. The principal programs for atmosphere, ocean, and terrestrial observations of climate include GCOS, GOOS, and GTOS, as well as their working groups.

U.S. scientists also serve on more operationally oriented international committees that support ongoing observational activities. Examples include the Commissions of the WMO (e.g., Basic Systems, Climatology, Hydrology, etc.) and the Joint Commission for Oceanography and Marine Meteorology sponsored by WMO and the Intergovernmental Oceanographic Commission. Individuals named to such groups serve as national representatives. These

groups establish observing requirements, develop protocols for data and information exchange, and obtain international commitments.

CCSP encourages U.S. scientists to continue their important roles in support of these diverse international programs.

Goal 6: Manage the observing system with an effective interagency structure.

The development of a national observing and monitoring system will require a coordinated management plan, which involves all federal agencies that conduct research and operational observations of the climate system. The management approach described in this goal follows the general guidelines outlined in Chapter 16. However, based upon the "best practices" learned over the history of USGCRP, a more detailed management structure is described for the observation system within this goal. Experience has shown that scientific oversight is key for the maintenance of climate-quality data from observing systems; consequently, the management plan will engage the science and user communities in key oversight and evaluation capacities. Finally, the observing system will be global and carried out with the nations of the world; therefore, cooperation at the international level for both operational and scientific oversight will need further development (see Goal 5).

At present, some of these management elements are in place, but a well-coordinated national management plan for observing the climate system is not, and will be addressed in the coming decade. At the federal level, coordination at the program level will be strengthened to address implementation, and oversight by agency principals will be established. Scientific oversight will occur through a layered pyramid of working groups starting at the most basic level of instrument teams and building up to an observing system science advisory council. This structure has been demonstrated to be an effective strategy over the lifetime of USGCRP and this goal expands the "best practices" learned to the observing system as a whole.

Objective 6.1: Provide coordinated management groups for the observing system elements

The management of the observing system elements will vary depending on size and complexity. Space systems, both operational and research, require focused project management groups that are dedicated to monitoring health and safety, providing command and control, receipt and quality control of downlinked data, fault recovery, and ground tracking, if necessary.

In situ systems can vary from control and maintenance of a single convenient site on land to maintenance of a large international network at sea. While the former can rely on a simple management program, the latter requires a much larger national management group and international coordination. CCSP will coordinate management groups of similar measurement techniques and logistics; coordinate with international management where applicable; work with international implementation coordination groups, such as the Joint Commission on Oceanography and Marine Meterology; and finally provide overall management and guidance for the national system as a whole through an interagency working group. The selection and oversight of management groups will depend on whether they are research or operational, and federal, commercial, or university.

Objective 6.2: Provide Climate Data Science Teams for climate data stewardship

The foundation of the management strategy is to obtain key CDRs through the use of CDSTs. Definition of the CDSTs is based on the last 2 decades of experience and lessons gained from previous Earth observation systems focused on climate measurements. A clear message from this experience is the need for CDSTs. These teams are composed of a group of scientists and engineers whose purpose is to convert raw instrument data into CDRs, including calibration, algorithm development, validation, error analysis, quality control, and data product design. If the data volume is small, the CDST may also produce and distribute the data products. If the data volume is large, the CDST may interface to a separate data center for production, archiving, and distribution.

Examples of effective CDSTs include the production of climate versions of the surface air temperature records; most NASA EOS satellite data products, as well as those of precursor activities such

as the Total Ozone Mapping Spectrometer, Stratospheric Aerosol and Gas Experiment, and Active Cavity Radiometer Irradiance Monitor; the international Argo ocean profiling float program; the NOAA-led Baseline Surface Radiation Network; and DOE's ARM Program. The reason that CDSTs are required for most, if not all, climate data records is the extreme accuracy needed for rigorous climate records (see Objective 3.1). Because the methods of measurement vary so greatly, effective CDSTs focus on just a few of the climate data records, and some only on a single CDR. A minimal list of variables needed as CDRs is given in Appendix 12.1.

Most CDSTs are chosen by scientific peer review. They consist of a principal investigator and a set of co-investigators. The co-investigators lead key instrument, algorithm, sampling, validation, and data management functions. CDSTs include members of the climate modeling or climate analysis community (i.e., data users). This is key to keeping the CDST focused on the most effective approach to meeting users' needs. Most CDSTs are currently funded by a single U.S. agency. CDSTs have their work and products peer-reviewed. CDSTs are often used in national and international assessments of the state of climate science and climate impacts. They document the

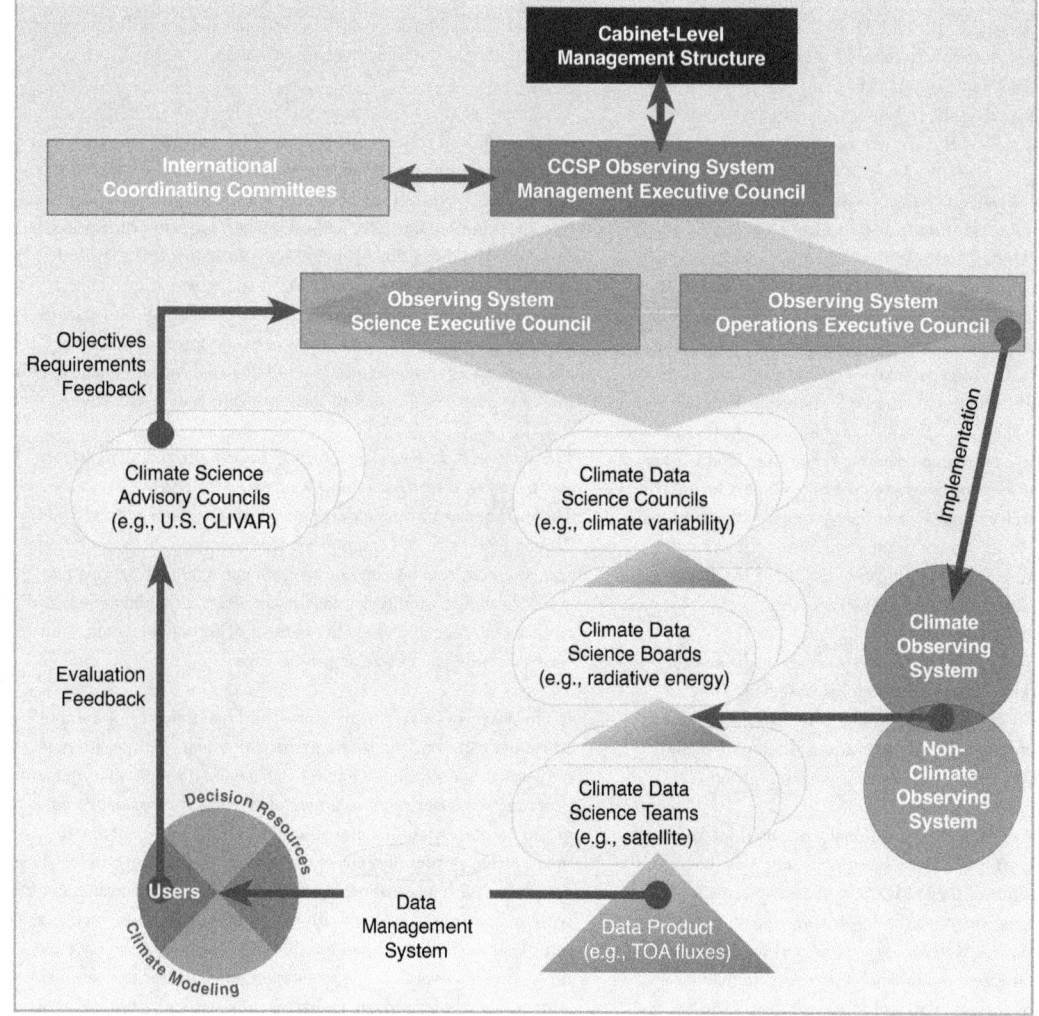

Figure 12-6: Schematic of the overall management structure for the observing system.

climate data products and their quality through web documents, conference papers, and journal papers.

Objective 6.3: Develop science and management advisory boards and councils to prioritize across climate system components and to guide system evolution

Observing system science direction will be provided by boards and councils. It is too long a step between a single CDST focusing on one or two climate variables and CCSP. A logical structure will be needed between these two extremes. The structure chosen must be able to handle the following set of changes and prioritizations:

- Evolving scientific understanding and changes in accuracy requirements
- Changing financial/human resources and delivery schedules
- Evolving technology for developing and implementing improvements
- Changing instrument, surface site, and spacecraft operations
- Changing data management interfaces and technology (e.g., production, distribution, and archival)
- User-driven changes in data product format requirements.

Addressing these trade-offs between resources, requirements, and climate variables will require a hierarchical science management committee structure. While there are many ways to provide such a structure, climate data and climate modelers will be best served by a structure that gives highest priority to tight integration of major climate system components (see Figure 12-6). Portions of such a system have evolved over the last decade, primarily at the international level as discussed under Goal 5. It is most fully evolved and effective for ocean observations, while the process for terrestrial observations is at a much earlier stage of development. More of this structure will be developed by CCSP for the United States.

The CCSP cabinet-level management structure is at the top level, responsible for resource commitments to the climate observing system.

The second level consists of a sub-group of CCSP—the Observing System Management Executive Council (OSMEC), composed of agency executives capable of committing agency funding. This group will provide a formal interface with international coordination groups, such as CEOS and IGOS-P.

Two groups comprise the third level. The first is an interagency program management group responsible for providing resources to implement the observations and to support science investigations— the Observing System Operations Executive Council (OSOEC). The second is the highest level science advisory group—the U.S. Observing System Science Executive Council (OSSEC). This science council makes recommendations to both OSMEC and OSOEC by prioritizing across the entire climate observing system. In addition, OSSEC will develop the objectives and requirements for the observing system, based on input from the Climate Science Advisory Councils of the research elements and decision support groups, and will evaluate the ability of the observing system to meet these objectives. These Climate Science Advisory Councils will also provide OSSEC with evaluation reports on the performance of the system in meeting the needs of its users.

At the fourth level, Climate Data Science Councils are responsible for climate observations to support each of the seven major CCSP research elements: Atmospheric Composition, Climate Variability and Change, Water Cycle, Land-Use/Land-Cover Change, Carbon Cycle, Ecosystems, and Human Contributions and Responses, as well as Decision Support activities. The Climate Data Science Councils could be a sub-group of the Climate Science Advisory Councils shown in Figure 12-6. Some of the Climate Science Advisory Councils already exist, such as the U.S. Climate Variability and Predictability (CLIVAR) team for the Climate Variability and Change theme. The Climate Data Science Councils address the complete range of climate variables within their theme.

A minimum set of climate variables is given in Appendix 12.1. Some variables will be relevant to several of the Climate Data Science Councils. Such overlaps are inevitable in any management structure given the tight coupling of processes in the climate system. It is the responsibility of OSSEC to assign primary responsibility for each variable to one of the Climate Data Science Councils. The Climate Data Science Councils are responsible for setting CDR requirements for absolute accuracy, stability, and space/time sampling.

At the fifth level, Climate Data Science Boards will address data parameters grouped into the most natural subsets for climate processes and/or for the type of instrumentation. For example, one of the Climate Variability and Change science boards would cover radiative energy from the top of atmosphere to the surface of the Earth. Some of these groups have already been formed by individual agencies (e.g., sea surface topography, ocean winds, sea surface temperature, and ocean color within NASA). Within each Science Board there are *in situ*, satellite, and field experiment observations. The advantage of this approach is that the linkages of surface, satellite, and *in situ* data types for key climate parameters are actively considered in any observing system trade-offs and will be valuable to the implementation decisions of OSOEC. Each CDST will report to the relevant Climate Data Science Board.

The Climate Data Science Boards and Executive Councils will meet as required to assess progress in the data system, nominally once per year. These meetings are expected to be indepth workshops sufficient to deal with the complex trade-offs and linkages between the observing system components. Every 5 years, a major assessment of the entire climate observing system would be produced, reviewed by the National Research Council, and coordinated with international assessments (e.g., the International Conference on Ocean Observing Systems for Climate, GCOS Adequacy Reports, IPCC assessments).

These boards and councils must include members of the climate data user community (e.g., climate modelers, climate analysts), as well as CDST representatives and data management representatives. In addition, members of the community who participate in international science or operations groups will be *ex officio* members of their appropriate counterpart board or council. The most efficient organization will be to maximize the extent that the data management and climate modeling management structures parallel that of climate observing and monitoring.

Objective 6.4: Provide a management structure that allows clear interagency responsibility, prioritization, peer review, and evolution of the observing system

Observations and monitoring of the climate system are carried out by a number of federal agencies. Success will require a series of steps to be initiated jointly by OSMEC, OSOEC, and OSSEC. These steps include:

- Assign agency responsibility for measurement contributions to each variable and measurement type. These contributions would typically be defined in terms of measurement system (satellite, *in situ*, field experiment) and sampling (e.g., satellite orbits, number/location of ground sites). OSSEC will be necessary in this step in order to ensure that the climate monitoring principles (see Appendix 12.4) are adhered to in assigning these responsibilities and evaluating the ability of the observations to meet the science objectives and requirements. If more than one agency is involved in measuring a single parameter, OSSEC must clearly define the responsibilities of each. Given the need for independent observations, it is recommended that the United States and the international community each provide at least one measurement of each variable. This would ensure the absolute minimum of two observations for each key climate variable.

- Assess the state of the climate observing system, including current agency plans. This assessment must include climate observing system requirements for each component as well as current and planned capabilities.

- Once shortcomings are identified, the Executive Councils (Science, Operations, and Management) and international groups must coordinate a plan for eliminating these shortcomings over time.

- Select the CDSTs for each climate variable, preferably through peer review. Typically, for any given variable, there will be separate CDSTs for surface and satellite observations. This is dictated by the large differences in instrumentation, calibration, operations, and sampling for these systems.

CHAPTER 12 AUTHORS

Lead Authors
Chester Koblinsky, NASA and CCSPO
Thomas Spence, NSF
Waleed Abdalati, NASA
Michael Dettinger, USGS
Howard Diamond, NOAA
Michael Johnson, NOAA
Thomas Karl, NOAA
James Mcguire, NASA
Lawrence Pettinger, USGS
Sushel Unninayer, NASA
Bruce Wielicki, NASA

Contributors
Jeff Amthor, DOE
James Andrews, ONR
James Butler, NOAA
Margarita Conkright, NOAA and CCSPO
Wanda Ferrell, DOE
Gerald Fraser, NIST
David Goodrich, NOAA
Carol Johnson, NIST
Jack Kaye, NASA
William Kirby, USGS
Laura Miner-Nordstrom, USEPA
Steve Piotrowicz, NOAA
Cynthia Rosenwieg, NASA
David Shultz, USGS
Brent Smith, NOAA
Sidney Thurston, NOAA

APPENDIX 12.1

EARTH CLIMATE SYSTEM OBSERVATIONS

This table provides a summary of "State" and "Forcing/Feedback" variables for the major components of the Earth system for which observations are required. In parentheses, "I" and "S" denote measurements made by *in situ* and space-based instruments, respectively. See Annex C for source information.

STATE VARIABLES	EXTERNAL FORCING OR FEEDBACK VARIABLES
(1) Atmosphere	
• wind (I/S) • upper air temperature (I/S) • surface air temperature (I/S) • sea-level pressure (I) • upper air water vapor (I/S) • surface air humidity/water vapor (I/S) • precipitation (I/S) • clouds (I/S) • liquid water content (I/S)	• sea surface temperature (I/S) • land surface soil moisture/temperature (I/S) • land surface structure and topography (I/S) • land surface vegetation (I/S) • CO_2 and other greenhouse gases, ozone and chemistry, aerosols (I/S) • evaporation and evapotranspiration (I/S) • snow/ice cover (I/S) • shortwave and longwave surface radiation budget (I/S) • solar irradiance and shortwave/longwave radiation budget (S)
(2) Ocean	
• upper ocean currents (I/S) • sea surface temperature (I/S) • sea-level/surface topography (I/S) • sea surface salinity (I/S) • sea ice (I/S) • wave characteristics (I/S) • mid- and deep-ocean currents (I) • subsurface thermal structure (I) • subsurface salinity structure (I) • ocean biomass/phytoplankton (I/S) • subsurface carbon (I), nutrients (I) • subsurface chemical tracers (I)	• ocean surface wind and wind stress (I/S) • incoming surface shortwave radiation (I/S) • downwelling longwave radiation (I/S) • surface air temperature/humidity (I/S) • precipitation (freshwater/salinity flux) (I/S) • evaporation (I/S) • freshwater flux from rivers and ice melt (I/S) • CO_2 flux across the air-sea interface (I) • geothermal heat flux—ocean bottom (I) • organic and inorganic effluents (into ocean) (I/S) • biomass and standing stock (I/S) • biodiversity (I) • human impacts—fishing (I) • coastal zones/margins (I/S)
(3) Terrestrial	
• topography/elevation (I/S) • land cover (I/S) • leaf area index (I) • soil moisture/wetness (I/S) • soil structure/type (I/S) • permafrost (I) • vegetation/biomass vigor (I/S) • water runoff (I/S) • surface ground temperature (I/S) • snow/ice cover (I/S) • subsurface temperature and moisture (I/S) • soil carbon, nitrogen, phosphorus, nutrients (I) • necromass (plant litter) (I) • subsurface biome/vigor (I) • land use (I/S) • groundwater and subterranean flow (I) • lakes and reservoirs (I/S) • rivers and river flow (I/S) • glaciers and ice sheets (I/S) • water turbidity, nitrogen, phosphorus, dissolved oxygen (I/S)	• incoming shortwave radiation (I/S) • net downwelling longwave radiation (I/S) • fraction of absorbed photosynthetically active radiation (I/S) • surface winds (I) • surface air temperature and humidity (I/S) • albedo (I/S) • evaporation and evapotranspiration (I/S) • precipitation (I/S) • land use and land-use practices (I/S) • deforestation (I/S) • human impacts—land degradation (I/S) • erosion, sediment transport (I/S) • fire occurrence (I/S) • volcanic effects (on surface) (I/S) • biodiversity (I/S) • chemical (fertilizer/pesticide and gas exchange) (I) • waste disposal and other contaminants (I) • earthquakes, tectonic motions (I/S) • nutrients and soil microbial activity (I) • coastal zones/margins (I/S)

APPENDIX 12.2

ILLUSTRATIVE RESEARCH MILESTONES FOR OBSERVATIONS AND MONITORING

This table provides examples of observing priorities highlighted in the research element chapters (Chapters 3-9) of this plan.

Atmospheric Composition

- Continue baseline observations of atmospheric composition over North America and globally.
- Improve description of the global distributions of aerosols and their properties.
- Develop and improve inventories of global emissions of methane, carbon monoxide, nitrous oxide, and nitrogen oxides (NO_x) from anthropogenic and natural sources.
- Monitor global distributions of tropospheric ozone and some of its precursors (e.g., NO_x).
- Continue monitoring trends in ultraviolet radiation.

Climate Variability and Change

- Improve effectiveness of observing systems, including deployment of new systems and re-deployment of existing systems, as well as the collection of targeted paleoclimatic data.
- Improve estimates of global air-sea-land fluxes of heat, moisture, and momentum.
- Regularly update and extend global climate reanalyses.
- Conduct process studies for needed observations of critical ocean mixing processes.
- Develop a paleoclimatic database to evaluate climate models.

Water Cycle

- Develop an integrated global observing strategy for water cycle variables.
- Characterize water vapor in the climate-critical area of the tropical tropopause.
- Monitor drought based on improved measurements of precipitation, soil moisture, and runoff.
- Test parameterizations for clouds and precipitation processes.
- Initialize and test boundary layers and other components in models.

Land-Use/Land-Cover Change

- Continue to acquire global calibrated coarse-, moderate-, and high-resolution remotely sensed data.
- Provide global maps of areas of rapid land-use and land-cover change, and location and extent of fires.
- Quantify rates of regional, national, and global land-use and land-cover change.
- Develop global high-resolution satellite land-cover databases.
- Provide operational global monitoring of land use and land-cover conditions.

Carbon Cycle

- Provide U.S. contributions to an international carbon observing system, including measurements of carbon storage, fluxes, and complementary environmental data.
- Assessment of the quality of measurements that support global carbon cycle science.
- Measure atmospheric carbon dioxide and methane concentrations and related tracers in under-sampled locations.
- Observe global air-sea fluxes of carbon dioxide, lateral ocean carbon transport, and delivery of carbon from the land to the ocean.
- Develop database of agricultural management effects on carbon emissions and sequestration in the United States.

Ecosystems

- Define requirements for ecosystem observations to quantify feedbacks to climate and atmospheric composition, to enhance existing observation systems, and to guide development of new capabilities.
- Quantify biomass, species composition, and community structure of terrestrial and aquatic ecosystems in relation to disturbance patterns.
- Maintain and enhance satellite terrestrial, atmospheric, and oceanic observing systems and networks, to monitor trends in ecosystem variables to parameterize models and verify model projections.
- Continue and enhance long-term observations to track changes in seasonal cycles of productivity, species distributions and abundance, and ecosystem structure.
- Provide data quantifying aboveground and belowground effects of elevated carbon dioxide concentration in combination with elevated ozone concentration on the structure and functioning of agricultural, forest, and aquatic ecosystems.

Human Contributions and Responses to Environmental Change

- Gridded world population database, including time series as far as possible into the past.
- Human footprint data set that depicts the geographic extent of human impacts on the environment.
- Produce elevation maps depicting areas vulnerable to sea-level rise and planning maps depicting how state and local governments plan to respond to sea-level rise.

APPENDIX 12.3

CLIMATE CHANGE RESEARCH INITIATIVE (CCRI) ACTIVITIES IN OBSERVATIONS AND MONITORING

CCRI Milestones

Ocean Observations

CCRI funds will be used to work toward the establishment of an ocean observing system that can accurately document climate-scale changes in ocean heat content, carbon uptake, and sea-level changes. Global tropical measurements will be augmented to improve seasonal forecasts. Ocean reference stations will be added to improve routine analyses of ocean-atmosphere fluxes at these stations to improve energy balance studies and coupled modeling parameterizations. In addition, key locations will be instrumented to improve understanding of abrupt climate change detection. The requirements for ocean observations for climate have been well documented, the relevant technology is available, and the international community is mobilized through GCOS and GOOS to implement key elements of the system.

Atmosphere Observations

CCRI funds will be used to work with other developed countries to reestablish the benchmark upper air network, emphasizing data-sparse regions; to upgrade the GCOS surface network for baseline variables; to deploy mobile Atmospheric Radiation Measurement program facilities; to begin planning for an early copy of an NPOESS aerosol instrument; and to place new Global Atmospheric Watch stations in priority sites to measure pollutant emissions, aerosols, and ozone, in specific regions.

Aerosol Observations

Aerosols and tropospheric ozone play unique, but poorly quantified, roles in the atmospheric radiation budget. CCRI investments will be used to begin implementation of plans developed by the interagency National Aerosol-Climate Interactions Program to define and evaluate the role of aerosols that absorb solar radiation, such as black carbon and mineral dust. Proposed activities include field campaigns, *in situ* monitoring stations to measure black carbon and aerosol precursors, global climatologies of tropospheric aerosols, and satellite algorithm development.

Carbon Cycle Observations

Research objectives for carbon cycle science include improved observations to address some of the field's greatest areas of uncertainty. CCRI funds will be targeted for the integrated North American Carbon Program (NACP), a priority of the *U.S. Carbon Cycle Science Plan*. NACP calls for implementing a North American Carbon Cycle Observing System consisting of a network of small aircraft stations and tall towers to obtain profiles of carbon gases for determining sources and sinks of carbon dioxide in the United States, expansion of the AmeriFlux sites, the development of automated carbon dioxide and methane sensors, improvements in ground-based measurements, and inventories of forest and agricultural lands. In addition, funds are provided for improved estimates of carbon dioxide over and into the ocean derived from ship-based instruments.

In addition, CCRI will select and award development assistance projects for climate monitoring in developing nations.

APPENDIX 12.4

GCOS CLIMATE MONITORING PRINCIPLES[4]

Effective monitoring systems for climate should adhere to the following principles:

1. The impact of new systems or changes to existing systems should be assessed prior to implementation.
2. A suitable period of overlap for new and old observing systems is required.
3. The details and history of local conditions, instruments, operating procedures, data processing algorithms, and other factors pertinent to interpreting data (i.e., metadata) should be documented and treated with the same care as the data themselves.
4. The quality and homogeneity of data should be regularly assessed as a part of routine operations.
5. Consideration of the needs for environmental and climate-monitoring products and assessments, such as IPCC assessments, should be integrated into national, regional, and global observing priorities.
6. Operation of historically uninterrupted stations and observing systems should be maintained.
7. High priority for additional observations should be focused on data-poor regions, poorly observed parameters, regions sensitive to change, and key measurements with inadequate temporal resolution.
8. Long-term requirements, including appropriate sampling frequencies, should be specified to network designers, operators, and instrument engineers at the outset of system design and implementation.
9. The conversion of research observing systems to long-term operations in a carefully planned manner should be promoted.
10. Data management systems that facilitate access, use, and interpretation of data and products should be included as essential elements of climate monitoring systems.

Furthermore, satellite systems for monitoring climate need to:
a. *Take steps to make radiance calibration, calibration monitoring, and satellite-to-satellite cross-calibration of the full operational constellation a part of the operational satellite system.*
b. *Take steps to sample the Earth system in such a way that climate-relevant (diurnal, seasonal, and long-term interannual) changes can be resolved.*

Thus, satellite systems for climate monitoring should adhere to the following specific principles:
11. Constant sampling within the diurnal cycle (minimizing the effects of orbital decay and orbit drift) should be maintained.
12. A suitable period of overlap for new and old satellite systems should be ensured for a period adequate to determine inter-satellite biases and maintain the homogeneity and consistency of time-series observations.
13. Continuity of satellite measurements (i.e., elimination of gaps in the long-term record) through appropriate launch and orbital strategies should be ensured.
14. Rigorous pre-launch instrument characterization and calibration, including radiance confirmation against an international radiance scale provided by a national metrology institute, should be ensured.
15. On-board calibration adequate for climate system observations should be ensured and associated instrument characteristics monitored.
16. Operational production of priority climate products should be sustained and peer-reviewed new products should be introduced as appropriate.
17. Data systems needed to facilitate user access to climate products, metadata, and raw data, including key data for delayed-mode analysis, should be established and maintained.
18. Use of functioning baseline instruments that meet the calibration and stability requirements stated above should be maintained for as long as possible, even when these exist on de-commissioned satellites.
19. Complementary *in situ* baseline observations for satellite measurements should be maintained through appropriate activities and cooperation.
20. Random errors and time-dependent biases in satellite observations and derived products should be identified.

---

[4] The ten basic principles were adopted (in paraphrased form) by the Conference of the Parties (COP) to the U.N. Framework Convention on Climate Change through Decision 5/CP.5 of COP-5 at Bonn in November 1999.

CHAPTER

CHAPTER CONTENTS

For each goal, this chapter introduces the objectives for data management to be addressed in the coming decade based upon current knowledge and infrastructure.

Goal 1: Collect and manage data in multiple locations.

Goal 2: Enable users to discover and access data and information via the Internet.

Goal 3: Develop integrated information data products for scientists and decisionmakers.

Goal 4: Preserve data.

Management Structure

National and International Partnerships

One of the goals of the U.S. Climate Change Research Initiative (CCRI) is to enhance and integrate observation, monitoring, and data management systems to support climate process and trend analyses. This chapter lays the strategy for managing integrated data and information for the next decade.

The nature of the concerted effort of the Climate Change Science Program (CCSP) calls for an overarching data policy that provides full and open access to Earth science-related data in a timely fashion. The terms and conditions of exchange and use for this purpose should be agreed to both nationally and internationally. In the early 1990s, the U.S. Global Change Research Program (USGCRP) agreed to data exchange principles that are still adhered to today (see Box 13-1). The governing law for U.S. Government agencies, OMB Circular A130, specifically states that the "open and efficient exchange of scientific and technical government information, subject to applicable national security controls and the proprietary rights of others, fosters excellence in scientific research and effective use of federal research and development funds." Office of Management and Budget (OMB) Circular A130 establishes agency user charges at the marginal cost of dissemination, including a provision that agencies can plan to "establish user charges at less than cost of dissemination because of a determination that higher charges would constitute a

significant barrier to properly performing the agency's functions, including reaching members of the public whom the agency has a responsibility to inform". This lofty standard should be emulated by all participants in the larger endeavor described by this plan.

The need to manage data as a shared national resource in a manner that focuses on the needs of end users has not previously been recognized, nor has the challenge been undertaken in a serious and systematic manner. Climate data are complex and variable as the data are obtained by diverse means, across a broad range of disciplines, for a variety of purposes, and by wide-ranging organizations—individual researchers; institutions; private industry; and federal, state, and local government organizations. These data come in different forms, from a single variable measured at a single point (e.g., a species name) to multi-variate, four-dimensional data sets that may be petabytes (10^{15} bytes) in size.

Although new data sets that integrate information from multiple sources are being developed, current efforts are limited in scope and a significant expansion is required to meet the needs of policymakers and scientists. The challenge is that data are often inconsistently calibrated in space or time—making scientifically sound integration of multiple data sets difficult. No simple data standard can be designed that all data providers will utilize. Moreover, the U.S. government has limited resources to support long-term electronic data management beyond the life of individual investigators' projects or programs. Currently, no interagency management structure

BOX 13-1

DATA MANAGEMENT FOR GLOBAL CHANGE RESEARCH POLICY STATEMENTS

- The USGCRP requires an early and continuing commitment to the establishment, maintenance, validation, description, accessibility, and distribution of high-quality, long-term data sets.
- Full and open sharing of the full suite of global data sets for all global change researchers is a fundamental objective.
- Preservation of all data needed for long-term global change research is required. For each and every global change data parameter, there should be at least one explicitly designated archive. Procedures and criteria for setting priorities for data acquisition, retention, and purging should be developed by participating agencies, both nationally and internationally. A clearinghouse process should be established to prevent the purging and loss of important data sets.
- Data archives must include easily accessible information about the data holdings, including quality assessments,

supporting ancillary information, and guidance and aids for locating and obtaining the data.
- National and international standards should be used to the greatest extent possible for media and for processing and communication of global data sets.
- Data should be provided at the lowest possible cost to global change researchers in the interest of full and open access to data. This cost should, as a first principle, be no more than the marginal cost of filling a specific user request. Agencies should act to streamline administrative arrangements for exchanging data among researchers.
- For those programs in which selected principal investigators have initial periods of exclusive data use, data should be made openly available as soon as they become widely useful. In each case, the funding agency should explicitly define the duration of any exclusive use period.

developed not to fill unmet measurement needs, but instead to improve the quality of existing measurements.

Fulfilling the need for climate and climate-related data that are useful for scientists, planners, and other end users will be a complex task. The overall challenge, then, is:

To provide seamless, platform-independent, timely, and open access to integrated data, products, information, and tools with sufficient accuracy and precision to address climate and associated global changes.

This challenge can be met through development of a system that efficiently links observations to data management and analysis, and ensures timely delivery of climate data and related information and their preservation for future generations (see Figure 13-1). This integration can be implemented using proven and emerging technologies such as the Internet and digital libraries. Specific goals in this effort are:

1) Collect and manage data in multiple locations
2) Enable users to discover and access data and information via the Internet
3) Develop integrated information data products for scientists and decisionmakers
4) Identify data to be preserved.

exists to develop and enforce adoption of a complex data management solution. Scientific data that are not institutionally managed are at serious risk of vanishing when the scientist or data collector turns to other projects or retires.

Traditional core activities within data management have been regarded to be data curation—quality control, context-setting (i.e., metadata), preservation, etc.—and distribution of data sets. In order to focus on the needs of the scientists who use the data, we must significantly expand this core to include data discovery (the ability to locate data that are distributed across multiple institutions and disciplines) and data interoperability (changes to how we conduct data management in ways that free users from the productivity losses associated with incompatible formats, unwieldy file sizes, and large non-aggregated collections).

In addition, many of the scientific and decision support needs of the CCSP require analysis and processing of data into specialized products. Even with a large number of measurement systems, there will always be quantities of interest that are either impractical or impossible to measure directly or routinely. Thus, physical models using instrument data as inputs are implemented and can help fill some of the unmet measurement needs of the program. Additionally, products are

These goals will be achieved through implementation of an effective management structure that will ensure interagency coordination of these efforts, scientific and technological guidance, and user input and requirements.

Researchers, planners, and decisionmakers need seamless access not only to information produced by CCSP efforts, but also to the larger scope of information produced by other federal, non-federal, regional, and international programs and activities. These users should be able to focus their attention on the information content of the data, rather than how to discover, access, and use it. The challenge for data management is a system where the user experience will change fundamentally from the current process of locating, downloading, reformatting, and displaying to one of accessing information, browsing, and comparing data with standard tools, such as web browsers, geographic information system (GIS) programs, and scientific visualization/analysis systems, without concern for data format, data location, or data volume.

The strategy for building this framework must be an evolutionary process with a development model based on ongoing interactions with users. In addition, modifications to existing systems and the development of new systems will require use of existing technologies

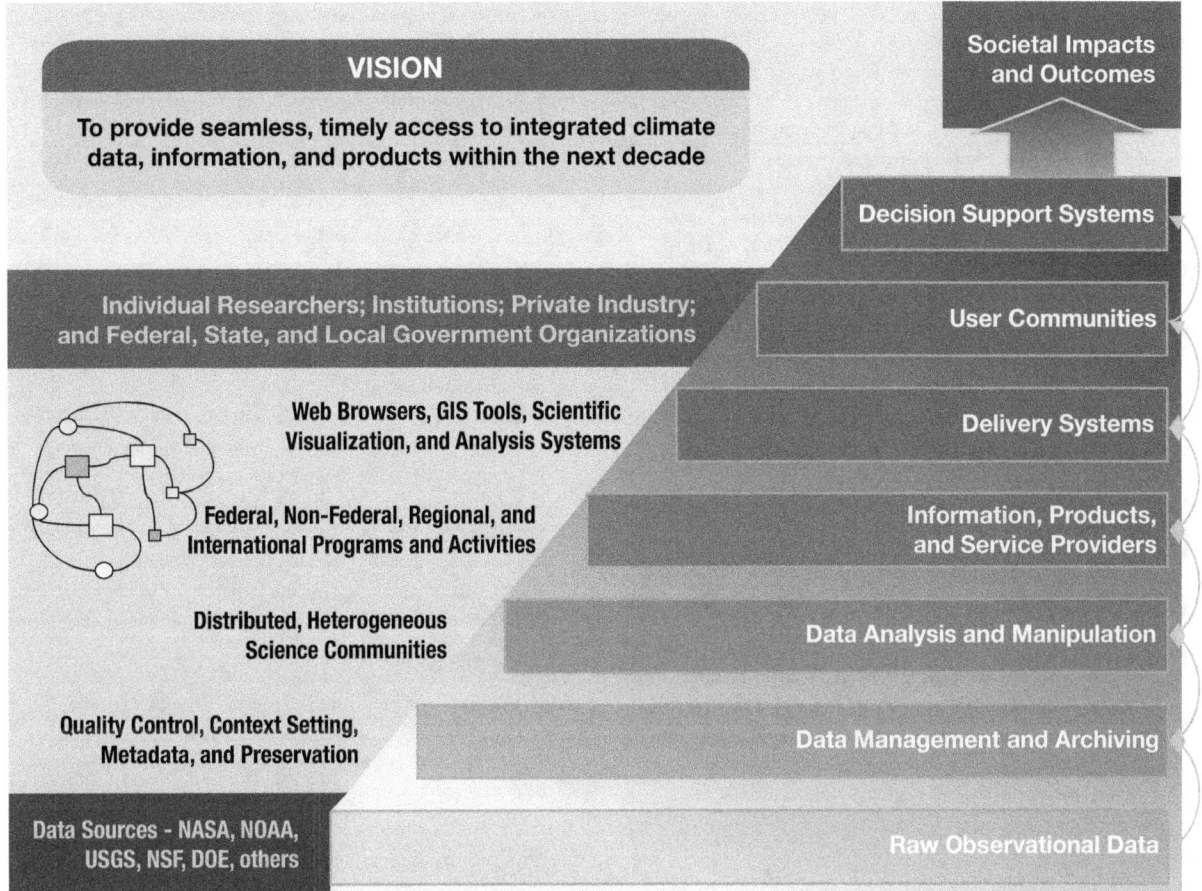

Figure 13-1: Roadmap from data collection to decision support. Source: NASA.

with the vision that the systems would be regularly updated with new technologies to respond to user requirements. Such a framework, with established metadata and quality control/quality assurance standards, mechanisms of transport, protocols, and requirements, will permit data and product providers to contribute their information as well as allow users to query and access the system for relevant information. The challenges to CCSP will be pursuing unprecedented levels of cooperation across current data management institutions and programs and a commitment to mapping the future development and execution of a suitable strategic plan.

The guiding principles for this CCSP data management plan are:
- The measure of success will be the ability of scientists and decisionmakers to access "stand-alone" or "integrated" data and information in a consistent and easily accessed format.
- The value added will be integration—many types of climate data from different suppliers will be available in a manner consistent with user requirements.
- The methods used by data suppliers to deliver data to their "customers" need will evolve with new technology.
- It will be easy for users to discover and access data (local, regional, national, and international).
- The system will be responsive to user feedback.
- The system will preserve irreplaceable data, making use of effective compression technologies where appropriate.

- There will be an open design and open standards process.
- Operations will be reliable, sustained, and efficient.

Goal 1: Collect and manage data in multiple locations.

A distributed system requires CCSP to exploit advances in information technology that enable the development of a distributed data and information system in which data will be collected and managed in multiple locations including federal, state, and local agencies; academic institutions; and non-governmental organizations. Our ability to provide Climate Data Records (CDRs; see Chapter 12, Objective 3.4) and climate information to the community will depend on the interoperability of the system and metadata standards. Long-term archiving and stewardship of the data will be the responsibility of accredited (typically federal) data centers.

Objective 1.1: Develop standard metadata guidelines
Under this objective, CCSP will provide additional specific community-based guidelines for scientific metadata content where and as appropriate. One approach will be to adopt the ISO 19115/TC211 Geographic Information/Geomatics standard (<http://www.isotc211.org>), which is built on the Federal Geospatial Data Clearinghouse (FGDC) core standards.

Objective 1.2: Expand collaboration among data providers

CCSP will expand the collaboration between the federal data centers and external (university, commercial, and non-profit) data service providers. This collaboration will build on the strong foundation provided by existing distributed systems, encompassing the data centers established by federal science agencies, such as the National Aeronautics and Space Administration (NASA), National Oceanic and Atmospheric Administration (NOAA), Department of Energy, U.S. Department of Agriculture, U.S. Geological Survey (USGS), and the National Science Foundation. The data management plan also calls for expanding partnerships with foreign governments, intergovernmental agencies, and international scientific bodies and data networks to provide data that are needed to address the international character of research and decisionmaking. These collaborations should improve access to regional, state, and local data.

Goal 2: Enable users to discover and access data and information via the Internet.

This goal requires a greater emphasis on the development of a framework to respond to the need for integration and communication of information across disciplines and among scientists and policymakers. Multi-agency and multidisciplinary institutional and data resources will need to be targeted to develop standards and processes for sound data management. System upgrades need to include the

implementation of tools to enable communication among multiple data locations. The process of identifying the data requirements of the program on a regular basis, including visualization, analysis, and modeling requirements, needs to be strengthened. Human resources will be required to perform these tasks, particularly individuals with the technical expertise to support user requirements. These needs will be addressed by CCSP (see Figure 13-2).

Objective 2.1: Improve access to data

Several activities are planned under this objective which will enable improved access to data:

- Expand the Global Change Master Directory (GCMD) to facilitate access to data. Agencies will provide descriptions in the format needed for this action.
- Ensure the provision of socioeconomic data collected by federal statistical agencies (e.g., the Census Bureau and the Bureau of Economic Affairs), by resource management agencies (e.g., the U.S. Department of Agriculture, U.S. Army Corps of Engineers, the U.S. Bureau of Reclamation, the U.S. Bureau of Land Management, and the U.S. Fish and Wildlife Service), by energy-related agencies (e.g., the Department of Energy and the Environmental Protection Agency), and by state and local agencies.

Objective 2.2: Management of biological data

Management of biological data will receive priority. Objective 2.3 of the Observing and Monitoring research element focuses on developing new capabilities for ecosystem observations. This is a CCRI priority and is a critical need for evaluating the effects of

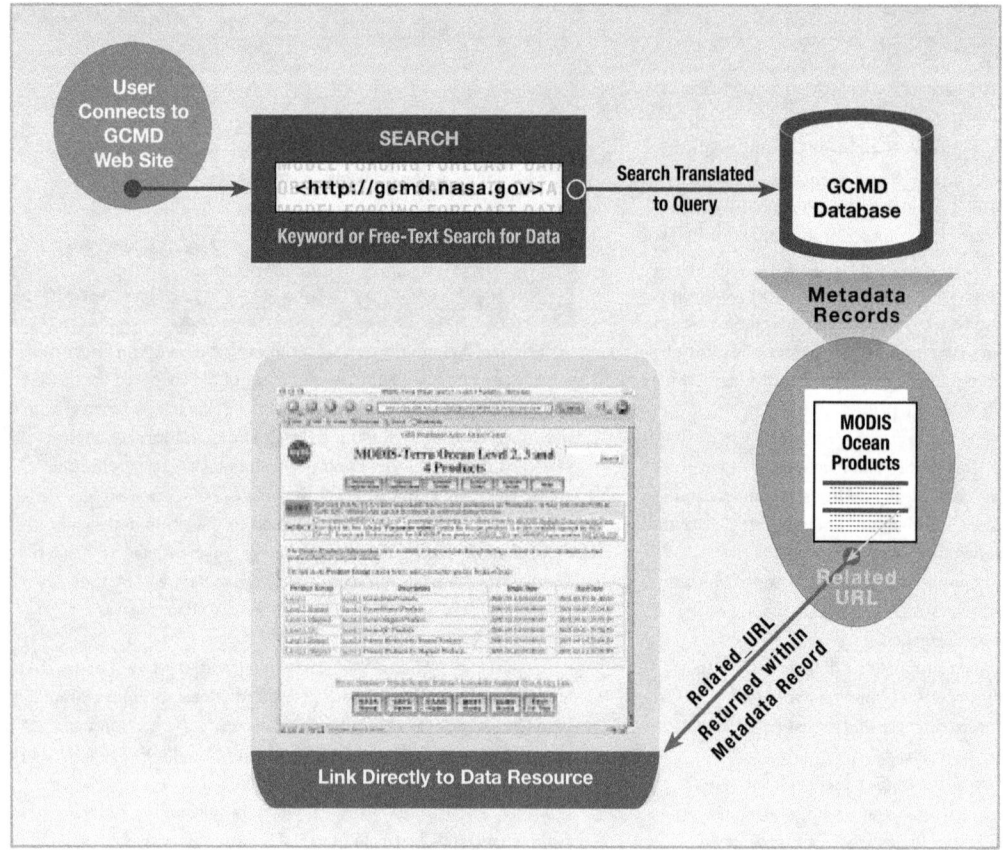

Figure 13-2: Search and direct retrieval of data set information from NASA's Global Change Master Directory. Source: NASA.

climate change on ecosystems. Biological data management is hampered by its requirement for extensive metadata, changes in named taxonomic species, and availability (at present mostly in non-electronic form and in the hands of individual investigators).

Objective 2.3: Data and information portals

Under this objective, CCSP will consolidate agency data information into one portal; that is, an agency home page would provide a mechanism for identifying all available data and information. CCSP will create special, tailored portals for data products of interest to the various CCSP working groups. These portals will use the emerging web metadata clearinghouse technology to allow researchers to locate and access coincident data of interest from various observation systems. This will require implementation of climate-quality data and metadata documentation, standards, and formatting policies that will make possible the combined use of targeted data products taken at different times, by different means, and for different purposes. Additionally, CCSP will work toward supporting the national climate observing system monitoring architecture described in Chapter 12 (Objective 1.7)

Goal 3: Develop integrated information data products for scientists and decisionmakers.

The goal of information analysis and interpretation is to incorporate the multidisciplinary science elements of the CCSP in order to integrate information and provide integrated products. This requires that links between scientists and data managers on one hand and data quality and data products on the other need to be enhanced to provide a more effective translation of user requirements into data products (see Chapter 12, Objective 3.4). Data managers must be able to understand, communicate, and work closely with scientists and others to ensure proper stewardship for the data archive and its distribution. Data managers must be included in scientific working groups and steering committees to guide the integration of data management and science and decision support. CCSP will ensure data quality and preservation by making data management an integral part of any observing or data collection program. Decision support needs will set the priorities for integrated products and help to define and address data management issues associated with the integrated products.

Objective 3.1: Establish links between data providers and decisionmakers

A dialogue needs to be established between data providers and decisionmakers to understand how scientific resources and knowledge are currently used by decisionmakers. Scientific discussion, planning, and implementation are needed for key data issues (e.g., gridding algorithms, gap-filling periods of missing observations) to permit assimilation of many CCSP data products. CCSP data analysis will draw on and promote further advances in data-processing automation, data visualization techniques, and web-based data delivery mechanisms. Activities under this objective include:

- Create a link on the CCSP website where decisionmakers can search for, locate, and link to data and information products identified by CCSP working groups as potentially of significant utility.

- Develop a prototype of the provision of support services for decision support systems. Provide an initial operational capability that interfaces one or more CCSP data systems to one or more decision support systems.
- Implement procedures to solicit climate information requirements from regions, sectors, and users who are using climate projections for management and policy decisions.

Objective 3.2: Application of data products and information

The emphasis on regional data, modeling, and decision support activities is increasing. Researchers and stakeholders are collaborating to develop applications based on research findings. An example of this effort is an experiment to integrate scientists and stakeholders to frame and apply El Niño-Southern Oscillation forecasts and other research products in a variety of regions and economic sectors. Regional environmental data products that can provide up-to-date information on environmental conditions to decisionmakers—and, if appropriate, allow an interactive "If…, then…" environment—must be anchored in considerations of input and process uncertainties and outcome accuracies. Decision support services must provide information about uncertainty to be of maximum utility to decisionmakers. This objective includes:

- Improve access to climate data and information for addressing regional concerns and issues
- Provide geo-referenced and spatially and temporally averaged socioeconomic data products for integrated studies
- Continually improve and clearly articulate the accuracy of regional data.

Objective 3.3: Harness emerging technologies

CCSP needs to take advantage of emerging information systems such as Digital Libraries (DLs). DL is a paradigm for investment by several agencies that has the potential of becoming the world's most vital environment for discourse and resources promoting excellence in science and education. Data management could be greatly informed and enabled by the DL technology. The guiding principles for the development of DL are to provide a spectrum of interoperability, to provide one library with many portals, and to leverage the energy and achievements of others. DL's effort will focus on building a comprehensive library of digital resources and this effort will enhance CCSP's successful implementation.

Goal 4: Preserve data.

One daunting challenge of the 21st century is the management of the large volume of highly diverse data describing the Earth's climate. These data are a result of comprehensive observing and monitoring systems and models producing new data sets from the climate observations. The size of the data archives is growing faster than we can derive information from them. For example, NASA's Earth science data holdings increased by a factor of six between 1994 and 1999; the total amount of data doubled between 1999 and 2000. It is estimated that by 2010, the size of a major U.S. archive for data from NOAA, NASA, and USGS will be 18,000 terabytes (10^{18} bytes). Lessons learned from NASA's pioneering efforts in handling their current holdings (more than 2,500 terabytes) must be used by

BOX 13-2

EXAMPLES OF DATA PRODUCTS

Chapter 3 –
Atmospheric Composition
- Improved description of the global distributions of aerosols and their properties.
- A 21st century chemical baseline for the Pacific region against which future changes can be assessed.

Chapter 4 –
Climate Variability and Change
- New and improved climate data products, including assimilated data from satellite retrievals and other remotely sensed and *in situ* data for model development and testing; consistent and regularly updated reanalysis data sets suitable for climate studies; centuries-long retrospective and projected climate system model data sets; high-resolution data sets for regional studies; and assimilated aerosol, radiation, and cloud microphysical data for areas with high air pollution.
- A paleoclimatic database designed to evaluate the ability of state-of-the-art climate models to simulate observed decadal to century-scale climate change, responses to large changes in climate forcing, and abrupt climate change.

Chapter 5 – Water Cycle
- Integrated long-term global and regional data sets of critical water cycle variables such as evapotranspiration, soil moisture, groundwater, clouds, etc., from satellite and *in situ* observations for monitoring climate trends and early detection of climate change.
- Enhanced data sets for feedback studies, including water cycle variables, aerosols, vegetation, and other related feedback variables, generated from a combination of satellite and ground-based data.

Chapter 6 –
Land-Use/Land-Cover Change
- Global high-resolution satellite remotely sensed data and land-cover databases.
- Operational global monitoring of land use and land-cover conditions.

Chapter 7 – Carbon Cycle
- Global, synoptic data products from satellite remote sensing documenting changes in terrestrial and marine primary productivity, biomass, vegetation structure, land cover, and atmospheric column CO_2.

- Landscape-scale estimates of carbon stocks in agricultural, forest, and range systems and unmanaged ecosystems from spatially resolved carbon inventory and remote-sensing data.

Chapter 8 – Ecosystems
- Data sets for examining effects of management and policy decisions on a wide range of ecosystems to predict the efficacy and tradeoffs of management strategies at varying scales.
- Synthesis of known effects of increasing CO_2, warming, and other factors (e.g., increasing tropospheric ozone) on terrestrial ecosystems based on multi-factor experiments.

Chapter 9 –
Human Contributions and Responses to Environmental Change
- Assessments of the potential economic impacts of climate change on the producers and consumers of food and fiber products.
- Elevation maps depicting areas vulnerable to sea-level rise.

the community. In addition, new technologies need to be developed that will enable us to keep all data needed for long-term global change research, reducing the need to prioritize which data will be archived. This endeavor would also consider lessons learned from communities that already handle this volume of data (e.g., defense intelligence, commercial video streaming).

Objective 4.1: Enhance the data management infrastructure
Telecommunications bandwidth capacity must be adequate to accommodate the movement of these larger data volumes as they progress through an information cycle including measurements, distributed scientific analyses, science models, predictions, decision support tools, assessments, and policy and management decisions. Increased levels of bandwidth will become available through government research, development, and funding; commercial availability and acquisition; and non-profit sector partnering. It is important to keep in mind that the evolutionary realization of this vitally needed infrastructure must be continually planned. Another critical area requiring enhancement is the

development of new technologies for storage of large volumes of data and information.

Objective 4.2: Preserve historical records
At the same time, many important heritage data sets face a growing risk of loss due to deterioration of paper records, obsolescence of electronic media and associated hardware and software, and the gradual loss of experienced personnel (see, e.g., Figure 13-3). We look to these historical records, from which we can derive long-term trends, to help provide the missing pieces of the overall climate puzzle. The primary focus under this objective will be to identify and rescue data that are at risk of being lost due to media deterioration, poor accessibility, or limited distribution.

Objective 4.3: Support an open data policy
Another data management challenge is data policy—described as the set of rules, regulations, laws, or agreements governing the access and use of data. Database protection legislation, enacted in Europe and proposed in the United States, has raised concerns that the flow of scientific information may become much more constrained.

Many of these policies are in conflict with each other and the challenge will be to understand these conflicts and chart a course that benefits all. This will necessitate the close interaction of and negotiation between the database rights holders and users, in order to strike a balance between protection and fair use (NRC, 1999f). Compiling long-term climate-quality data sets from which long-term climate trends can be derived will be greatly affected by the future data policies of national and international bodies. CCSP will develop and implement guidelines for when and under what conditions data will be made available to users other than those who collected them.

Management Structure

Working in partnership with members and representatives of the research community in federal agencies and academia, and with appropriate committees of the National Research Council, CCSP will seek to identify the data requirements of the program on a regular basis, including visualization, analysis, and modeling requirements. Priority attention will be given to those observations and data that are central to a specific research element but for which requirements are not currently being met, or that exist but are not part of a publicly available data system. Accomplishment of these goals will require an integrated management structure that involves the CCSP agencies with oversight by members of the external community. A Data Management Steering Committee composed of federal, state, academic, and industry managers and decisionmakers will provide oversight, priorities, coordination, and recommendations to the CCSP Data Management Working Group (DMWG). DMWG will be responsible for the preparation, implementation, and periodic review of data management activities, and publication of annual reports describing milestones achieved and future activities. Close links via shared membership will be maintained with the Observing and Monitoring boards and councils as described in Chapter 12.

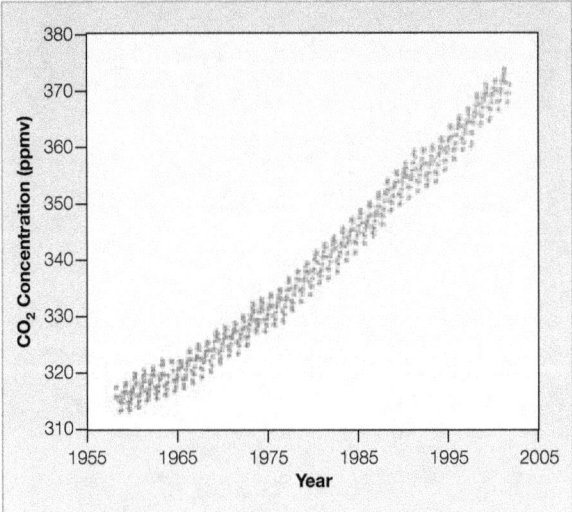

Figure 13-3: Changes in atmospheric carbon dioxide (CO_2) concentration at Mauna Loa, Hawaii, over time. This figure illustrates the critical need to preserve historical data. Source: Dave Keeling and Tim Whorf, Scripps Institution of Oceanography.

National and International Partnerships

CCSP will facilitate access to the data and information required and generated as part of this program. A critical need for observations and data are identified throughout this plan. Box 13-2 provides examples from each research chapter that illustrates the type of data products to be generated; Box 13-3 provides examples of the type of information products. Note that this latter box does not reflect output from the various modeling activities described in each chapter.

BOX 13-3

EXAMPLES OF INFORMATION PRODUCTS

Chapter 3 – Atmospheric Composition
- A *State of the Atmosphere 2006* report that describes and interprets the annual status of the characteristics and trends associated with atmospheric composition, ozone layer depletion, temperature, rainfall, and ecosystem exposure.

Chapter 4 – Climate Variability and Change
- Documented impacts of climate extremes on regions and sectors, and evaluations (both positive and negative) of the implications should climate change in the future.

Chapter 5 – Water Cycle
- Assessment reports on the status and trends of water flows, water uses, and storage changes for use in analyses of water availability.

Chapter 6 – Land-Use/Land-Cover Change
- Report on the regional and national impacts of different scenarios of land use and land cover on water quality and quantity.

Chapter 7 – Carbon Cycle
- *State of the Carbon Cycle* report focused on North America.

Chapter 8 – Ecosystems
- Reports describing the potential consequences of global and climatic changes on selected arctic, alpine, wetland, riverine, and estuarine ecosystems; selected forest and rangeland ecosystems; selected desert ecosystems; and the Great Lakes.

Chapter 9 – Human Contributions and Responses to Environmental Change
- Assessments of the potential health effects of combined exposures to climatic and other environmental factors (e.g., air pollution).

The generation of U.S. and global data products will require cooperation with national and international data centers and institutions. The CCSP will utilize and participate in the development of the data discovery and data interoperability framework being advanced by other programs such as the U.S. Integrated Ocean Observing System effort. CCSP will coordinate its activities with international programs to take advantage of emerging data management and information tools and technologies and sharing of climate change data and information. Examples of international programs that actively engage in data management include the World Data Center system, which functions under the guidance of the International Council of Scientific Unions (ICSU) and facilitates international exchange of scientific data; the World Climate Research Programme, which sponsors multiple major projects involving international cooperation and data collection with guidance by a Joint Scientific Committee; the International Human Dimensions Programme on Global Environmental Change Data and Information System (IHDP/DIS), which links social science data centers and scientists; and the Data Management Coordination Group of the Joint World Meteorological Organization/

Intergovernmental Oceanographic Commission Technical Commission for Oceanography and Marine Meteorology, which is currently developing an Oceans Information Technology Pilot Project.

CHAPTER 13 AUTHORS

Lead Authors
Margarita Conkright, NOAA and CCSPO
Wanda Ferrell, DOE
Clifford Jacobs, NSF
Martha Maiden, NASA

Contributors
Vanessa Griffin, NASA
Steve Hankin, NOAA
Thomas Karl, NOAA
Chet Koblinsky, NASA

CHAPTER

CHAPTER CONTENTS

The Need for Effective Communication

Dissemination of Credible and Reliable Research Findings

Communication with Diverse Audiences

Coordination and Implementation

This chapter describes how the Climate Change Science Program (CCSP) will improve communication of federal government climate change science research. It highlights current CCSP communication activities and outlines a series of future steps.

The Need for Effective Communication

The CCSP has a responsibility for credible and effective communications with stakeholders in the United States and throughout the world on issues related to climate variability and climate change science. Effective communication is not always straightforward, however, as climate science is complex and rapidly evolving, and respected scientists can sometimes disagree in their interpretations. The economic and policy dimensions of the issue can give rise to even greater disputes among individuals with different policy views. Professional, institutional, and cultural perspectives can also impede the flow of information.

At the same time, the public investment in understanding, mitigating, and adapting to climate variability and change is substantial. As an essential part of its mission and responsibilities, the CCSP will enhance the quality of public discussion by stressing openness and transparency in its

scientific research processes and results, and ensuring the widespread availability of credible, science-based information.

The CCSP and individual federal agencies generate substantial amounts of authoritative scientific information on climate variability and change. While public and private sector interests have made progress in generating valuable climate science information, efforts to improve public access to this information have not always kept pace. Research findings are generally well reported in the scientific literature, but relevant aspects of these findings need to be reported in formats suitable for use by diverse audiences whose understanding and familiarity with climate change science issues vary.

To further its commitment to the effective communication of climate change science information, the CCSP has established the following goals:
- Disseminate the results of CCSP activities credibly and effectively
- Make CCSP science findings and products easily available to a diverse set of audiences.

In addition to CCSP efforts, individual federal agencies have relationships with stakeholders, and they disseminate science-based climate information. Efforts include activities in which agencies respond directly to inquiries from the public and other stakeholders; maintain websites and listservs; produce and distribute hardcopy documents and multimedia products; conduct or sponsor briefings, lectures, and press conferences; testify before Congress or other government bodies; finance scholarships, fellowships, and internships; support museum exhibits and other public displays; sponsor, participate, or otherwise contribute to meetings attended by stakeholders; provide scientifically sound content for K-12 education activities; and fund communication efforts managed outside the federal government. Working with the agencies' climate science communications activities will be a key component of CCSP communications efforts.

The CCSP acknowledges that in seeking to strengthen its ongoing communications with diverse constituencies—ranging from international, national, and regional policymakers to academic researchers, adaptive management technical experts, the media, and concerned citizens—it faces the challenges of constrained resources and institutional fragmentation. CCSP's ongoing communications efforts will build on existing resources and explore new communications and information-sharing opportunities.

Dissemination of Credible and Reliable Research Findings

Responding to the direction of President Bush that the best available scientific information be developed to support decisionmaking on global climate change issues, CCSP has developed its strategic planning and public review processes to facilitate "credible fact finding" on (a) key climate science issues, (b) comprehensive climate and ecosystem observing and data management systems, and (c) the development of decision support resources.

Transparency and Comprehensiveness

CCSP will undertake a number of approaches for ensuring that research plans, reports, and related activities are prepared in an open fashion. Approaches will include early and ongoing review of draft and final work products by stakeholder communities; reporting of key assumptions, methodologies, data, and uncertainties; the use of a variety of decision support tools discussed in Chapter 11 that relate research findings to policymakers and technical experts involved in adaptive management decisions; and web-based information that is freely available to all users.

To maintain credibility among users of the CCSP analyses and projections, CCSP draft and final plans, reports of findings, and projections of future outcomes will be posted on publicly accessible websites, and all comments communicated by interested stakeholders also will be posted for public review. CCSP will aim to make its analyses comprehensive (i.e., covering a range of plausible policy options) within the limits of the resources available for analysis. Moreover, whenever possible and practical, CCSP will facilitate comparisons with other relevant studies.

The CCSP public workshop held in December 2002 set a high standard of open proceedings. The workshop presented the current state of climate change science and gathered comments from both scientists and public stakeholders on defining a strategy for climate observations and research. All elements of the strategic

planning process, including the *Discussion Draft Strategic Plan*, all of the workshop proceedings, and all public comments received after the workshop are openly available at the CCSP website, <www.climatescience.gov>.

Reports on the Basis for Findings and Levels of Confidence

CCSP aims to describe the basis for each of its key findings and projections, with sufficient detail to allow independent reviewers to replicate the underlying analyses. Evaluation and communication of uncertainty and levels of confidence is also a crucial issue. Uncertainties can arise from lack of knowledge; from problems with data, models, terminology, or assumptions; and from other sources, creating room for misunderstanding.

Because uncertainty can never be completely eliminated, CCSP will develop systematic approaches for evaluating levels of confidence and uncertainty and communicating this information. This will enable users of the information to understand the uses and limits of the information they are seeking to apply. Where appropriate, "confidence level" descriptions will be used to communicate these characterizations. The introduction of uncertainty is not intended to imply a basis for inaction. In cases where the uncertainty of analyses or projections is so large as to make the discrimination between options impractical, this finding will be reported directly.

Communication with Diverse Audiences

The Audience

Section 102 of the 1990 Global Change Research Act (P.L. 101-606) requires CCSP to "consult with actual and potential users of the results of the Program to ensure that such results are useful in developing national and international policy responses to global change." As described in Chapter 11, CCSP is responding to this charge within the framework of three decisionmaking categories: (1) public discussions and evaluations based on available state-of-science syntheses; (2) operational adaptive management decisions undertaken by resource managers; and (3) support for policy formulation.

Each of these decisionmaking categories has a unique set of stakeholders. Stakeholders are individuals or groups whose interests (financial, cultural, value-based, or other) are affected by climate variability, climate change, or options for adapting to or mitigating these phenomena. CCSP stakeholders range from members of Congress to those involved in the development of state and regional drought action plans, to those managers of agricultural operations, water resource systems, fire, and fisheries, to the general public.

Interactive Communication

CCSP will actively seek to *learn from* its constituencies through ongoing dialogue and feedback mechanisms. The CCSP will aim to improve dialogue with public and private sector constituencies with

Climate Change Science Program Strategic Plan   Chapter 14. Communications

Figure 14-1: Two-way communications. The CCSP plan encourages input from specific constituent groups and from the public at large. Here, Janine Bloomfield of Environmental Defense, a participant in the CCSP Planning Workshop for Scientists and Stakeholders, comments in a session on Resource Management Decision Support (4 December 2002). Source: Nick Sundt, CCSPO.

the end result of providing stakeholders with adequate opportunities to help frame important scientific research activities. This dialogue is an essential component of the development of decision support discussed in Chapter 11 (see Figure 14-1).

Where practical and effective, CCSP will communicate through professional, civic, and other membership organizations representing key constituency groups, and through the news media. In this way, CCSP can leverage those organizations' and institutions' information dissemination capabilities.

CCSP will initiate a series of pilot projects that will serve as a basis for longer term sustained efforts to improve two-way communications with a wide range of interests. This effort will reflect a commitment to learn from those same interests, and incorporate feedback received regarding successes and shortcomings. The initial pilots will be modest in scope and will likely involve expansion of existing partnerships—for instance, an initiative to coordinate information access and linkages among existing climate change science-related websites. The pilot projects, as is the case with ongoing communications efforts, will include an evaluation phase to determine the design and implementation of subsequent communication activities.

Accessibility of Science Information for Diverse Audiences

The American public is making a sizeable financial commitment to improving scientific understanding of climate variability and change. In turn, the American public has a right to expect ongoing access to high-quality scientific assessments and evaluations.

There are many channels of communication that are available to CCSP, including:
- Peer-reviewed scientific publications and science summaries for diverse audiences
- The CCSP website
- The media
- Other outreach materials.

Publications
Most of the information produced under CCSP sponsorship is in the form of peer-reviewed, published scientific articles and reviews.

Figure 14-2: *Our Changing Planet*. Issued annually by CCSP, this report summarizes the program's recent accomplishments and near-term plans. Published in hardcopy and in digital format, *Our Changing Planet* is a primary source of information about federally funded research on global change.

These articles and reviews are the products of specific agency program efforts and are the working base of scientific knowledge. In addition to the core research publications, CCSP will produce interagency reports that integrate and synthesize information from nationally and internationally sponsored climate science activities. One of these reports will address climate change issues framed in a practical, real-life perspective for the concerned, general public.

As described in Chapter 2, CCSP will focus interagency attention on the ongoing development of synthesis and assessment products. These products will support both policymaking and adaptive management. The synthesis and assessment products take account of the need for assessments on the full range of issues spanning all CCSP objectives and will provide a "snapshot" of knowledge of the climate, environmental, and socioeconomic aspects of climate variability and change. The products will support specific groups or decision contexts across the full range of issues addressed by CCSP, and where appropriate, the Climate Change Technology Program (CCTP).

Another CCSP publication is the *Our Changing Planet* report (see Figure 14-2), which has established itself as the authoritative guide to ongoing climate science research by federal agencies. This annual report describes the activities, budgets, and plans of the U.S. Global Change Research Program (USGCRP). The 2003 report, a supplement to the President's Fiscal Year 2003 Budget, also describes the start-up activities of the U.S. Climate Change Research Initiative (CCRI).

The CCSP Website

CCSP's web initiatives are a valuable tool for public education and communications. Combined, the CCSP and related websites—
<www.climatescience.gov>,
<www.usgcrp.gov>, and
<www.gcrio.org>—attract roughly
1 million visitors annually (see Box 14-1).
These websites provide a comprehensive reference for users seeking information on climate science structured around scientific disciplines as well as by sponsoring agency. In addition, scientists can find information on funding sources and announcements of research opportunities.

The CCSP will implement a web development strategy to improve content and delivery to diverse audiences to:
• Continue to update and improve related interagency sites
• Increase target audience familiarity with the CCSP websites and the resources available through them
• Improve processes for preparing documents for the web to ensure full accessibility

• Strengthen ties and links among CCSP interagency and agency websites
• Expand online support for interagency working groups
• Better meet the growing demand for digital images
• Foster timely interactive dialogues with stakeholders
• Solicit from users reactions on usefulness of websites and ways to make them more effective
• Produce periodic and annual summary reports documenting progress and near-term objectives.

The Media

Americans obtain most of their news and information, and therefore their impressions, about climate science through the mass-circulation print and broadcast news media (see Figure 14-3). As they do in other important public policy/science issues, the news media play an essential role in informing Americans about climate change and climate variability. CCSP will help ensure that the news media are well equipped to meet the needs of their audiences by supporting distribution of timely, credible, and useful information to news media through printed and electronic means, and through participation with and support for climate science journalism training activities as appropriate.

Other Outreach Materials

CCSP will develop outreach materials on climate change science fundamentals geared to varied audiences, using formats such as Frequently Asked Questions (FAQs) and educational fact sheets.

BOX 14-1

'E' GOVERNMENT AND CLIMATE SCIENCE

The CCSP's initial interagency efforts to improve communications via the web (primarily through <www.usgcrp.gov>, <www.climatescience.gov>, and <www.gcrio.org>) have attracted nearly 1 million visitors annually, a number that has continued to steadily increase. Additional users, including scientists, resource managers, planners, business executives, journalists, students, teachers, and others are increasing their use of the Internet. By providing outstanding content, well-designed websites can serve these diverse users.

The E-Government Strategy presented by the Office of Management and Budget (OMB) in February 2002 emphasized that "the President has made 'Expanding E-Government' integral to a five-part Management Agenda for making government more focused on citizens and results." One example of a low-cost,

high-payoff opportunity for expansion is the extension of the CCSP's web strategy to the new website of the Climate Change Technology Program (CCTP), <www.climatetechnology.gov>. Accordingly, the CCSP is cooperating with its partner program in developing the new site, and the resulting close web collaboration will continue to be an ongoing high priority of both programs. This cooperative effort, in addition to efficiently using existing interagency resources, will provide early and tangible evidence of the fundamental linkage between climate science and technology—and between the people and organizations working on different facets of the climate change issue.

US Climate Change Science Program
www.climatescience.gov

Figure 14-3: A pivotal role for journalists. Because of their capability to inform large and diverse audiences, the news media are crucial to CCSP's communications efforts. Shown here is Dr. John H. Marburger III, Assistant to the President for Science and Technology and Director of the Office of Science and Technology Policy, responding to journalists' questions at the CCSP Planning Workshop for Scientists and Stakeholders (3 December 2002). Source: Nick Sundt, CCSPO.

These FAQs will address particular climate science issues, including model studies, state-of-scientific-understanding reports, "If..., then..." scenario analyses, and decision support resources. Each will be coordinated and produced with the involvement of the appropriate CCSP technical working groups.

CCSP also recognizes the value of collaborating with the traditional formal educational community. Working with a network of federal departments and agencies—and their own networks of partner educational interests—CCSP will develop a series of "Climate in the Classroom" educational fact sheets on climate science. These "white papers" will be written to provide the lay public with information on topics such as climate models, the role of aerosols, or the influence of oceans in shaping climate. While concise and easily accessible to a broad audience, these fact sheets will also point to additional resources that will allow individuals to get additional information. They will be disseminated electronically via the web and also will be targeted to key educational institutions and to key resource organizations for distribution to specialized constituencies. The number of fact sheets for the coming year will be determined by resource availability. At a minimum, four fact sheets will be developed and distributed over the first year.

Coordination and Implementation

In order to accomplish the goals of the CCSP communications efforts, new mechanisms for interagency coordination and implementation are needed. CCSP will:
* Formalize the interagency climate Outreach and Communications working group so that it can participate actively in the development of the CCSP master communications plan and

carry out CCSP-sponsored information dissemination and outreach activities. This group will work with the CCSP science-based and decision support working groups to ensure that the program's communications needs are met.
* Develop a comprehensive communications implementation plan and funding plan, based on an inventory of existing communications activities and building on experience from several short-term pilot projects to determine and shape ongoing effective communication strategies.

Improved Interagency Working Relationships

CCSP will formalize and engage the existing informal interagency climate Outreach and Communications group. Membership will consist of key agency professionals whose responsibilities routinely involve work with climate change science as part of their communications activities—typically through agency communication and education programs.

The Outreach and Communications working group will work with the seven CCSP science-based and separate decision support, data management, and observations and monitoring groups. It will also work with stakeholders to ensure that the information needs of public and private sector interests are understood. Efforts to improve coordination among CCSP participating agencies will capitalize on existing distribution systems and the strengths of participating agencies.

Comprehensive Communications Plan

With interagency input, the CCSP will develop a comprehensive communications implementation plan and the corresponding communications/education budget using the results of the climate science outreach and education activities inventory now underway. The inventory will describe ongoing federal agency communications and educational outreach activities related to climate variability and change science. The interagency communications plan will be based on the inventory and agency expertise and will include several short-term communications pilot projects in order to explore the effectiveness of new communication strategies.

The implementation plan will contain specific benchmarks and time tables to allow tracking of the plan's progress. It will be developed in cooperation with agency representatives and other stakeholder interests and will be completed by the end of 2003.

CHAPTER 14 AUTHORS

Lead Authors
Bud Ward, CCSPO
Nick Sundt, CCSPO
Susan Avery, NOAA and CCSPO
Stephanie Harrington, NOAA and CCSPO
James Mahoney, DOC
Richard Moss, CCSPO
Kathryn Parker, USEPA

CHAPTER

CHAPTER CONTENTS

Goals for International Cooperation in Climate Science

The International Framework

International Assessments and Applications

Engaging the International Global Change Research Community

Bilateral Discussions

Research

Observing Systems

Data Management

Capacity Building

U.S. Plans and Objectives for Future International Cooperation

Goals for International Cooperation in Climate Science

Climate varies over a wide range of geographic scales that transcend national boundaries. Climate change and its impacts are therefore intrinsically international in scope. To study climate change and variability on appropriate scales thus requires international cooperation—cooperation among scientists and research institutions and governmental agencies. U.S. scientists, institutions, and agencies are at the forefront of such international cooperation, reflecting the leadership role of U.S. climate science. The United States, at the governmental level, similarly leads in efforts to develop and maintain an intergovernmental framework within which climate change science, including research and observational programs, can be planned and implemented.

The overarching goals of U.S. efforts to promote international cooperation in support of the U.S. Climate Change Science Program (CCSP) are therefore to:
• Actively promote and encourage cooperation between U.S. scientists and scientific institutions and agencies

and their counterparts around the globe so that they can aggregate the scientific and financial resources necessary to undertake research on change at all relevant scales, including both the regional and global.
• Expand observing systems in order to provide global observational coverage of change in the atmosphere and oceans and on land, especially as needed to underpin the research effort.
• Ensure that the data collected are of the highest quality possible and suitable for both research and forecasting, and that these data are exchanged and archived on a timely and effective basis among all interested scientists and end users.
• Support development of scientific capabilities and the application of results in developing countries in order to promote the fullest possible participation by scientists and scientific institutions in these countries in the above research, observational, and data management efforts.

To achieve these goals, CCSP has assigned high priority to development of international cooperation to support the research elements of the program in the realization of programmatic goals. This includes working closely with U.S. scientists involved in the various CCSP components to ensure effective advocacy of their programmatic needs and interests in the complex international framework that has been established for global change research. CCSP supports U.S.

scientists in interfacing with international organizations and other countries on the many broad issues related to global change and global climate change.

CCSP remains dedicated to supporting and participating in existing cutting-edge international climate change science and technology research and assessments—for example, by assisting scientists in the planning and implementation of international collaborative projects. In doing so, CCSP continues to advocate maintaining the flexible international framework that permits U.S. scientists and agencies to select the approach that best suits their needs from various avenues for international cooperation.

CCSP will continue to support improvement of data sharing across international boundaries, thus contributing to the most effective development of scientific results and knowledge; to seek reductions in administrative and political barriers to international collaborative research and scientific exchange; to support expansion of specific observation systems and associated scientific infrastructure; and to support research in areas such as the mitigation of and adaptation to change—areas essential to economic development and its sustainability. CCSP will also continue to support programs that are building capacity through the training of competent scientists to address global change issues, the education of stakeholders, and the strengthening of institutions.

The International Framework

CCSP is a leader within a global network of active and engaged international research scientists and institutions. This network has developed an extensive framework to address both research and observational requirements for addressing global climate change issues. This framework includes a series of global-scale research programs; non-governmental and intergovernmental international organizations at both the regional and global level; various networks for coordination of observing systems—both *in situ* and remote sensing—and data exchange and management; international assessments; and organizations that focus on education, training, and capacity building.

The International Geosphere-Biosphere Programme (IGBP), the International Human Dimensions Programme (IHDP), the World Climate Research Programme (WCRP), Diversitas, and the Earth System Science Partnership (ESS-P) are among the most important of these organizations. CCSP has and will continue to interact with these organizations directly when appropriate, and by supporting U.S. scientists to participate in and provide dynamic scientific leadership for them through financial support, provision of data products, and scientific input.

National agencies that conduct and fund global change research, satellite remote-sensing systems, agricultural and forest research, and development projects also coordinate their efforts with their counterpart agencies in other countries through a number of organizations and networks. These include the International Group of Funding Agencies for Global Change Research (IGFA), the Committee on Earth Observation Satellites (CEOS), and the

Consultative Group on International Agricultural Research (CGIAR). The United States, for example, is one of the largest donors to CGIAR, which sponsors 16 international agricultural research centers devoted to improving food security, alleviating poverty, and improving the management of natural resources in developing nations. These centers are engaged in biological research that is intended to increase production of basic food crops and livestock and to maintain and enhance the natural resource base relating to soil, water, aquatic resources, agroforestry, and forestry.

The United States interacts at the intergovernmental level with partner countries in United Nations organizations that support global change research, both directly and indirectly. Preeminent among these are the World Meteorological Organization (WMO); the Intergovernmental Oceanographic Commission (IOC) of the United Nations Educational, Scientific, and Cultural Organization (UNESCO); the United Nations Environment Programme (UNEP); the Food and Agriculture Organization (FAO); the United Nations Development Programme (UNDP); and the World Health Organization (WHO). The anticipated re-entry of the United States into UNESCO should provide additional opportunities for U.S. scientists to interact with and participate in the full range of UNESCO science activities. Through its participation in UNEP, UNDP, and the World Bank, the United States also participates actively in and supports the Global Environment Facility (GEF), the primary international institution for the exchange of energy and carbon sequestration technologies with the developing world.

International Assessments and Applications

As a leader in climate change science, the United States assumes responsibility to participate in and provide data to international assessments such as those concerning ozone, biodiversity, ecosystems, and climate. In addition to the role that the United States plays in assessments, the federal government also provides support for and data to a number of programs that apply climate change information to provide critical decision support resources such as the U.S. Agency for International Development's (USAID) Famine Early Warning System (FEWS NET).

The 1987 Montreal Protocol on Substances that Deplete the Ozone Layer stands as one of the best examples of effective international cooperation on an environmental issue. In response to growing evidence of stratospheric ozone depletion, the nations of the world were able to develop the Montreal Protocol. Through global cooperation and compliance and the development of new non-depleting technologies, the increase in atmospheric concentrations of ozone-depleting gases has slowed and in some cases started to decline. International assessment has been a critical part of this process from the very beginning, providing the scientific underpinnings for decisionmaking on this issue. The series of Scientific Assessments of Ozone Depletion has provided decision support resources in the form of scientific information about the global state of ozone-depleting compounds.

The Millennium Ecosystem Assessment (MA) is building a new framework to link ecosystem services (e.g. food, fuel, clean water)

with human well-being. MA takes an integrated and multi-scale (local, regional, to global) approach and focuses on ecosystem services, the consequences of changes in ecosystems for human well-being, and the consequences of changes in ecosystems for other life on Earth. It addresses questions like: "What are the potential impacts, both positive and negative, of economic growth and globalization on ecosystems?" and "What policies and actions concerning ecosystems can best contribute to the alleviation of poverty?" In particular, it assesses the current and historical trends in ecosystems and their contribution to human health; choices for conserving and sustainably managing ecosystems to augment their contribution to human well-being; and scenarios for changes in ecosystems and their impacts on human well-being. By taking an integrated multi-scale approach that brings the latest scientific data, MA will provide information that will allow decisionmakers to review the trade-offs associated with choices they need to make.

WMO and UNEP established the Intergovernmental Panel on Climate Change (IPCC) in 1988 to conduct multidisciplinary reviews of current scientific and socioeconomic information pertaining to climate change. IPCC performs its critical task through an organizational structure composed of three working groups and a task force: Working Group I assesses the science of the climate system and climate change; Working Group II assesses the vulnerability of socioeconomic and natural systems to climate change; Working Group III assesses options for limiting greenhouse gas emissions and mitigating climate change; and the Task Force focuses on inventories of national greenhouse gases. CCSP participates extensively in the IPCC process, including support of the Working Group I Technical Support Unit, support for the IPCC Task Group on Scenarios for Climate and Impact Assessment, and support for scientist participation in IPCC activities.

The Famine Early Warning System Network is an example of how climate information can and is directly supporting decisionmaking in areas that are highly susceptible to both man-made and natural hazards, including change. The network combines food economy analysis, which analyzes the structure of poor urban and rural households in Africa, with a variety of data including climate to determine the risk of food shortage. These analyses are conducted on a seasonal basis and in response to predicted and observed hazards as well as other conditions that may threaten food security (e.g. natural disasters, market data).

Engaging the International Global Change Research Community

The United States and U.S. scientists engage the international global change research community in a variety of ways—for example, by sponsoring and contributing to climate change science-related workshops that foster cutting-edge science collaborations.

The Interagency Working Group for International Research and Cooperation (IWG-IRC) brings together the U.S. federal agencies

interested in international cooperation in climate change science. This group traces its beginnings to the International Cooperation in Global Change Research Act of 1990 (Title II). The primary purpose of the working group is to provide international affairs support to the research programs of CCSP. It consists of representatives of 11 U.S. agencies that have major interests in global climate change science and technology. Meeting regularly, member agencies exchange information on international issues pertaining to climate change research and develop interagency approaches directed at resolving these issues. The working group coordinates U.S. interagency support for the international infrastructure that provides coordination for the major international global change research programs.

U.S. participation in IGFA aids in efforts to engage the national funding agencies of over 20 member countries. This group ensures that national funding agencies are regularly informed about national global change research programs, supporting initiatives, facilities, and related issues. The agencies then identify topics of mutual interest that they address through their respective national processes and, in some cases, coordinated international efforts. IGFA is also concerned with the reduction of barriers to international collaborative research at all levels—for example, constraints on the exchange of scientists and equipment.

The United States actively promotes global change research in the Antarctic and Arctic. Regarding the Antarctic, the United States works through cooperation with parties to the Antarctic Treaty and the Scientific Committee on Antarctic Research (SCAR). Regarding the Arctic, the United States works through the Arctic Council, the International Arctic Sciences Committee (IASC), and the Arctic Ocean Sciences Board (AOSB). Work with these organizations will advance fundamental knowledge of the polar regions as well as provide observations that are critical to our understanding of climate. The Arctic Climate Impact Assessment (ACIA), conducted under the auspices of the Arctic Council and IASC, is assessing the consequences of climate change on the circum-Arctic environment, its resources, economy, and peoples.

An especially important example of U.S. efforts to engage scientists in other countries in cooperative research involves Japan. The United States is involved in a focused cooperative effort with Japan in the geosciences and environment. Within this effort, a series of 10 annual workshops have been conducted on global change research. This year, the United States hosted the 10th U.S.–Japan workshop in this series, on the topic of "Water and Climate." Workshop topics for the two previous years were "Health and the Environment" and "Carbon Cycle Management in Terrestrial Ecosystems." These meetings have proven to be an excellent forum for the exchange of ideas and information and are serving to stimulate international scientific collaborations.

A new series of bilateral discussions has recently been added to the United States' international affairs repertoire as a result of the Climate Change Research Initiative (CCRI) of 2001. This initiative provides yet another means to engage other countries in scientific collaboration on climate change science and technology. These activities are described in detail in the following section.

Bilateral Discussions

President Bush's climate change policy announcements on 11 June 2001 and 14 February 2002 highlighted the importance of international cooperation to develop an effective and efficient global response to the complex and long-term challenge of climate change. Under the leadership of the Department of State, the United States adopted a Bilateral Climate Change Strategy, focusing on countries or regional entities that are responsible for nearly 75% of the world's greenhouse gas emissions. Through this important network of bilateral and regional partnerships, the United States is advancing the science of climate change, enhancing the technology to monitor and reduce greenhouse gases, and assisting developing countries through capacity building and technology transfer.

Working with a mix of developed and developing countries, the Bilateral Climate Change Cooperation effort builds upon and supplements the four U.S. international research and cooperation activity goals: research, observations, data management and distribution, and capacity building. Examples from the range of activities included under the bilateral climate change effort are: global climate observing systems; evaluation of climate systems models; observations and data exchange; research on polar regions, aerosols, and clouds; ocean and atmospheric research; research on greenhouse gas sinks including land use, land-use change, and forestry; advancement of clean and renewable energy technologies; and energy efficiency. Capacity building and technology transfer activities with developing countries also include cooperation in economic and environmental/climate modeling, carbon cycle measurements, monitoring and measurement of greenhouse gases,

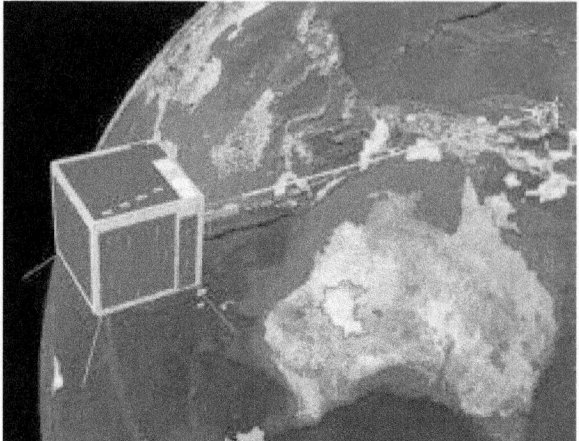

Figure 15-1: Australia's Federation Satellite carrying NASA's Global Positioning System receiver.

development of adaptive capacity to improve resource management, implementation of integrated environmental strategies, investments in climate observing systems, and creating partnerships for energy efficiency.

The United States has made significant progress on many aspects of the President's international climate change agenda through the establishment of results-oriented "action plans" with bilateral and regional partners. These partners include Australia, Japan, the seven Central American countries (Belize, Costa Rica, El Salvador, Guatemala, Honduras, Nicaragua, and Panama), the People's Republic of China, the European Union (EU), India, Italy, New Zealand, Canada, the Russian Federation, the Republic of Korea, Mexico, and South Africa. Specific examples follow.

Under the U.S.–Australia Climate Action Partnership (CAP) the focus is on practical approaches towards dealing with climate change. Specific priority areas included under CAP are climate change science and monitoring; renewable and reduced emission stationary energy technologies; engagement with business on technology development; policy design and implementation; capacity building in developing countries; and greenhouse gas accounting in the forestry and agriculture sectors (see Figure 15-1).

Figure 15-2: The NOAA Climate Monitoring and Diagnostics Laboratory (CMDL) Carbon Cycle Greenhouse Gases (CCGG) Group's Cooperative Air Sampling Network at Syowa, Antarctica, is a collaboration with Japan's National Institute of Polar Research. The flasks collected here are analyzed for 12 chemical species including greenhouse gases and related isotopes.

The U.S.-Japan High Level Consultations on Climate Change (HLC) Working Group on Science and Technology focuses on seven priority areas: improvements of climate models; impacts and adaptation/mitigation policy assessment employing emission-climate impact integrated models; observation and international data exchange and quality control (see, for example, Figure 15-2); research on greenhouse gas sinks including land use, land-use change, and forestry; research on polar regions; and development of mitigation and prevention technologies. Experts also are collaborating on issues relating to developing countries and market-based approaches.

The United States and seven Central American countries (Belize, Costa Rica, El Salvador, Guatemala, Honduras, Nicaragua, and Panama) agreed to enhance climate change collaboration under the auspices of the Central American-United States of America Joint Accord, known by its Spanish acronym CONCAUSA. The partnership emphasizes the need for intensified cooperative efforts to address climate change through scientific research, estimating and monitoring greenhouse gases, investing in forestry conservation, enhancing energy efficiency, utilizing new environmental technologies, enhancing capacity to adapt to climate change, and collaborating to better understand its regional impacts. In March 2002, the State Department announced a CONCAUSA Action Plan that included $2 million in new funds from USAID, with supplemental project support from EPA and NASA.

The United States and the People's Republic of China have agreed under an approved charter to collaborate on a broad suite of climate change science and technology issues, including cooperative research on and analysis of gases other than carbon dioxide, economic/environmental modeling, integrated assessments and evaluation strategies for adaptation, capture and sequestration, observations and measurements, institutional partnerships, energy and environment follow-up to the World Summit on Sustainable Development, and existing clean energy protocols and annexes.

Under the agreement of representatives to the U.S.–EU High Level Dialogue on Climate Change on 23 April 2002, a "U.S.–EU Joint Meeting on Climate Change Science and Technology Research" convened in Washington on 5-6 February 2003, to enhance cooperation on climate-related science and research. The two sides identified cooperative research activities in six areas: (1) carbon cycle research; (2) aerosol-climate interactions; (3) feedbacks, water vapor, and thermohaline circulation; (4) integrated observation systems and data; (5) carbon capture and storage; and (6) hydrogen technology and infrastructure.

The United States and Italy have developed a bilateral partnership encompassing a wide range of cooperative science and technology projects and activities, including climate change modeling, atmospheric processes, the carbon cycle, remote sensing, human and ecosystem health, and ocean observations and the ocean ecosystem. On the technology side, the partnership is advancing cooperative efforts on hydrogen infrastructure and energy technologies, including fuel cells, renewable energy, advanced power systems, and advanced

Figure 15-3: NASA/NASDA Tropical Rainfall Measuring Mission (TRMM) microwave imager data.

energy technologies including carbon capture and sequestration. The two countries have announced their intention to promote the exchange of graduate students, young scientists, and senior scientists in the area of climate change science and technology.

Most recently, the United States and India identified a broad range of cooperative programs, including science and technology activities. The two countries have identified collaborative efforts in the area of climate change science—for example, the NASA/Japanese Space Agency (NASDA) Tropical Rainfall Measuring Mission (see Figure 15-3). Additional anticipated areas of collaboration are adaptation, land-use change and forestry, climate modeling, ocean observations for the Indian Ocean, economic and environmental modeling, source-level measurement and monitoring, and a broad range of energy issues.

Research

As discussed in the opening section, the United States has a number of goals with respect to climate change science research. International cooperation in research has an important role in focusing the world's scientific resources on the highest priority global change research issues, in helping to reduce scientific redundancy in a world of limited financial resources, and in improving the exchange of information internationally. U.S. research efforts are significantly enhanced through international cooperation. By developing both conceptual and research frameworks, international research programs provide models that aid U.S. program managers to plan and coordinate their efforts. CCSP will to continue to facilitate cutting-edge climate change research by providing appropriate venues and

resources for scientists to meet, to identify research areas of common interest, and to plan and implement joint projects.

Much of the research conducted and sponsored through CCSP benefits from and contributes to projects sponsored by the four major international research programs: IGBP, WCRP, IHDP, and the Diversitas program (see Table 15-1). The Diversitas program is dedicated to fostering global research on biodiversity in order to provide information to be integrated into conservation and sustainable management of biodiversity.

Examples of the links between the CCSP research elements and international programs include the Climate Variability and Predictability (CLIVAR) program and the Global Energy and Water Cycle Experiment (GEWEX), both core programs of WCRP. The CCSP Climate Variability and Change research element will continue to work with CLIVAR, Stratospheric Processes and their Role in Climate (SPARC), Past Global Changes (PAGES), and Climate and Cryosphere (CliC) programs. Both the CCSP Climate Variability and Change and Water Cycle research elements will continue their work in close coordination with the GEWEX program.

The CCSP research elements maintain regular exchanges with the various international projects to which they are related. The U.S. scientists leading CCSP projects have offered substantial input to the design and implementation of many of these projects that are now underway. In addition to programs and projects that operate under the auspices of the international global change research programs, the CCSP research elements interact with many others (see Table 15-2). For example, the scientists involved in the U.S. North American Carbon Program (NACP) will, in order to realize the research goals of the program, expand their interaction with

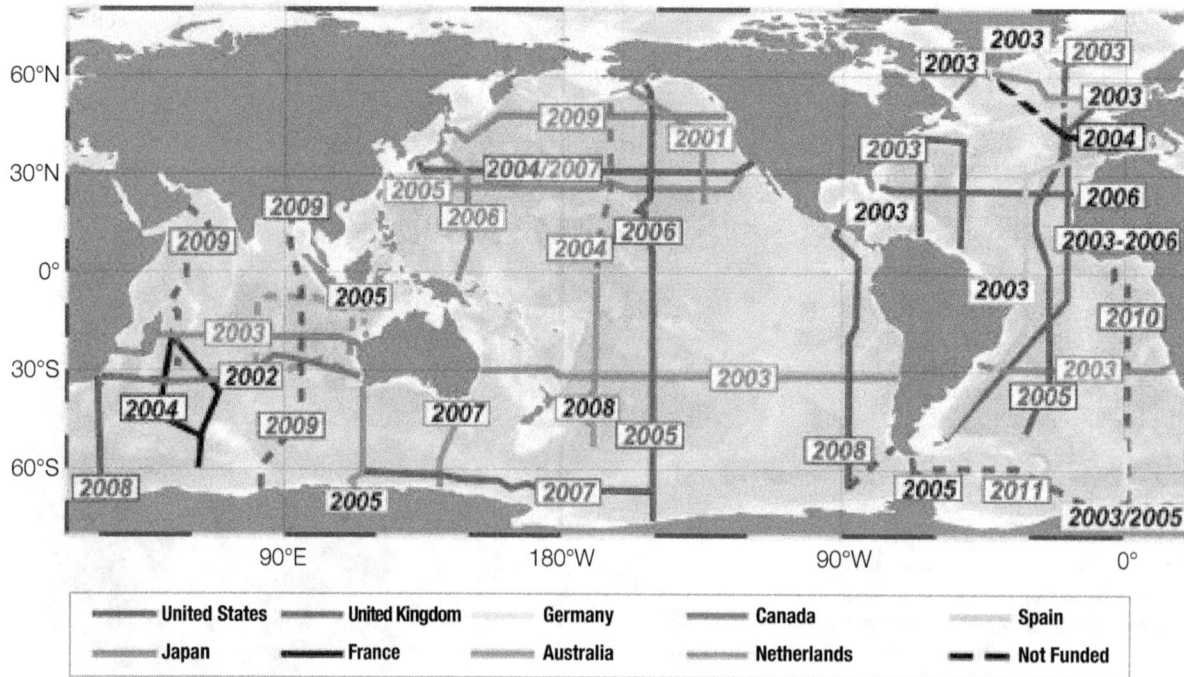

Figure 15-4: Map of planned deep-ocean, repeat hydrography cruises to address scientific objectives of the U.S. CLIVAR and Carbon Cycle Science Program objectives. Source: Chris Sabine, NOAA-PMEL.

colleagues in Canada and Mexico and with the Carboeurope cluster, the International Global Observing Strategy Partnership (IGOS-P), and the Global Carbon Project as well (see Chapter 7). The programs described in these two tables represent the primary international interactions of the research programs within CCSP.

The synergies and coordination developed by such parallel programs are instrumental in addressing global and regional issues. For example, representatives of the world's major climate modeling centers form the WCRP/CLIVAR Working Group on Coupled Modeling (WGCM) to collectively address needed advancements in coupled

TABLE 15-1

Examples of key programs and projects with which the CCSP research elements cooperate, coordinated through the four international research programs

Project	Description
International Geosphere-Biosphere Programme (IGBP)	
International Global Atmospheric Chemistry (IGAC)	IGAC is dedicated to understanding the processes that determine atmospheric composition. This includes detailed examination of the interactions between atmospheric and chemical composition, physical, biospheric, and climatic processes. They also investigate the prediction of impacts on atmospheric composition associated with natural and anthropogenic forcing (see Chapter 3).
Past Global Changes (PAGES)	The primary objective of PAGES is to facilitate international cooperation in paleo-environmental research. PAGES coordination embraces paleoclimatic aspects of the interactive physical, chemical, and biological processes that regulate the Earth system, concentrating on those aspects which best inform our understanding of potential future changes of relevance to human concerns (see Chapter 4).
Joint Global Ocean Flux Study (JGOFS)	The JGOFS program was designed to advance our understanding of the processes that control carbon exchanges between the atmosphere, surface ocean, ocean interior, and continental margins, and the sensitivity of these fluxes to climate change.
GLOBal ocean ECosystems dynamics (GLOBEC)	The GLOBEC program is designed to investigate and advance our knowledge of the global ocean ecosystem, its major subsystems, and its response to physical forcing, with emphasis on natural variability (see Chapter 8).
Global Climate and Terrestrial Ecosystems (GCTE)	The GCTE program investigates the effects of climate changes, atmospheric composition, and land use on the structure and function of terrestrial ecosystems and how they lead to feedbacks to the atmosphere and the climate system (see Chapter 8).
Land-Ocean Interface in the Coastal Zone (LOICZ)	LOICZ studies biogeochemical processes and changes in the fluxes of materials from within river catchments to coastal shelf boundaries, the influence of human activities on these changes, and the impact of flux changes on human welfare. LOICZ also provides science information to the global community, especially decisionmakers and coastal zone managers (see Chapter 8).
Biospheric Aspects of the Hydrological Cycle (BAHC)	BAHC is an interdisciplinary project that fosters and promotes research to better understand the role that the terrestrial biosphere plays in the Earth system, and the increasing human-induced changes (see Chapter 8).
World Climate Research Programme (WCRP)	
Climate Variability and Change (CLIVAR)	The CLIVAR program is one of the major research efforts of WCRP concerned with climate variability, extending predictions of climate variation and refining estimates of anthropogenic climate change. CLIVAR will advance the findings of the successfully completed Tropical Ocean and Global Atmosphere (TOGA) project, and continues its valuable work with the World Ocean Circulation Experiment (WOCE) (see Chapter 4).
Stratospheric Processes And their Role in Climate (SPARC)	SPARC concentrates on the interaction of dynamic, radiative, and chemical processes in the stratosphere and what role they play in climate. SPARC is constructing a stratospheric reference climatology; improving the understanding of trends in temperature, ozone, and water vapor in the stratosphere; and improving the understanding of gravity waves (see Chapter 4).
The Global Energy and Water Cycle Experiment (GEWEX)	GEWEX focuses on the study of atmospheric and thermodynamic processes that determine the hydrological cycle and water budget as well as their adjustments as a function of climate change (see Chapters 4 and 5).
Climate and Cryosphere, Arctic Climate System Study (CliC, ACSYS)	ACSYS—a regional project studying climate of the Arctic region including its atmosphere, ocean, sea ice, and hydrological regime—is being expanded into a global project, CliC, to investigate the role of the entire cryosphere in global climate.
World Ocean Circulation Experiment (WOCE)	The WOCE program was designed to understand and predict changes in the world's ocean circulation, volume, and heat storage that would result from changes in atmospheric climate and net radiation with a variety of *in situ*, remote-sensing, and modeling methodologies.
International Human Dimensions Programme (IHDP)	
Land-Use and Land- Cover Change (LUCC)	LUCC, co-sponsored by IGBP, focuses on the understanding of the interrelationship between land use and land-cover change, biogeochemistry, and climate. It addresses these issues with an interdisciplinary approach utilizing case studies, models, and integrative analyses. Topics such as climate change, food production, health, urbanization, coastal zone management, transboundary migration, and the availability and quality of water are addressed in LUCC research (see Chapter 6).

TABLE 15-1 (CONTINUED)

Examples of key programs and projects with which the CCSP research elements cooperate, coordinated through the four international research programs

Project	Description
Jointly Sponsored IGBP/WCRP/IHDP Projects	
Global Environmental Change and Food Systems (GECAFS)	GECAFS tries to address the suite of issues associated with population growth and globalization of economies as they relate to food stability in the context of global environmental change. This includes consideration of the potential impacts of global change on both individuals and societies (see Chapter 8).
The Global Carbon Project (GCP)	GCP integrates atmospheric, oceanic, terrestrial, and human dimension aspects of the carbon cycle with commitment and balanced input from the global environmental change programs. It has developed an international framework for carbon research and investigates system-wide questions of interactions between humans and the carbon cycle. National and regional studies contribute to the implementation of the project (see Chapter 7).
The Global Water System Project (GWS)	A joint water project, sponsored by all four global environmental change programs, is in the planning phase and will take an integrative look at the global water system. GWS aims at understanding impacts of global change on local and regional coupled water-human systems, and how local and regional anthropogenic activities in turn affect global environmental change. Approaches to establish more sustainable water systems will be identified (see Chapter 5).
Global Change SysTem for Analysis, Research, and Training (START)	START focuses on regional interdisciplinary global change research. Its objective is to build scientific capacity in developing countries through regional research and training to address the science and policy issues relevant to environmental change and sustainable development. START has regional offices in East Asia, South Asia, Southeast Asia, Oceania, Africa, and the Mediterranean.
Surface Ocean-Lower Atmosphere Study (SOLAS)	SOLAS is a new international research initiative to carry out research at the interface between the oceans and the atmosphere in order to achieve a quantitative understanding of the key biogeochemical-physical interactions and feedbacks between the ocean and the atmosphere, and how this coupled system affects and is affected by climate and environmental change (see Chapter 8).

modeling of the global physical climate system. Research activities coordinated by WCRP elements have led to major advancements in the understanding and modeling of the North Atlantic Oscillation/Northern Hemisphere Annular Mode (NAO/NAM), improved precipitation and solar insulation products, and advanced representations of the Arctic in models.

The CLIVAR Variability of the American Monsoon (VAMOS) program has been instrumental in enlisting the participation of South and Central American scientists to focus on observing, modeling, and assessing variability of the American monsoon system and our ability to predict its changes. A new era of cooperation between participating scientists and key organizations responsible for national meteorological and oceanographic observational systems is not only entraining key scientists, but is also enhancing existing observational systems, initiating new observation capabilities (e.g., rain gauge and sounding networks) for climate and application research, and providing improved practical knowledge for decisionmakers.

U.S. scientists play an important role in the research programs carried out by the Inter-American Institute for Global Change Research (IAI), and in bringing together the more than 200 research universities and government institutions in the Western Hemisphere that make up its research network. Research programs sponsored by IAI have aided in the development of new decision and management tools in diverse areas, ranging from the incorporation of long-range forecasts into dam management for hydropower and irrigation to the establishment of a tri-national sardine fishery forum that regularly brings together regulatory agencies, resource managers, fishermen, and researchers from Canada, Mexico, and

the United States. In addition, IAI research enabled the first rigorous scientific ranking of the drivers of global change, based on scenarios of changes in global biodiversity.

Some of the most productive international cooperative interactions are developed through international research expeditions and field experiments. The Ocean Drilling Program (ODP) is an outstanding example of a long-term field study that has obtained cores of sediments and crustal rock from the oceans to improve understanding of the history of changes in the oceans and climate. The National Science Foundation (NSF) and Japan's Ministry of Education, Culture, Sport, Science, and Technology (MEXT) signed a memorandum of understanding in April of 2003 to further advance scientific ocean drilling through a new Integrated Ocean Drilling Program (IODP—see Figure 15-5). IODP will develop ocean sediment records from a global array of sites to allow a sophisticated and detailed analysis of the causes, rates, and severity of changes in the Earth's climate system and their relation to major pulses in biologic evolution on time scales ranging from a few hundred years to more than 500,000 years.

Other recent examples of successful field campaigns include the Indoex campaign off of India and over the Indian Ocean to study atmospheric aerosols; the ACE-Asia and TRACE-P field campaigns to study atmospheric trace constituent and aerosol observations over East Asia and the Western Pacific Ocean; and the Large Scale Biosphere-Atmosphere Experiment in Amazonia (LBA) campaign to study the ecology and hydrology of the Amazon region. Such field campaigns and others have demonstrated the great value of having good relationships with host countries to ensure the highest quality scientific observations.

Observing Systems

Although climate modeling capabilities have progressed in recent years, sustained improvements will require substantial expansion of Earth observing systems, both remote and *in situ*, in order to fill gaps in existing databases, especially in those areas of the world for which existing data is sparse. Such data-sparse areas include remote

regions, especially those with harsh environments, and areas where existing capabilities to make observations and collect data are limited, such as the oceanic and interior land areas of the Southern Hemisphere and both polar regions.

The United States hosted an Earth Observation Summit in July 2003 to engage the international community in a continuing dialogue on the issues associated with building a comprehensive, integrated Earth observation system. The Earth Observation Summit had four

TABLE 15-2

Additional examples of key international and regional programs and projects with which CCSP research elements cooperate and/or coordinate

Organization	Project	Description
Intergovernmental Organizations		
World Meteorological Organization (WMO)	Hydrology and Water Resources Programme	This group focuses on the application of hydrology to sustainable development, the mitigation of water-related disasters, and effective environmental management at national and international levels (see Chapter 5).
United Nations Educational, Scientific, and Cultural Organization (UNESCO)	International Hydrological Programme	This is an intergovernmental capacity-building program and information center for information pertaining to water resources. It is dedicated to building the capacity of member states to better manage and develop their water resources (see Chapter 5).
WMO/UNESCO	Hydrology for Environment, Life, and Policy (HELP) Program	HELP is a joint initiative of UNESCO and WMO, led by the International Hydrological Programme. HELP is creating a new approach to integrated catchment management through the creation of a framework for water law and policy experts, water resource managers, and water scientists to work together on water-related problems (see Chapter 5).
Northern Eurasia Earth Science Partnership Initiative (NEESPI)		NEESPI will identify the critical science questions and establish a program of coordinated research on the state and dynamics of terrestrial ecosystems in northern Eurasia and their interactions with the Earth's climate system to enhance scientific knowledge and develop predictive capabilities to support informed decisionmaking and practical applications (see Chapter 6).
Intergovernmental Oceanographic Commission (IOC)/ Scientific Committee on Ocean Research (SCOR)	Global Ecology and Oceanography of Harmful Algal Bloom (GEOHAB)	The GEOHAB program is an international program designed to improve international capabilities for effective management and mitigation of harmful algal blooms. This research program utilizes a comparative approach at a variety of scales from the cellular to the ecosystem level. This research effort will improve understanding of the distribution of and trends in harmful algal blooms and the influences of anthropogenic and climate-related factors (see Chapter 8).
Intergovernmental and Similar Organizations and Activities		
International Research Institute for Climate Prediction (IRI)		IRI is dedicated to improving and providing climate predictions to help societies worldwide cope with climate fluctuations, particularly extreme events that have great impacts on both human populations and the environment.
European Commission	Carboeurope	The Carboeurope cluster is a research program dedicated to understanding the terrestrial carbon cycle for a range of environments in Europe, including carbon fixation and carbon sources and sinks. These issues are considered in the context of climate variability, the availability of nutrients, nitrogen deposition, and management. Their research effort is complemented by research in the Amazon forests.
Inter-American Institute for Global Change Research (IAI)		IAI is an intergovernmental research organization supported by 19 countries in the Americas. It is dedicated to the understanding of global change and its socioeconomic implications. This includes study of the natural and social sciences, the full and open exchange of data relevant to global climate change, augmenting the capacity of countries to conduct scientific research, and the provision of information in a timely manner and in a useful form for policymakers (see Chapter 9).
Asia-Pacific Network for Global Change Research (APN)		APN fosters global environmental change research in the Asia-Pacific region, increases the participation of developing countries in that research, and strengthens the connection between the science community and policymakers.
International Long-Term Ecological Research (ILTER)		ILTER is an international organization dedicated to ecological research on long temporal scales and large spatial scales (see Chapter 8).

Figure 15-5: The January 2002 launch of the Japanese riser drill ship Chikyu (Earth) in Kobe, Japan. This vessel and a U.S.-provided non-riser drill ship will be the primary global scientific ocean drilling platforms for the new Integrated Ocean Drilling Program. IODP will address many fundamental and societally important questions in Earth and ocean science including improving our understanding of past climate and climate change.

major goals: to promote the international exchange of *in situ*, aircraft, and satellite observations in a full and open manner and in a timely fashion with minimal cost; to gain international agreement on the concept of an international, comprehensive, integrated, and sustained Earth observation system that will meet collective requirements for observations, minimize data gaps, and maximize the utility of the system; to establish an intergovernmental ad hoc working group to develop a 10-year implementation plan for the system; and to help improve observing systems and advance the scientific capacity in developing countries.

The United States supports a number of international observing systems and networks with which the CCSP research programs work closely (see Chapters 5, 6, 7, and 8). The International Global Observing Strategy Partnership (IGOS-P) incorporates a number of large-scale observing systems that are designed to cover land, through the Global Terrestrial Observing System (GTOS); the ocean, through the Global Ocean Observing System (GOOS); and climate, through the Global Climate Observing System (GCOS) and the Global Observing System/Global Atmosphere Watch (GOS/GAW). The United States not only supports these programs directly, but also encourages the international global change research programs to ensure that the observational needs of research scientists are fed into IGOS-P and that the partners also benefit from the observational expertise developed in research projects.

In the Arctic, the United States is also involved in a number of international observing projects such as the Arctic and Subarctic Ocean Fluxes Study (ASOF) and the North Pole Environmental Observatory (NPEO). The ASOF program, which the European Commission also provides support for, will monitor and advance understanding of the oceanic fluxes of heat, salt, and freshwater and their effect on global ocean circulation. NPEO, coordinated with Japan and Canada, is a series of drifting buoys, oceanographic moorings, and hydrographic casts that serve collectively as an observing system.

The CCSP research programs will need to continue to work closely with these observational program groups in order to benefit from and contribute to them. CCSP assigns very high priority to the Argo ocean observations program, a global array of free-drifting profiling floats that measures the temperature and salinity of the upper 2,000

meters of the ocean. The United States is actively seeking, through a wide range of multilateral and bilateral cooperative efforts, to encourage and promote a wide range of international participation to complement the U.S. efforts. Cooperation in Argo has been initiated successfully with Japan, Australia, the European Commission, China, the Russian Federation, India, the Republic of Korea, Canada, Spain, Norway, Germany, Denmark, France, the United Kingdom, and New Zealand. Another important example of a focused observing system of importance to CCSP is the program for the Global Observation of Forest Cover and Global Observations of Land Cover Dynamics (GOFC-GOLD—see Chapter 6).

Satellite remote-sensing systems are continually being developed to take advantage of new technologies to collect an ever-widening range of data on both global and regional scales. Successful implementation of such systems requires development of collaborative international ground-based networks, maintenance of these networks, and assurance of calibration relative to widely recognized standards. This can be accomplished only through collaboration among scientists and national and international agencies from many nations. NASA and NOAA work closely with their counterpart agencies in other countries, such as the European Space Agency (ESA) and the Japanese Space Development Agency (NASDA) to coordinate planning and implementation of new satellite remote-sensing systems. These agencies interact directly with one another regularly and as a group in CEOS.

NASA also partners on a bilateral basis with its counterpart agencies in other countries in algorithm development [e.g., with Canada, Measurement of Pollution in the Troposphere (MOPITT)], instrument development [e.g., with the United Kingdom on the High-Resolution Dynamics Limb Sounder (HRDLS) and Brazil on the Humidity Sounder for Brazil (HSB)], and mission implementation (e.g., with Japan on the TRMM project and instrumentation for the AQUA spacecraft). The United States and Japan are now actively discussing the possibility of leading a multi-national Global Precipitation Measurement (GPM) mission in the future. Similarly, the United States and France have jointly provided for two satellites measuring ocean surface height (TOPEX/Poseidon in 1992, Jason-1 in 2002), and are actively exploring further investments in this area as part of a strategy to transition these research measurements to operations.

Data Management

Climate change science depends critically on the full and open exchange of scientific observations and data both within and across national borders. It is essential that scientists have the widest possible access to scientific data in order to conduct effective research and to develop reliable climate forecasts. Scientists involved in both research and forecasting must also be able to make the data that they use available for others to test, to replicate their research results and forecasts, and to indicate lines for future research and forecasting. Furthermore, these data must also be made available in a timely and useful manner to policymakers and decisionmakers in order to assist in efforts to mitigate the impacts of climate variability and change.

As a leader in and major supporter of climate change research, the United States has the opportunity to influence the rest of the world on issues pertaining to data management and thus to continue to advocate the full and open exchange of scientific data across international boundaries. It is also important to develop and promote wide application of common methodologies and protocols in order that collected data are of high quality, comparable, and easily accessible, especially through an effective international network of data archival and exchange systems.

A number of international efforts to manage and exchange data have been underway for many years. The World Data Center System (WDCS), established by the International Council of Scientific Unions (ICSU), has successfully collected, maintained, archived, and distributed scientific data, including climate change data. Since it was founded, WDCS has expanded into a number of worldwide independent sites. The Data and Information System (DIS) is the data management portion of IHDP and links social science data to scientists and institutions researching climate change.

CCSP expects to continue to interact closely with data management and dissemination efforts such as NASA's Global Change Master Directory (GCMD) and the International Directory Network (IDN). GCMD is the set of data and information systems from individual agencies that support climate change science and serves as the U.S. coordinating node of CEOS. The GCMD system utilizes a standard format, known as the Directory Interchange Format (DIF), that provides a simple user interface allowing access to over 1,800 data sets. Using the DIF system, CEOS developed IDN, which allows worldwide access to data and information about climate change. The Global Observation Information Network (GOIN) project is an example of U.S.-Japan bilateral cooperation in Earth observation information networks including satellite and *in situ* data. All of these programs are examples of the ideals of data sharing the United States continues to work toward internationally.

Capacity Building

CCSP and the major international research programs have recognized that to address the myriad of issues associated with understanding and responding to global change will require a truly global research effort. They have and will continue to provide resources in order to build scientific capacity in developing countries and thus to improve the ability of developing countries to undertake global change research and to benefit more fully from the results of such research. CCSP therefore supports a number of efforts to build scientific capacity in the developing world.

A key component of this effort is the System for Analysis, Research, and Training (START) that is sponsored jointly by IGBP, WCRP, IHDP, and Diversitas. Its mission is to build capacity in developing countries in order that they can conduct research on global change and be better prepared to understand and thus mitigate the potential impacts those changes may have for human health, agriculture, natural resources, water, and food security.

The START program is hosted by the United States, and supported in large part by CCSP, while also receiving direct support from USAID and EPA. START is undertaking capacity building on a regional basis in Africa, Southeast Asia, temperate East Asia, the Mediterranean, and Oceania. The START program also sponsors fellowships and training workshops, and is currently developing a young global scientists' network that will link young scientists in the developing world with their peers in the United States.

Another example of a science project with a strong capacity-building component is the Large-Scale Biosphere-Atmosphere Experiment in Amazonia. LBA is designed to promote understanding of the effects of climate, land-use change, ecology, and biogeochemical and hydrological cycling in Amazonia. In the LBA campaign, capacity-building issues are addressed by requiring the involvement of Brazilian scientists and students, and has become a critical criterion in the selection of participants for the program.

Established in 1991, the Global Environment Facility funds environmental projects in developing countries. Climate change projects constitute approximately 40% of GEF's activities—its largest single focus. The United States has pledged $500 million over the next 4 years for GEF, a 16% increase that leveraged $2.2 billion and helped bring about the largest funding commitment in GEF's history. A significant portion of this funding is to help build capacity by supporting the transfer of advanced energy and sequestration technologies to the developing world. The Administration has also pledged to pay U.S. arrears that had built up over the course of the previous replenishment period.

The United States also supports two regional efforts that grew out of the "White House Conference on Scientific and Economic Research Related to Global Change" hosted by President George Bush in 1990: the Inter-American Institute for Global Change Research, and the Asia-Pacific Network for Global Environmental Change Research (APN). The United States is the primary supporter of IAI, and has joined with Japan in supporting APN. A goal of both IAI and APN is to increase participation of developing countries in global climate change research and to strengthen links between the science community and policymakers. Both institutions, independently and in collaboration with organizations like START, have strong capacity-building programs. The United States seeks to complement support of IAI and APN regions by advancing regional cooperation in Africa. These efforts will not only contribute to improving the capabilities of African countries to conduct climate change science, but in the long term also will result in improved databases for this data-sparse region that, in turn, can improve models of climate change around the world.

U.S. Plans and Objectives for Future International Cooperation

The overall framework for international cooperation in global change research and observations has been responsive to the needs of U.S. global change science. However, this framework should be broadened and strengthened to keep pace with the evolving needs of this science with respect to both research and observations.

To expand cooperation internationally, the President has announced that the United States intends to:
- Commit $25 million to support the implementation of climate observation and response systems in developing countries
- Expand funding of GEF
- Support the transfer of advanced energy and sequestration technologies to developing countries in order to limit their greenhouse gas emissions growth
- Expand cooperation in climate change research and technology with a number of key countries and regional organizations
- Work with IAI and other institutions to better understand regional aspects of climate change.

CCSP also will:
- Continue to support and advance regional cooperation in Africa, possibly through a workshop that could lead to improved and expanded hemisphere-scale regional cooperation in global change research in Africa
- Expand international cooperation by continuing to support bilateral discussions in climate change science and technology
- Support further development and expansion of global observing systems through IGOS-P and three of its partners—GCOS, GOOS, and GTOS—and the Argo program for ocean

observations, through further multilateral and bilateral cooperative efforts analogous to those already initiated
- Expand cooperation in biodiversity research, especially through the Diversitas program
- Enhance efforts to bring science and technology to bear on increasingly complex problems of natural resource development (e.g., the application of climate information for improved adaptation and disaster preparedness)
- Work with the international global change research programs—WCRP, IGBP, IHDP, and Diversitas—to promote effective transition of a number of their present focused programs to cross-cutting programs (such as the new programs on water, carbon, and global change and food security) that are intended to relate global change research more directly to major societal and economic factors
- Continue to work with and be supportive of international assessments such as those of IPCC.

CHAPTER 15 AUTHORS

Lead Authors
Louis B. Brown, NSF
Toral Patel-Weynand, DOS
Jim Buizer, NOAA
David Allen, CCSPO

Contributors
Christo Artusio, DOS
Ko Barrett, USAID
Garik Gutman, NASA
Michael Hales, NOAA
Jack Kaye, NASA
Kate Maliga, NASA
Linda Moodie, NOAA
Duane Muller, USAID
Carrie Stokes, USAID
Lisa F. Vaughan, NOAA

Background

Several circumstances define the unique management
environment of the Climate Change Science Program
(CCSP). Fundamentally, CCSP integrates federal research
on global change and climate change, as sponsored by 13
federal departments and agencies (the Departments of
Agriculture, Commerce, Defense, Energy, Health and
Human Services, the Interior, State, and Transportation;
together with the Environmental Protection Agency, the
National Aeronautics and Space Administration, the
National Science Foundation, the Agency for International
Development, and the Smithsonian Institution). The Office
of Science and Technology Policy, the Council on
Environmental Quality, the National Economic Council , and
the Office of Management and Budget provide oversight.
Planning and implementation must be coordinated across
the participating departments and agencies because the
capabilities required for comprehensive scientific inquiries
and synthesis extend beyond the mission, resources, and
expertise of any single agency (see Box 16-1 for the roles
and responsibilities of CCSP's participating agencies).

As a federal program, CCSP is implemented within the
context of the federal budget cycle. Budget requests are
coordinated through interagency research working groups and other
mechanisms, but ultimate budget accountability resides with the
participating agencies and departments. As a result of its interagency
composition, activities in the CCSP budget are funded by the U.S.
Congress through six separate Appropriations bills. There is also
oversight provided by a number of Authorization committees, making
the relationship between CCSP budgeting and the appropriations
process complex.

CCSP partners closely with the independent national and international
research community, which has partitioned research on the Earth
system and climate change into discrete and manageable research
issues and questions. CCSP has developed management mechanisms
to ensure that data and information needs are coordinated across
disciplines and research areas so that synthesis and model development
can proceed as rapidly as scientific advances allow. In addition,
CCSP must manage its research portfolio to facilitate discovery-
driven investigation on a broad range of global change issues while
also providing sufficient focus to achieve its climate objectives.

Under the President's direction, CCSP is placing increasing emphasis
on realizing returns on society's investment in scientific discovery
by developing resources that apply this information in support of
climate change policymaking and adaptive management/planning.
This requires the program to improve interactions between
researchers and users of scientific information, and aid in the transfer
of capabilities from research and development to operational services.

BOX 16-1

PRINCIPAL AREAS OF FOCUS FOR THE CCSP AGENCIES

Department of Agriculture (USDA)

USDA-sponsored research supports long-term studies to improve our understanding of the roles that terrestrial systems play in influencing climate change, and the potential effects of global change (including water balance, atmospheric deposition, vegetative quality, and ultraviolet-B radiation) on food, fiber, and forestry production in agricultural, forest, and range ecosystems. USDA's research program is strengthening efforts to determine the significance of terrestrial systems in the global carbon cycle, and to identify agricultural and forestry activities that can contribute to a reduction in greenhouse gas concentrations. USDA's research agencies will support the Department in responding to the President's directive to develop accounting rules and guidelines for carbon sequestration projects. Contributions from USDA's research program include the development of improved emission and sequestration coefficients, new tools for accurately measuring carbon and other greenhouse gases, and the development of improved sequestration methodologies.

Department of Commerce (DOC)

DOC's National Oceanic and Atmospheric Administation (NOAA) mission is: "To understand and predict changes in the Earth's environment and conserve and manage coastal and marine resources to meet the nation's economic, social, and environmental needs." The long-term global change efforts of NOAA are designed to develop a predictive understanding of variability and change in the global climate system, and to advance the application of this information in climate-sensitive sectors through a suite of process research, observations and modeling, and application and assessment activities. Specifically, NOAA's research program includes ongoing efforts in operational *in situ* and satellite observations with an emphasis on oceanic and atmospheric dynamics, circulation, and chemistry; understanding and predicting ocean-land-atmosphere

interactions, the global water cycle, and the role of global transfers of carbon dioxide among the atmosphere, ocean, and terrestrial biosphere in climate change; improvements in climate modeling, prediction, and information management capabilities; the projection and assessment of variability across multiple time scales; the study of the relationship between the natural climate system and society and the development of methodologies for applying climate information to problems of social and economic consequences; and archiving, managing, and disseminating data and information useful for global change research. DOC's National Institute of Standards and Technology (NIST) provides measurements and standards that support accurate and reliable climate observations. NIST also performs calibrations and special tests of a wide range of instruments and measurement techniques for accurate measurements. NIST provides a wide array of data and modeling tools that provide key support to developers and users of complex climate prediction models.

Department of Defense (DOD)

DOD does not support dedicated global change research, but continues a history of participation in CCSP through sponsored research that concurrently satisfies national security requirements and stated CCSP goals. All data and research results are routinely made available to the civil science community. DOD science and technology investments are coordinated and reviewed through the Defense Reliance process and published annually in the Defense Science and Technology Strategy, the Basic Research Plan, the Defense Technology Area Research Plan, and the Joint Warfighting Science and Technology Plan.

Department of Energy (DOE)

Research supported by DOE's Office of Biological and Environmental Research (BER) is focused on the effects of energy production and use on the global Earth

system, primarily through studies of climate response. Research includes climate modeling, aerosol and cloud properties and processes affecting the Earth's radiation balance, and sources and sinks of energy-related greenhouse gases (primarily carbon dioxide). It also includes research on the consequences of climatic and atmospheric changes for ecological systems and resources, the development of improved methods and models for conducting integrated economic and environmental assessments of climate change and of options for mitigating climate change, and education and training of scientists for climate change research.

Department of Health and Human Services (HHS)

Four National Institutes of Health (NIH) institutes support research on the health effects of ultraviolet (UV) and near-UV radiation. Their principal objectives include an increased understanding of the effects of UV and near-UV radiation exposure on target organs (e.g., eyes, skin, immune system) and of the molecular changes that lead to these effects, and the development of strategies to prevent the initiation or promotion of disease before it is clinically defined. In addition, the National Institute of Environmental Health Sciences (NIEHS) supports research on the health effects of chlorofluorocarbon replacement chemicals, including studies on the metabolism and toxicity of hydrofluorocarbons and halogenated hydrocarbons. HHS (NIH and the Centers for Disease Control and Prevention) also conducts research related to other impacts of global change on human health, including renewed concern about infectious diseases whose incidence could be affected by environmental change. In addition, NIH sponsors a program to assess the impact of population change on the physical environment and to account for effects of the physical environment on population change.

BOX 16-1 (CONTINUED)

PRINCIPAL AREAS OF FOCUS FOR THE CCSP AGENCIES

Department of the Interior (DOI)
Research at DOI's U.S. Geological
Survey (USGS) contributes directly to
the CCSP's intellectual framework of a
whole-system understanding of global
change (i.e., the interrelationships
among climate, ecological systems, and
human behavior). USGS examines
terrestrial and marine processes and the
natural history of global change,
including the interactions between
climate and the hydrologic system.
Studies seek to understand the character
of past and present environments and
the geological, biological, hydrological,
and geochemical processes involved in
environmental change. USGS supports a
broad area of global change research,
with a focus on understanding the
sensitivity of natural systems and impacts
of climate change and variability,
surficial processes, and other global
change phenomena on the nation's lands
and environments at the regional scale.
Specific goals of the program are to
improve the utility of global change
research results to land management
agencies; to emphasize monitoring the
landscape and developing technical
approaches to identifying and analyzing
changes that will take advantage of a
burgeoning archive of remotely sensed
and *in situ* data; and to emphasize the
response of biogeographic regions and
features, particularly montane, coastal,
and inland wetland ecosystems.

Department of State (DOS)
Through DOS annual funding, the
United States is the world's leading
financial contributor to the United Nations
Framework Convention on Climate
Change and to the Intergovernmental
Panel on Climate Change (IPCC), a
major organization for the assessment of
scientific, technical, and socioeconomic
information relevant to the understanding
of climate change, its potential impacts,
and options for adaptation and mitigation.
Recent DOS contributions to the IPCC
provide substantial support for the
Global Climate Observing System,
among other activities.

**Department of
Transportation (DOT)**
DOT utilizes existing science to
improve decisionmaking tools in three
primary areas: (1) impact of climate
variability and change on transportation
(research to examine the effects that
climate change and variability may have
on transportation infrastructure and
services, and to identify potential
adaptation strategies for use by
transportation decisionmakers, operators,
state and local planners, and infrastructure
builders); (2) increasing energy efficiency
and reducing greenhouse gases (research
on reducing energy use will cover
mitigation of transportation's
environmental impacts both through
conservation and through the application
of new technology); and (3) modeling
(research to develop and improve
analytical tools for transportation energy
use to support decisionmaking throughout
government and in the private sector).

**Agency for International
Development (USAID)**
USAID provides decisionmakers with
the information to effectively respond
to drought and food insecurity through
the Famine Early Warning System
Network (FEWS NET). FEWS NET
analyzes remote-sensing data and
ground-based meteorological, crop, and
rangeland observations to track progress
of rainy seasons in semi-arid regions of
Africa in order to identify early
indications of potential famine.

**Environmental Protection
Agency (EPA)**
EPA's Global Change Research Program
is an assessment-oriented program with
primary emphasis on understanding the
potential consequences of climate
variability and change on human health,
ecosystems, and socioeconomic systems
in the United States. This entails: (1)
improving the scientific basis for
evaluating effects of global change on air
quality, water quality, ecosystems, and
human health in the context of other
stressors and in light of human dimensions

(as humans are catalysts of and respond
to global change); (2) conducting
assessments of the risks and opportunities
presented by global change; and (3)
assessing adaptation options to increase
resiliency to change and improve society's
ability to effectively respond to the risks
and opportunities presented by global
change. EPA's program emphasizes the
integration of the concepts, methods,
and results of the physical, biological,
and social sciences into decision support
frameworks.

**National Aeronautics and Space
Administration (NASA)**
The mission of NASA's Earth Science
Enterprise is to understand and protect
our home planet by using our view from
space to study the Earth system and
improve prediction of Earth system
change. NASA programs are aimed at
understanding the Earth system and
applying Earth system science to
improve prediction of climate, weather,
and natural hazards in partnership with
other federal agencies and international
space and research programs. Its
Research Strategy orchestrates observing
and modeling programs to address these
essential questions:
*   How is the Earth changing, and what
    are the consequences for life on
    Earth?
*   How is the global Earth system
    changing?
*   What are the primary causes of
    change in the Earth system?
*   How does the Earth system respond
    to natural and human-induced
    change?
*   What are the consequences of
    change in the Earth system for
    human civilization?
*   How well can we predict future
    changes in the Earth system?
NASA's portfolio includes observations,
research, analysis, modeling, and
advanced technology development, in
order to answer select science questions,
and benchmarking decision support
resources to ensure society receives the
benefits of this research.

BOX 16-1 (CONTINUED)

PRINCIPAL AREAS OF FOCUS FOR THE CCSP AGENCIES

National Science Foundation (NSF)

NSF programs address global change issues through investments in challenging ideas, creative people, and effective tools. In particular, NSF global change research programs support research and related activities to advance the fundamental understanding of physical, chemical, biological, and human systems and the interactions among them. The programs encourage interdisciplinary activities and focus particularly on Earth system processes and the consequences of change. NSF programs facilitate data acquisition and information management activities necessary for fundamental research on global change, and promote the enhancement of models designed to improve understanding of Earth system processes and interactions and to develop advanced analytic methods to facilitate basic research. NSF also supports fundamental research on the general processes used by organizations to identify and evaluate policies for mitigation, adaptation, and other responses to the challenge of varying environmental conditions.

Smithsonian Institution

Within the Smithsonian Institution, global change research is conducted at the Smithsonian Astrophysical Observatory, the National Air and Space Museum, the Smithsonian Environmental Research Center, the National Museum of Natural History, the Smithsonian Tropical Research Institute, and the National Zoological Park. Research is organized around themes of atmospheric processes, ecosystem dynamics, observing natural and anthropogenic environmental change on daily to decadal time scales, and defining longer term climate proxies present in the historical artifacts and records of the museums as well as in the geologic record at field sites. The Smithsonian Institution program strives to improve knowledge of the natural processes involved in global climate change, provide a long-term repository of climate-relevant research materials for present and future studies, and to bring this knowledge to various audiences, ranging from scholarly to the lay public. The unique contribution of the Smithsonian Institution is a long-term perspective—for example, undertaking investigations that may require extended study before producing useful results and conducting observations on sufficiently long (e.g., decadal) time scales to resolve human-caused modification of natural variability.

Mission agencies that can benefit from observations, methods, and information developed through CCSP need to have ready access to these resources (see Chapter 11).

After more than a decade of experience, the U.S. Global Change Research Program (USGCRP) provides the foundation for managing an interagency research program on complex climate and global change issues. The new cabinet-level management structure instituted by President Bush in 2002 helps to focus efforts addressing the challenges of improving science-based information to manage the risks and opportunities of variability and change in climate and related systems during the coming decade. Among the new management techniques to be employed are approaches for addressing gaps in research and integration capacity that do not fit easily into the activities of any particular agency. In addition, the program coordinates with the Climate Change Technology Program (CCTP) to address issues at the intersection of science and technology, such as evaluating approaches to sequestration, monitoring of anthropogenic greenhouse gas emissions, and energy technology development and market penetration scenarios.

Program Criteria

CCSP uses a problem-driven rather than a disciplinary approach in setting priorities and sequencing investments, identifying for early action and support those projects and activities that meet the following agreed-upon criteria:

- *Scientific or technical quality*. The proposed work must be scientifically rigorous as determined by peer review. Implementation plans will include periodic review by external advisory groups (both researchers and users).
- *Relevance to reducing uncertainties and improving decision support tools in priority areas*. Programs must substantially address one or more of the CCSP goals. Programs must respond to needs for scientific information and enhance informed discussion by all relevant stakeholders.
- *Track record of consistently good past performance and identified metrics for evaluating future progress*. Programs addressing priorities with good track records of past performance will be favored for continued investment to the extent that time tables and metrics for evaluating future progress are provided. Proposed programs that identify clear milestones for periodic assessment and documentation of progress will be favorably considered for new investment.
- *Cost and value*. Research should address Climate Change Research Initiative (CCRI)/USGCRP goals in a cost-effective way. Research should also be coordinated with and leverage other national and international efforts. Programs that provide value-added products to improve decision support resources will be favored.

The Climate Change Science Program Management Strategy

Management of CCSP involves five mechanisms:
- Executive direction by the cabinet-based management structure, including priority setting, management review, and accountability
- Program implementation by CCSP-participating agencies

- Coordinated planning and program implementation through interagency working groups
- External interactions for guidance, evaluation, and feedback
- Coordination and management support from an interagency office accountable to the CCSP interagency governing committee.

Interactions among those responsible for these five mechanisms are critical for improving the scientific planning, the effectiveness of interagency management, and the focus of climate and global change research to support governmental and non-governmental needs. CCSP will also employ guidance from the President's Management agenda to strengthen the implementation of this plan.

Executive Direction by the Cabinet-Based Management Structure

CCSP's executive direction is provided by the cabinet-based management structure introduced in Chapter 1 and shown in Figure 16-1. At the highest level, this structure includes the

Executive Office of the President, with policy review provided by a combined National Security Council (NSC), Domestic Policy Council (DPC), and National Economic Council (NEC) panel. The Committee on Climate Change Science and Technology Integration (CCCSTI), consisting of cabinet secretaries and agency heads, was developed to provide management oversight to the federal climate change science and technology programs. The Interagency Working Group on Climate Change Science and Technology (IWGCCST) reports to CCCSTI and consists of the Deputy/Under Secretaries (or the counterparts of these positions in non-cabinet agencies and offices). The working group provides oversight for both CCSP and CCTP (which develops and reviews climate technology programs within the federal government), and makes recommendations to CCCSTI about funding and program allocations, in order to implement a coordinated climate change science and technology program that will better support policy development.

Membership on CCSP's interagency governing body, which is chaired by the CCSP Director (a Department of Commerce

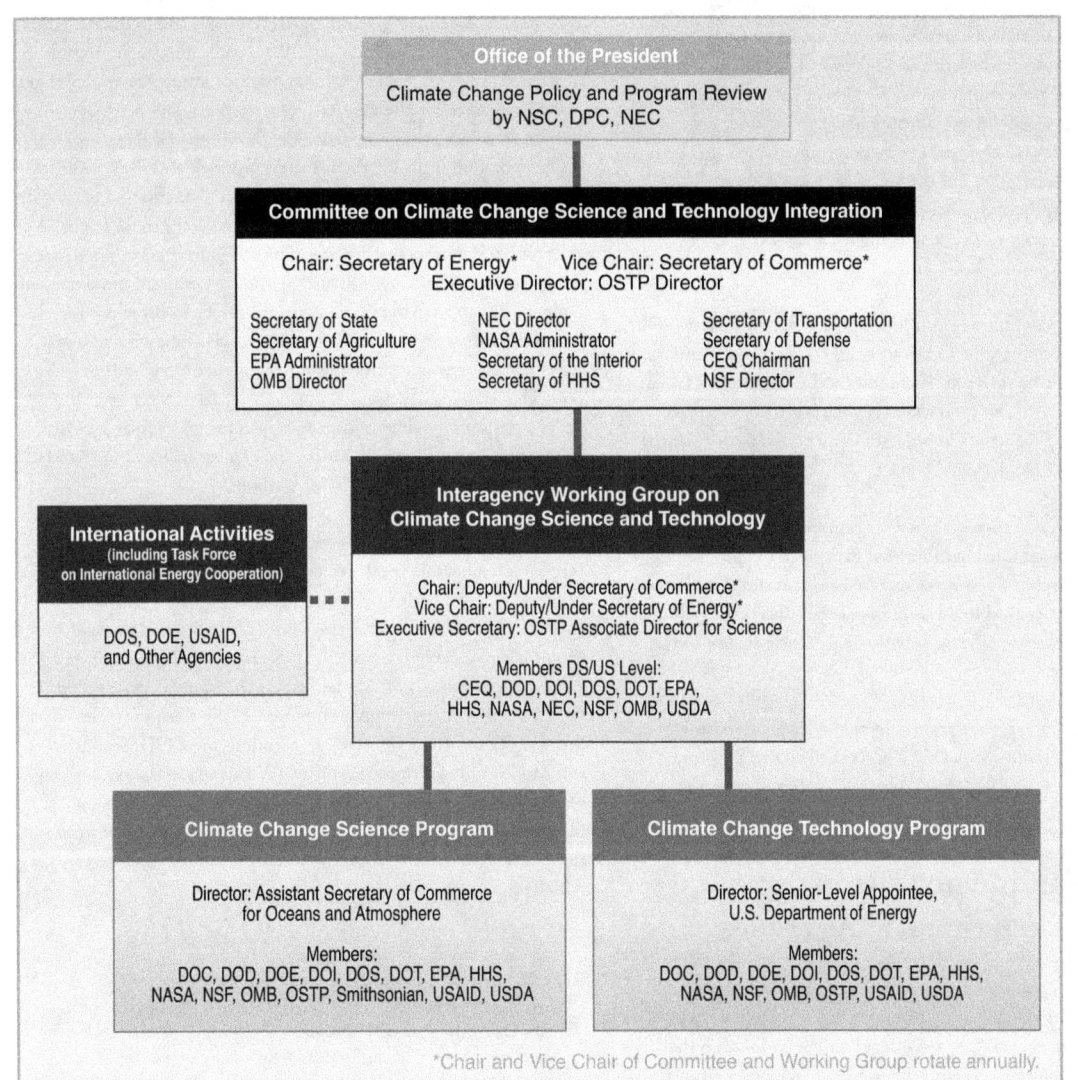

Figure 16-1: The Climate Change Science and Technology Programs are overseen by a cabinet-level management structure of the U.S. government.

appointee), is joint with the Subcommittee on Global Change Research (SGCR), the interagency committee that coordinates USGCRP. Its membership includes representatives from agencies that have mission responsibilities and/or funding in climate and global change research. USGCRP and CCRI are integrated within CCSP and responsibility for compliance with the requirements of the Global Change Research Act of 1990—including its provisions for annual reporting of finding and short-term plans, scientific reviews by the National Academy of Sciences/National Research Council, and periodic publication of a strategic plan for the program—rests with the CCSP's interagency governing body.

The CCSP interagency governing body provides overall management direction and is responsible for ensuring the development and implementation of an integrated interagency program. It oversees and directs all aspects of the program, including setting top-level goals for the program and determining what products will be developed and produced to meet those goals.

Agency representatives on the CCSP's interagency committee coordinate development of an integrated, interagency budget as a component of the President's annual budget request to Congress. They conduct periodic inventories and reviews of agency or departmental programs within the CCSP budget cross-cut (e.g., for the FY03 inventory, see <www.climatescience.gov>) and, in cooperation with IWGCCST and OMB, determine which agency programs are considered part of the CCSP budget cross-cut. They ensure that agency/departmental climate and global change research programs are prioritized and aligned with CCSP's interagency goals.

CCSP's governing committee ensures periodic program reviews and evaluations involving both the CCSP agencies and external partners, including the scientific research community and users of global change information. It also conducts periodic reviews of progress toward interagency objectives in order to evaluate the performance and effectiveness of the allocated budget.

The CCSP interagency governing body is responsible for ensuring the availability of scientific inputs needed to achieve the program's mission. It will develop and oversee mechanisms that support crucial research that is not central to the core missions of the participating agencies and that is most effectively and efficiently carried out in an interagency setting.

Finally, the CCSP interagency governing committee is responsible for coordinating activities with CCTP and other related programs; several members serve in leadership roles in coordinating committees for both CCSP and CCTP.

Program Implementation by CCSP-Participating Agencies

The goals and objectives of CCSP and the plans designed by the interagency working groups are carried out by the participating agencies, either individually or collaboratively. Each agency has its respective mission, capabilities, and budget authorizations, on the basis of which it commits to conduct research, make observations, run models, and generate products contributing to CCSP objectives.

Agency managers serve as chairs and members of the CCSP interagency working groups, facilitating the alignment of agency plans and CCSP implementing strategies. Agency managers pursue collaborative agreements with other CCSP agencies and other organizations as required to generate the products they commit to in the CCSP context.

Coordinated Planning and Implementation through Interagency Working Groups

At the implementation level, CCSP draws on the strengths of many agencies and departments. It requires a significant degree of coordination to ensure that the individual agency research conducted under the umbrella of CCSP supports program scientific objectives and that developments are effectively and efficiently synthesized and transferred into operational and sustained societal benefits.

Interagency Working Groups (IWGs) of program managers who have budget authority within their agencies to implement programs will oversee development of integrated science and implementation plans for each of the CCSP IWGs.[5] IWGs will commission preparation of science plans (see next section on external interactions) to provide overall guidance for the research, observations, and modeling needed to achieve program goals. IWGs will themselves prepare implementation plans for the research program elements, which will provide performance objectives. The implementation plans will describe the priority scientific programs necessary to meet goals, the roles of the participating agencies, and joint agency initiatives, where needed. The plans will also provide generalized timelines and budget estimates for the investments necessary to carry out the activities, noting any critical dependencies. The implementation plans will also include priorities for the observations and/or observing systems necessary to meet the goals of the research as well as critical modeling efforts and/or information management issues. These priorities will inform the choices that will need to be made by agencies and by CCSP as a whole.

Appropriate CCTP working groups will coordinate with CCSP through membership on the relevant CCSP interagency working group [e.g., CCTP Measurement, Monitoring, and Verification (MMV) Working Group members on CCSP Observations and Monitoring Working Group, CCTP Scenarios and Modeling Task Force member on CCSP Modeling and Decision Support groups, CCTP Sequestration group member on CCSP Carbon Cycle Working Group] and *vice versa*. In addition, the CCTP Scenarios and Modeling Task Force and the MMV and sequestration working groups will meet jointly on occasion with the CCSP Decision Support, Observations and Monitoring, and Carbon Cycle groups, respectively, to review program directions and potential joint initiatives.

---

[5] Currently, CCSP has research-oriented interagency working groups focusing on Atmospheric Composition, Climate Variability and Change, Global Water Cycle, Land-Use/Land-Cover Change, Global Carbon Cycle, Ecosystems, Human Contributions and Responses to Environmental Change, Decision Support, Modeling, Observations and Monitoring, International, and Data Management.

External Interactions for Guidance, Evaluation, and Feedback

The science community brings essential expertise to CCSP activities. CCSP recognizes the need to develop and utilize a variety of processes and mechanisms to provide an open and transparent process, program evaluation, and feedback. Scientific input will be sought from individuals from universities, federal research agencies, non-governmental organizations, and industry.

Relevant committees and boards of the National Academy of Sciences and the National Research Council (NRC) will continue to be asked to provide scientific guidance as appropriate. Since the establishment of USGCRP, NRC has provided a wealth of valued advice to USGCRP, leading to such documents as *Global Environmental Change: Research Pathways for the Next Decade* (NRC, 1999a), *Our Common Journey* (NRC, 1999c), *The Science of Regional and Global Change: Putting Knowledge to Work* (NRC, 2001e), and *Climate Change Science: An Analysis of Some Key Questions* (NRC, 2001a). Most recently, NRC has engaged in the review of the development of this strategic plan by participating in the CCSP Workshop in December 2002, and providing a review of the CCSP draft strategic plan. The NRC will provide an overall review of the CCSP Strategic Plan and its development process after the plan is published.

CCSP has considered the NRC recommendation that a permanent external advisory group be created to oversee CCSP performance. After careful review, CCSP believes that essential program oversight is better provided by the use of a number of external advisory mechanisms, including periodic overall program reviews by NRC or other groups, rather than a single body. Additional mechanisms to seek external scientific input—such as workshops, steering committees, ad hoc working groups, and review boards—will be employed as needed. CCSP will continue to consider creation of a permanent overall advisory group as program implementation proceeds.

The research community, in cooperation with users, will develop science plans for each research element of the program. These plans will describe in greater detail than is possible in this Strategic Plan the research that is required to address the questions in the research program elements. An example of a detailed science plan is the *U.S. Carbon Cycle Science Plan* (CCWG, 1999). This science plan was requested by several agencies participating in USGCRP and was developed by a Carbon and Climate Working Group that drew on the expertise of the U.S. carbon cycle science community through workshops. Scientific plans have been developed or are being developed for the other CCSP research program elements as well to guide research efforts.

Coordination on an ad hoc basis across advisory bodies of the participating agencies involved in CCSP and related programs, such as CCTP, will be encouraged by the CCSP's interagency governing body. This will help ensure that the respective advisory bodies of participating CCSP agencies are aware of the roles and responsibilities of each of the other agencies in the program. It will also enable the interagency governing body for CCSP to consult with an independent group of individuals who also serve as members of scientific advisory committees for the individual CCSP agencies.

Coordination and Management Support by an Interagency Office

The agencies participating in CCSP fund and supervise an interagency office—the Climate Change Science Program Office (CCSPO)—which fosters program development and coordination by supporting:

- Research coordination and integration
- Development, coordination, and integration of decision support resources
- Integration of agency activities in the areas of observing, monitoring, and data management
- Development and implementation of an interagency communications plan
- Secretariat support for the CCSP Director, the interagency governing body of CCSP, and related bodies.

Research Coordination and Integration. CCSPO staff will support interagency working groups responsible for coordinating major CCSP program elements. Specific responsibilities will include (1) helping to develop scientific strategies and implementation plans; (2) assisting in the coordination of element-specific planning and implementation; (3) assisting in supporting and managing interactions with scientific advisory groups; (4) assisting in development of program milestones/deliverables and approaches for monitoring progress toward objectives; and (5) providing general support including coordinating meetings, acting as a central point of contact, and responding to queries. In addition, CCSPO staff will take responsibility for helping to achieve better integration across the major research elements. Strides have been made in science coordination and integration in recent years in several areas of CCSP/USGCRP, including climate variability and change, the carbon cycle, and the water cycle.

Development, coordination, and integration of decision support resources. CCSPO staff will support the CCSP agencies/departments and interagency groups responsible for oversight and implementation of CCSP decision support activities. This will foster evaluation of differing approaches to decision support analyses. CCSP focuses on solutions that integrate climate change research results into decision support resources based on national and international priorities. CCSP coordinates agency activities employing peer-review solicitations to identify best-in-class decision support solutions. Interagency working groups benchmark the performance of decision support solutions associated with specific local, regional, national, and international areas of emphasis. CCSPO staff will also support interagency oversight and review processes, ensuring a feedback link between decision support and research planning.

Integration of agency activities in the areas of observing, monitoring, and data management. CCSPO will coordinate and facilitate the work of the interagency CCSP Observations and Monitoring and Data Management working groups. CCSPO staff for observing and monitoring and data management will work with the agencies to: (1) develop a strategy for the U.S. contributions to a global climate observation and monitoring system; (2) coordinate an interagency design of U.S. contributions to the global system; (3) coordinate and ensure interagency implementation of the system (i.e., moving from design to operations); and (4) help evaluate the development and performance of the system in meeting the needs of the science and decision support communities.

Development and Implementation of an Interagency Communications Plan. CCSPO will facilitate program reporting and communications. CCSPO staff coordinate preparation of the annual *Our Changing Planet* report to Congress and other stakeholders, as well as other reports as requested. They are responsible for designing and maintaining the program's websites (including the USGCRP and CCSP websites, and the web presence of related programs in areas such as the carbon cycle). CCSPO will also take responsibility for maintaining an inventory of agency communications efforts, and working with agency CCSP representatives to develop an interagency communications plan.

CCSPO also supports the Working Group on International Research and Cooperation.

Secretariat Support for the CCSP Director, the Interagency Governing Body of CCSP, and Related Bodies. Activities include:

- Providing support to the CCSP Director and program staff
- Preparing and staffing the meetings of CCSP/SGCR, including timely distribution of needed materials, preparation of agenda, recording decisions, and supporting follow-up actions
- Supporting the annual production of *Our Changing Planet*
- Preparing and compiling budgets and supporting materials for the office and other activities supported through the distributed cost mechanism
- Establishing and maintaining office infrastructure (computer network and peripherals; databases; records of CCSP and CCSPO activity)
- Coordinating U.S. government review of Intergovernmental Panel on Climate Change and other assessment reports.

CHAPTER 16 AUTHORS

Lead Authors
James R. Mahoney, DOC
Richard H. Moss, CCSPO
Chester Koblinsky, NASA and CCSPO

Contributors
Ghassem Asrar, NASA
Margaret S. Leinen, NSF
James Andrews, ONR
Mary Glackin, NOAA
Charles Groat, USGS
William Hohenstein, USDA
Linda Lawson, DOT
Melinda Moore, HHS
Patrick Neale, Smithsonian Institution
Aristides Patrinos, DOE
Emmy B. Simmons, USAID
Michael Slimak, USEPA
Harlan Watson, DOS

ANNEXES

Authors, Reviewers, and Workshop Participants

Mark R. Abbott	Oregon State University
Waleed Abdalati	National Aeronautics and Space Administration
Thomas P. Ackerman	Pacific Northwest National Laboratory
Larry Adams	U.S. Department of Agriculture
Thomas C. Adang	The Aerospace Corporation
Jeremy A. Adler	Canadian Embassy
Robert F. Adler	NASA Goddard Space Flight Center
Syun Akasofu	University of Alaska Fairbanks
Saromi Adedeji Akeem	Compotech Nigeria Limited
Marryl Alber	University of Georgia
Steven R Albersheim	Federal Aviation Administration
Bruce Alberts	National Academy of Sciences
Mariann Albjerg	NASA Goddard Space Flight Center
Daniel L. Albritton	NOAA Aeronomy Laboratory
Stephen C. Aldrich	Bio Economic Research Associates
Latonya Alexander	National Aeronautics and Space Administration
Arthur G. Alexiou	UNESCO/Intergovernmental Oceanographic Commission
David M. Allen	Climate Change Science Program Office
Douglas R. Allen	Naval Research Laboratory
Anthony D. Amato	University of Maryland
Stephen D. Ambrose	National Aeronautics and Space Administration
P. Foerd Ames	Ocean Wave Energy Company
Jeff Amthor	U.S. Department of Energy
Paul Anastas	Office of Science and Technology Policy
Cort Anastasio	University of California, Davis
Karen Andersen	National Aeronautics and Space Administration
Stephen Andersen	U.S. Environmental Protection Agency
Donald E. Anderson	National Aeronautics and Space Administration
James Anderson	Harvard University
Lee Anderson	University of Delaware
Ken Andrasko	U.S. Environmental Protection Agency
Jean-Claude Andre	NASA Earth System Modeling Framework Advisory Board
James K. Andreasen	U.S. Environmental Protection Agency
Jeffrey A. Andresen	Great Lakes Regional Climate Change Assessment/Michigan State University
James E. Andrews	Office of Naval Research
James R. Angel	Illinois State Water Survey
Richard A. Anthes	University Corporation for Atmospheric Research
Behrooz Arastoo	Natural Resources Higher Education of Semnan
Hugo Arevalo	Insituto del Mar del Perú
Richard Arimoto	New Mexico State University
Phillip A. Arkin	University of Maryland
Joan L. Aron	Science Communication Studies
Vicki Arroyo	Pew Center on Global Climate Change
Christo Artusio	U.S. Department of State
Dan E. Arvizu	CH2M HILL
Richard Aspinall	National Science Foundation
Ghassem R. Asrar	National Aeronautics and Space Administration
Cynthia Atherton	Lawrence Livermore National Laboratory
Carol L. Auer	National Oceanic and Atmospheric Administration
Stephen J. Auer	National Oceanic and Atmospheric Administration/OAR
Susan K. Avery	University of Colorado
Roni Avissar	Duke University
Peter W. Backlund	National Center for Atmospheric Research
David Bader	U.S. Department of Energy
Stephen Baenziger	Crop Science Society of America
Mitchell T. Baer	U.S. Department of Energy
William L. Baer	Space Imaging
Thomas J. Baerwald	National Science Foundation
Rick Bailey	Commonwealth Scientific and Industrial Research Organisation
Ronald Bailey	Reason Magazine
Ronald C. Baird	National Oceanic and Atmospheric Administration/OAR
Marcia Baker	University of Washington
Wayman Baker	NOAA/Nat'l Centers for Environmental Prediction
Dennis Baldocchi	University of California, Berkeley
Mark P. Baldwin	Northwest Research Associates, Inc.
Todd W. Baldwin	Island Press
Virgil C. Baldwin, Jr.	USDA Forest Service
Roger C. Bales	University of Arizona
Peter J. Balint	George Mason University
Roberta Balstad Miller	Center for Int'l Earth Science Information Network
Seema Balwani	Office of Senator Daniel K. Akaka
Gil Bamford	
Anjuli S. Bamzai	National Oceanic and Atmospheric Administration
Raymond J. Ban	The Weather Channel
Mary Barber	Ecological Society of America
Fairley J. Barnes	Los Alamos National Laboratory
Christopher D. Barnet	NASA/Goddard Space Flight Center
Tim P. Barnett	Scripps Institution of Oceanography
Thomas O. Barnwell, Jr.	U.S. Environmental Protection Agency
Jill Baron	U.S. Geological Survey
Enriqueta Barrera	National Science Foundation
Jennifer H. Barrett	Office of Representative Mark Udall
Ko Barrett	U.S. Agency for International Development
Eric J. Barron	Pennsylvania State University
Ana P. Barros	Harvard University
Roger Barry	University of Colorado, Boulder (CIRES)
Patrick J. Bartlein	University of Oregon
Gerald S. Barton	NOAA (retired)
Joseph L. Bast	The Heartland Institute
John J. Bates	NOAA/National Environmental Satellite, Data, and Information Service
Robert Bauer	NASA Goddard Space Flight Center
Eric J. Bayler	National Oceanic and Atmospheric Administration
Reginald A. Beach	Consortium for Oceanographic, Research and Education (CORE)
William H. Beasley	University of Oklahoma
Joe M. Becerra	U.S. Navy
Glynda A. Becker	U.S. Department of Commerce
Markus E. Becker	RWE
David Beecy	U.S. Department of Energy
Carolyn C. Bell	U.S. Geological Survey
Nancy Beller-Simms	University Corporation for Atmospheric Research/NOAA
Deborah R. Belvedere	NASA Goddard Space Flight Center
Michael Bender	Princeton University
Richard E. Benedick	Battelle Pacific Northwest National Laboratory
Timothy C. Benner	U.S. Environmental Protection Agency
Richard A. Benson	Los Alamos National Laboratory
Kathleen A. Beres	Orbital Sciences Corporation
Jonas Berg	Humboldt State University
Brian Berger	Space News
Wolfgang Berger	Scripps Institution of Oceanography, UCSD
Belle Bergner	University of Pennsylvania
Chris Bernabo	National Council for Science and the Environment
James Berner	Alaska Native Tribal Health Consortium
Leonard S. Bernstein	L. S. Bernstein & Associates, L.L.C.
Joe Berry	Carnegie Institution of Washington
Uma Bhatt	University of Alaska Fairbanks
Kathryn A. Bickel	U.S. Department of Agriculture
Eugene W. Bierly	American Geophysical Union
Robert Bindschadler	National Aeronautics and Space Administration
Richard A. Birdsey	USDA Forest Service
Jennifer L. Biringer	World Wildlife Fund
Ronald J. Birk	National Aeronautics and Space Administration
Charon M. Birkett	University of Maryland College Park
Jeanne M. Bisanz	Great Lakes Regional Climate Change Assessment
Suzanne S. Bishop	Arctic Research Consortium of the United States
Maurice L. Blackmon	National Center for Atmospheric Research
Rainer Bleck	Los Alamos National Laboratory
Paul Bledsoe	National Commission on Energy Policy
Jack Blevins	City of Fairfax, Virginia
Janine Bloomfield	Environmental Defense
Richard A. Blum	Integrity in Science Project
Richard Blumenthal	State of Connecticut
Tom A. Boden	Oak Ridge National Laboratory
Samuel W. Bodman	U.S. Department of Commerce
Bo Bodvarssi	Lawrence Berkley National Laboratory
Kerry D. Bolognese	National Association of State Universities and Land Grant Colleges
George D. Bolton	AgCert International, LLC
Suzanne H. Bolton	National Oceanic and Atmospheric Administration/NMFS
John Bonitz	Independent Reviewer
Brian Bonlender	Congressman Jay Inslee
James V. Bonta	USDA/Agricultural Research Service
Paula Bontempi	National Aeronautics and Space Administration
Davis B. Bookhart	Consumer Energy Council of America
James E. Bossert	Los Alamos National Laboratory
Lisa D. Botluk	National Oceanic and Atmospheric Administration
Sara R. Bowden	U.S. Global Change Research Program/International Program
Michael Bowers	National Science Foundation
Dan Braden	National Council for Science and the Environment
Raymond S. Bradley	University of Massachusetts, Amherst
Russell Bradley	PRBO Conservation Science
Raymond J. Braitsch	U.S. Department of Energy
Sarah K. Brandel	U.S. Department of State
Stephen B. Brandt	NOAA/Great Lakes Environmental Research Laboratory
Lee E. Branscome	Climatological Consulting Corporation
Marcia Branstetter	Oak Ridge National Laboratory
Garrett W. Brass	U.S. Arctic Research Commission
Richard C. Bratcher	Electric Power Research Institute

Hans G. Brauch	Free University of Berlin/AFES-PRESS
Troy Bredenkamp	American Farm Bureau Federation
Sean T. Brennan	U.S. Geological Survey
William J. Brennan	U.S. Department of Commerce
David D. Breshears	Los Alamos National Laboratory
Francis P. Bretherton	University of Wisconsin-Madison
Michael J. Brewer	National Weather Service
Lawson W. Brigham	U.S. Arctic Research Commission
Julie Brigham-Grette	IGBP-PAGES
Anthony J. Broccoli	NOAA/Geophysical Fluid Dynamics Laboratory
Eduardo Brondizio	Indiana University
Priscilla Brooks	Conservation Law Foundation
Dail W. Brown	National Marine Fisheries Service
Joel Brown	New Mexico State University
Judy P. Brown	Rio Tinto North American Group Companies
Louis B. Brown	National Science Foundation
Sandra Brown	Winrock International
W. Steven Brown	National Institute of Standards and Technology
William H. Brune	Pennsylvania State University
David Bttisti	University of Washington
Claire L. Buchanan	Interstate Commission on the Potomac River Basin
Katherine F. Buckley	U.S. Environmental Protection Agency
Shannon H. Buckley	Hart Downstream Energy Services
Janice L. Buckner	NASA Goddard Space Flight Center
Marilyn A. Buford	USDA Forest Service Research and Development
James L. Buizer	NOAA Office of Global Programs
Stas Burgiel	Defenders of Wildlife
Virginia R. Burkett	U.S. Geological Survey
Mark A. Burnham	Lewis-Burke Associates
George H. Burrows	Autonomous Undersea Systems Institute
Robert C. Burruss	U.S. Geological Survey
Antonio J. Busalacchi	Earth System Science Interdisciplinary Center (ESSIC)
James H. Butler	NOAA Climate Monitoring and Diagnostics Laboratory
Carsten Butsch	German National Committee on Global Change Research
Lee Hayes Byron	U.S. Climate Action Network
Daewon W. Byun	University of Houston
Terry Cacek	NASA Earth System Modeling Framework Advisory Board/National Park Service
Ming Cai	University of Maryland
Ken Caldeira	Lawrence Livermore National Laboratory
John A. Calder	National Oceanic and Atmospheric Administration
Francisco J. Calderon	USDA/Agricultural Research Service
William Russell Callender	DOC/NOAA/OAR/Office of Scientific Support
Stephen E. Calopedis	Energy Information Administration
David B. Campbell	National Science Foundation
Julie K. Campbell	The Campbell Marketing Group, Inc.
Penelope Canan	University of Denver
Mark A. Cane	Lamont Doherty Earth Observatory
Ralph D. Cantral	NOAA Ocean Service
Linda A. Capuano	Honeywell Inc.
Robert Card	U.S. Department of Energy
Ann B. Carlson	Executive Office of the President
Peter S. Carlson	MHPC Corporation
R. Jared Carpenter	CREA
David J. Carson	World Meteorological Organization
Lynne M. Carter	Environmental Consultant
Thomas B. Carter	Portland Cement Association
Eileen M. Casabianca	DOC/NOAA/OAR
Ben Cash	Center for Ocean-Land-Atmosphere Studies
Michael Caughey	Illinois State Water Survey
Billy D. Causey	NOAA/Florida Keys National Marine Sanctuary
Nancy Cavallaro	USDA/Cooperative State Research, Education, and Extension Service
Daniel R. Cayan	Scripps Institution of Oceanography, UCSD
Richard P. Cebula	Science Systems and Applications, Inc.
Christine Celata	Lawrence Berkeley National Laboratory
Kelly Chance	Smithsonian Astrophysical Observatory
Stanley A. Changnon	Illinois State Water Survey
Elaine G. Chapman	Pacific Northwest National Laboratory
Thomas W. Chapman	National Science Foundation
Chris Charles	Scripps Institution of Oceanography
Robert J. Charlson	University of Washington
Michael L. Charney	Massachusetts Climate Action Network
Eric Chassignet	NASA Earth System Modeling Framework Advisory Board
Keya Chatterjee	National Aeronautics and Space Administration
Chen-Tung A. Chen	National Sun Yat-sen University
R. Douglas Chiappetta, M.A.	National Alternative Fuels Association
Gary Chilefone	
Anne F. Choate	ICF Consulting
Elizabeth Chornesky	University of California, Santa Cruz
Bob Christian	East Carolina University
John R. Christy	University of Alabama in Huntsville
Margaret M. L. Chu	U.S. Environmental Protection Agency
Pamela M. Chu	National Institute of Standards and Technology

William P. Chu	NASA Langley Research Center
Thomas Church	University of Delaware
Alice T. Cialella	Brookhaven National Laboratory
Noreen Clancy	RAND Corporation
Candyce E. Clark	National Oceanic and Atmospheric Administration
Cathy Clark	University Corporation for Atmospheric Research
Martyn Clark	University of Colorado
Rebecca Clark	National Institutes of Health
William C. Clark	Harvard University
Eileen Claussen	Pew Center on Global Climate Change
Mary L. Cleave	National Aeronautics and Space Administration
Pablo Clemente-Colón	NOAA/National Environmental Satellite, Data, and Information Service
Nicholas L. Clesceri	National Science Foundation and RPI
James A. Coakley	Oregon State University
Stella Melugin Coakley	Oregon State University
Roz E. Cohen	National Oceanic and Atmospheric Administration
Kenneth A. Colburn	Northeast States for Coordinated Air Use Mgmt
Alessandro Coletti	TITAN Systems Corp.
Kimberly N. Collell	American Soybean Association
Nancy S. Colleton	Institute for Global Environmental Strategies
Jackson D. Collier	National Aeronautics and Space Administration
Stan Coloff	U.S. Geological Survey
Rita Colwell	National Science Foundation
Josefino Comiso	NASA Goddard Space Flight Center
Susan G. Conard	USDA Forest Service
Margarita Conkright Gregg	NOAA/National Oceanographic Data Center
Joseph M. Conny	National Institute of Standards and Technology
Dawn M. Conway	Canadian Foundation for Climate and Atmospheric Sciences
Gary A. Cook	Greenpeace
Todd Phillip Cook	Lakeview Terrace Sanitarium
Charli Coon	The Heritage Foundation
Philip A. Cooney	White House Council on Environmental Quality
James N. Cooper	Earth Satellite Corporation
Ellen J. Cooter	U.S. Environmental Protection Agency
Rob Coppoch	
Robert W. Corell	American Meteorological Society and Harvard University
John G. Cormier	National Institute of Standards and Technology
Peter Cornillon	University of Rhode Island
Roman Michail Corobov	Institute of Geography, Moldavian Academy of Sciences
Christine Corwin	Bluewater Network
Ellis Cowling	North Carolina State University
Paul Craig	Sierra Club Global Warming and Energy Committee
Rachael G. Craig	National Science Foundation
Jonah B. Crawford	Stirling Energy Systems
Ted S. Cress	DOE/Pacific Northwest Laboratory
David Crisp	Jet Propulsion Laboratory/California Institute of Technology
Thomas M. Cronin	U.S. Geological Survey
Thomas J. Crowley	Duke University
Richard M. Cruse	Iowa State University
Ivan A. Csiszar	University of Maryland
Bill J. Cunningham	Unions for Jobs and the Environment (UJAE)
Thomas A. Curran	Sippican, Inc.
William B. Curry	Woods Hole Oceanographic Institution
Michael R. Curtis	U.S. Department of Energy
Robert M. Cushman	Oak Ridge National Laboratory
Ned Cyr	NOAA Fisheries
Harvey D. Dahljelm	ITT Industries
Roger C. Dahlman	U.S. Department of Energy
H. Daily	Independent Researcher
Virginia H. Dale	Oak Ridge National Laboratory
Marie Chantale Damas	National Science Foundation
William P. Dannevik	Lawrence Livermore National Laboratory
H. Lee Dantzler	DOC/NOAA/NESDIS
Margaret A Davidson	NOAA Ocean Service
Kert Davies	Greenpeace
Anthony B. Davis	Los Alamos National Laboratory
Joseph A. Davis	Society of Environmental Journalists
Danny M. Day	Eprda
Barbara M. De Rosa-Joynt	U.S. Department of State
Gitane De Silva	Canadian Embassy
Russell De Young	NASA Langley
Helvecio De-Polli	Embrapa Labex/ARS
Lois Dean	U.S. Department of Housing and Urban Development
Robert T. Deck	University of Toledo
Cynthia J. Decker	Office of the Oceanographer of the Navy
Philip L. DeCola	National Aeronautics and Space Administration
Ruth DeFries	University of Maryland
Victor E. Delnore	NASA Langley Research Center
John J. DeLuisi	National Oceanic and Atmospheric Administration
Thomas L. Delworth	NOAA/Geophysical Fluid Dynamics Laboratory
Scott Denning	Colorado State University
John G. Dennis	NASA Earth System Modeling Framework Advisory Board/National Park Service

Lawrence M Denton	The Weather Channel
Matthew S. DeRosier	Northrop Grumman
Paul V. Desanker	University of Virginia
Floyd DesChamps	U.S. Senate Committee on Commerce, Science, and Transportation
Mike Dettinger	U.S. Geological Survey
Russell DeYoung	NASA Langley
Atulya Dhungana	Independent
Howard J. Diamond	NOAA/National Environmental Satellite, Data, and Information Service
Hilda Diaz-Soltero	U.S. Department of Agriculture Forest Service
Robert Dickinson	Georgia Institute of Technology, American Geophysical Union
Nancy M. Dickson	Harvard University
Thomas Dietz	George Mason University
Lisa Dilling	National Center for Atmospheric Research
John Dilustro	University of Georgia
Joseph P. DiPietro	Energetics, Incorporated
Paul Dirmeyer	Center for Ocean, Land and Atmosphere
Gerald J. Dittberner	NOAA/National Environmental Satellite, Data, and Information Service
Keith W. Dixon	U.S. Department of Commerce / NOAA / GFDL
Larisa E. Dobriansky	U.S. Department of Energy
Richard D. Doctor	Argonne National Laboratory
David J. Dokken	U.S. Global Change Research Program/CCSP
Randall M. Dole	NOAA Climate Diagnostics Center
Dennis C. Doll	U.S. Environmental Protection Agency
J.R. M. Donald	AETC Inc.
Scott Doney	Woods Hole Oceanographic Institution
Martin J. Donohoe	Independent
John W. Doran	USDA-ARS/Soil Science Society of America
Craig Dorman	University of Alaska Fairbanks
Paul Doskey	Argonne National Laboratory
Bruce C. Douglas	Florida International University
David H. Douglass	University of Rochester
Harry J. Dowsett	U.S. Geological Survey
JR Drabick	Environmental and Energy Study Institute
Hank Drahos	NOAA/National Environmental Satellite, Data, and Information Service
John B. Drake	Oak Ridge National Laboratory
Robert S. Drozdzal	Computer Sciences Corporation
Charles J. Drummond	U.S. Department of Energy
Qingyun Duan	NOAA/National Weather Service
Manvendra K. Dubey	Los Alamos National Laboratory
Melissa L. Dubinsky	Stone and Webster Management Consultants, Inc.
Robert Duce	Texas A&M University
Claude E. Duchon	University of Oklahoma
James L. Duda	IPO/NPOESS
Chris Duffy	Pennsylvania State University
Philip B. Duffy	Lawrence Livermore National Laboratory
J.C. Duh	National Aeronautics and Space Administration
Robert J. Dumont	National Oceanic and Atmospheric Administration
Stephen V. Dunn	U.S. Environmental Protection Agency
Lisa F. Duriancik	USDA/CSREES
Margaret E. Dushin	University of Minnesota
Ellsworth Dutton	NOAA Climate Monitoring and Diagnostics Laboratory
Jon D. Dykstra	Earth Satellite Corporation
C. Mark Eakin	NOAA/National Climate Data Center
Richard C. Easter	Pacific Northwest National Laboratory
David R. Easterling	NOAA/National Environmental Satellite, Data, and Information Service
William E. Easterling III	The Pennsylvania State University
Cheryl Eavey	National Science Foundation
Myron Ebell	Competitive Enterprise Institute
Kristie L. Ebi	Electric Power Research Institute
Frank Eden	Eden Consulting
Susanna Eden	Climate Change Science Program Office
N. Terence Edgar	U.S. Geological Survey
Kodjovi Sidéra Edjame	University of Lomé
James A. Edmonds	Pacific Northwest National Laboratory
Marta Edwards	Compliance Solutions Corporation
Steven Edwards	National Oceanic and Atmospheric Administration
Kenneth Eggert	Los Alamos National Laboratory
Franco Einaudi	NASA Goddard Space Flight Center
Bernard P. Elero	Harris Corporation
Chris Elfring	The National Academies
Scott M. Elliott	Los Alamos National Laboratory
Ann L. Elsen	Montgomery County Department of Environmental Protection
Chris Elvidge	NOAA-NESDIS National Geophysical Data Center
Jerry W. Elwood	U.S. Department of Energy
William R. Emanuel	University of Virginia
David B. Enfield	NOAA AOML
Jürgen Engelhard	RWE
Jared K. Entin	National Aeronautics and Space Administration
A. Christine Eppstein	Environmental Council of the States
Paul R. Epstein	Center for Health and the Global Environment
David J. Erickson	Oak Ridge National Laboratory
Mary C. Erickson	National Oceanic and Atmospheric Administration
Jaime Esper	National Aeronautics and Space Administration

Robert Etkins	National Oceanic and Atmospheric Administration
Diane L. Evans	NASA/Jet Propulsion Laboratory
John T. Everett	UN Atlas of the Oceans
David Fahey	NOAA Aeronomy Laboratory
William L. Fang	Edison Electric Institute
Mike Farrell	Oak Ridge National Laboratory
Sherry O. Farwell	South Dakota School of Mines and Technology
Stephen P. Faulkner	U.S. Geological Survey
Kevin Fay	International Climate Change Partnership
Richard A. Feely	National Oceanic and Atmospheric Administration
Robert W. Fegley	EPA/Research and Development
Jay S. Fein	National Science Foundation
Howard J. Feldman	American Petroleum Institute
John C. Felmy	American Petroleum Institute
Murray Felsher	Washington Remote Sensing Letter
Benjamin Felzer	Marine Biological Laboratory
Ed J. Fendley	U.S. Department of State
George R. Fenske	Argonne National Laboratory
Ron Ferek	Office of Naval Research
Frank M. Ferrell	U.S. Department of Energy
Wanda R. Ferrell	U.S. Department of Energy
Anton C. Ferrer	Earth Scout International
Florence M. Fetterer	National Snow and Ice Data Center
Christopher B. Field	Carnegie Institute
Ann Fisher	Pennsylvania State University
Genene M. Fisher	Climate Change Science Program
Matt Fladeland	National Aeronautics and Space Administration
Brian P. Flannery	Exxon Mobil Corporation
Paul G. Flikkema	Northern Arizona University
Ruxandra M. Floroiu	The National Academies
Lawrence E. Flynn	U.S. Department of Commerce/NOAA
Gary Foley	U.S. Environmental Protection Agency
Jonathan A. Foley	Institute for Environmental Studies, University of Wisconsin
Peter F. Folger	American Geophysical Union
Jennifer E. Folta	International City/County Management Association
Kathleen S. Fontaine	NASA Goddard Space Flight Center
Keith J. Forbes	Fundação da Faculdade de Ciências da Universidade da Lisboa
Ned Ford	Sierra Club
Michael A. Fortune	Climate Science Forum
Ariele Foster	SustainUS
Ian Foster	NASA Earth System Modeling Framework Advisory Board
Joshua G. Foster	University Corporation for Atmospheric Research/NOAA
Guido F. Franco	California Energy Commission
Gerald T. Fraser	National Institute of Standards and Technology
Edward N. Frazier	TRW
Melissa P. Free	National Oceanic and Atmospheric Administration
J. Randall Freed	ICF Consulting
Andrew C. Freedman	Congressional Quarterly
Kenneth P. Freeman	Caelum Research Corporation
Amy J. Freise	International START Secretariat
Lars Friberg	Johns Hopkins University, SAIS
Elbert (Joe) W. Friday	National Academies, National Research Council
Julie Friddell	U.S. Army Corps of Engineers
Lawrence Friedl	National Aeronautics and Space Administration
Robert M. Friedman	The Heinz Center
Deborah Frisch	NSG
Jeffrey B. Frithsen	U.S. Environmental Protection Agency
J. Fruci	United States House of Representatives
Peter C. Frumhoff	Union of Concerned Scientists
Roland J. Fuchs	International START Secretariat
William Fulkerson	LERDWG
Inez Fung	University of California, Berkeley
John Furlow	U.S. Environmental Protection Agency
Clifford J. Gabriel	Office of Science and Technology Policy
Najwa Abbas Gadaheldam	UNIDO/ Energy Efficiency Branch
Mary Gade	Sonnenschein, Nath, and Rosenthal
Michael Gaffen	Datum International
Jeffrey S. Gaffney	Argonne National Laboratory
Peter T. Gage	World Resources Institute
William B. Gail	Ball Aerospace and Technologies Corporation
Gerardo V. Galamay	Volunteers Movement For Green Concerns
James Galloway	University of Virginia
L. Peter Galusky	Marathon Ashland Petroleum LLC
Janet L. Gamble	U.S. Environmental Protection Agency
Susan M. Gander	U.S. Environmental Protection Agency
Mary M. Gant	NIH/National Institute of Environmental Health Sciences
Yuan Gao	Princeton University
Jurgen D. Garbrecht	U.S. Department of Agriculture
Hernan E. Garcia	National Oceanic and Atmospheric Administration
Martha N. Garcia	U.S. Geological Survey
Edward P. Gardiner	American Museum of Natural History
David K. Garman	U.S. Department of Energy
Charles T. Garten, Jr.	Oak Ridge National Laboratory

Silvia L. Garzoli	National Oceanic and Atmospheric Administration
Lawrence W. Gates	Lawrence Livermore National Laboratory
Lydia Dumenil Gates	Lawrence Berkeley National Laboratory
Stephen D. Gawtry	University of Virginia
Gerald L. Geernaert	Los Alamos National Laboratory
Charlette A. Geffen	Pacific Northwest National Laboratory
Robert P. Gehri	Southern Company
Melvyn E. Gelman	National Oceanic and Atmospheric Administration
Bruce M. Gentry	National Aeronautics and Space Administration
Dylan George	National Science Foundation
Paul J. Georgia	Competitive Enterprise Institute
Janette C. Gervin	NASA Goddard Space Flight Center
Steven Ghan	Pacific Northwest National Laboratory
Anver Ghazi	European Commission
Michael Ghil	University of California, Los Angeles
Henry Gholz	National Science Foundation/Division of Environmental Biology
Anne Giblin	Marine Biological Laboratory
Lewis E. Gilbert	Columbia University
Ron S. Gird	NOAA National Weather Service
Anatoly A. Gitelson	University of Nebraska Lincoln
Mary M. Glackin	NOAA/National Environmental Satellite, Data, and Information Service
Karl M. Glasener	Soil Science Society of America
James F. Gleason	National Aeronautics and Space Administration
Peter Gleick	Pacific Institute for Studies in Development, Environment, and Security
Patricia Glick	National Wildlife Federation
Anand Gnanadesikan	NOAA/Geophysical Fluid Dynamics Lab
David S. Godwin	U.S. Environmental Protection Agency
Jeff Goebel	U.S. Department of Agriculture
Scott J. Goetz	University of Maryland
Roberto Goidanich	Brazilian Embassy in Washington
Alan M. Goldberg	The MITRE Corp.
Donald M. Goldberg	Center for International Environmental Law
Mitchell D. Goldberg	NOAA/National Environmental Satellite, Data, and Information Service
Stephen M. Goldberg	Argonne National Laboratory
Edward S. Goldstein	National Aeronautics and Space Administration
Rachel Goldstein	George Washington University Graduate Program
David J. Goldston	United States House of Representatives
Janet R. Gomon	Smithsonian Institution
Patrick Gonzalez	U.S. Geological Survey
Margaret G. Goodbody	District of Columbia Public Library
David C. Goodrich	USDA/Agricultural Research Service
David M. Goodrich	NOAA Climate Office
Steven M. Goralczyk	TRW
William D. Goran	U.S. Army Corps of Engineers
Thomas E. Graedel	Yale University
Thomas J. Grahame	U.S. Department of Energy
Anne E. Grambsch	U.S. Environmental Protection Agency
Mads Greauer	Resources for the Future
Kenneth Green	Fraser Institute
Richard S. Greenfield	American Meteorological Society
Suzie E. Greenhalgh	World Resources Institute
Greg Greenwood	California Resources Agency
Gregory B. Greenwood	Resources Agency of California
Robert A. Griffin	University of Alabama
Vanessa L. Griffin	NASA Goddard Space Flight Center
Roger B. Griffis	National Oceanic and Atmospheric Administration
David J. Griggs	Hadley Centre for Climate Prediction and Research
Stanley H Grigsby	Virginia Tech
William F. Griswold	University of Georgia
Charles Groat	U.S. Geological Survey
Robert Grossman	NorthWest Research Associates (NWRA)
Wolf D. Grossmann	Center for Environmental Research
Arnold Gruber	NOAA/National Environmental Satellite, Data, and Information Service
Hermann Gucinski	USDA Forest Service
Scott B. Gudes	National Oceanic and Atmospheric Administration
Franklin R. Guenther	National Institute of Standards and Technology
Manoj K. Guha	American Electric Power
Anthony R. Guillory	National Aeronautics and Space Administration
Mohan Gupta	NASA/Goddard Space Flight Center
Guy G. Guthridge	National Science Foundation
Garik Gutman	National Aeronautics and Space Administration
Shahid Habib	NASA/Goddard Space Flight Center
James Hack	National Center for Atmospheric Research
Peter Hacker	University of Hawaii
Charles S. Hakkarinen	Consultant
Ida M. Hakkarinen	NOAA/National Environmental Satellite, Data, and Information Service
Sirpa M.A. Hakkinen	NASA Goddard Space Flight Center
Robert C. Hale	University of Oklahoma
Michael B. Hales	NOAA/National Environmental Satellite, Data, and Information Service
Dorothy K. Hall	NASA Goddard Space Flight Center
Forrest G. Hall	University of Maryland, Baltimore County
Colin K. Haller	International City/County Management Association (ICMA)
Richard E. Hallgren	American Meteorological Society

David Halpern	Office of Science and Technology Policy
Kevin Hamilton	University of Hawaii
Annette P. Hanada	George Mason University
Steve Hankin	NOAA/PMEL
Lisa J. Hanle	U.S. Department of Energy
Lee Hannah	Center for Applied Biodiversity Science
Bryan J. Hannegan	United States Senate
James M. Hannon	Sippican Inc.
Lara J. Hansen	World Wildlife Fund
James E. Hansen	NASA Goddard Institute for Space Studies
Penelope Hansen	Environment and Energy Study Institute (EESI)
Robert C. Hansen	National Oceanic and Atmospheric Administration
Howard P. Hanson	Los Alamos National Laboratory
Paul J. Hanson	Oak Ridge National Laboratory
Penny Hanson	Environmental and Energy Study Institute
Singy Hanyona	Green Living Movement
Pat Harcourt	Waquoit Bay National Estuarine Research Reserve
Douglas R. Hardy	University of Massachusetts
Lowry A. Harper	USDA-Agricultural Research Service
Stephanie Harrington	National Oceanic and Atmospheric Administration
Susan W Harris	National Oceanic and Atmospheric Administration
Ed Harrison	NOAA/PMEL
Edwin F. Harrison	NASA Langley Research Center
Jake Haselswerdt	Lewis-Burke Associates
John C. Haskins	Moss Landing Marine Laboratories
Jerry L. Hatfield	USDA/Agricultural Research Service
John R. Haugh	Bureau of Land Management
Karl Hausker	PA Consulting Group
Katharine Hayhoe	ATMOS Research and Consulting
Anthony D. Haymet	CSIRO
John A Haynes	National Aeronautics & Space Administration
Badra Sriyani Hearath	University of Sri Jayawardhanapura
Sarah V. Heath	GF Energy, LLC
Michael A. Heeb	National Oceanic and Atmospheric Administration
Steve Heibein	
BertJan Heij	Netherlands Climate Change Research Programme
Douglas W. Heim	NEDO
Isaac M. Held	NOAA/Geophysical Fluid Dynamics Laboratory
Edward Helminski	National Carbon Sequestration Conference
George R. Hendrey	Brookhaven National Laboratory
Blair C. Henry	University of North Dakota
Robert T. Henson	University Corporation for Atmospheric Research
Einar-Arne Herland	European Space Agency - ESTEC
Charles N. Herrick	Stratus Consulting Inc.
Antonia Herzog	Natural Resources Defense Council
Shannon D. Heyck-Williams	Senate Environment & Public Works Committee
Joseph S. Hezir	EOP Group, Inc.
Thomas J. Hickey	Raytheon
Clifford A. Hickman	USDA Forest Service
Winston H. Hickox	California Environmental Protection Agency
Bruce B. Hicks	National Oceanic and Atmospheric Administration
Jennifer Hicks	Chesapeake Climate Action Network
Wayne Higgins	Climate Prediction Center/NOAA/NWS/NCEP
Steven D. Hilberg	Illinois State Water Survey
Peter H. Hildebrand	NASA Goddard Space Flight Center
Steve G. Hildebrand	Oak Ridge National Laboratory
Harvey S. Hill	National Aeronautics and Space Administration
Dexter Hinckley	Retired
Brian R. Hinman	U.S. Climate Action Network
Larry Hinzman	University of Alaska Fairbanks
Leonard P. Hirsch	Smithsonian Institute
Porter Hoagland	WHOI
Stacey Hockett Sherlock	University of Maryland
Robert Hoeft	American Society of Agronomy
Martin I. Hoffert	New York University
David J. Hofmann	NOAA Climate Monitoring and Diagnostics Laboratory
William G. Hohenstein	U.S. Department of Agriculture
Patrick W. Holden	Waterborne Environmental, Inc.
Eric K. Holdsworth	Edison Electric Institute
Elisabeth A. Holland	National Center for Atmospheric Research
David Y. Hollinger	USDA Forest Service
Steven E. Hollinger	Illinois State Water Survey
Tony Hollingsworth	European Centre for Medium-Range Weather Forecasts
Connie D. Holmes	National Mining Association
Susan H. Holte	U.S. Department of Energy
John L. Hom	USDA Forest Service
Maria Hood	Intergovernmental Oceanographic Commission - UNESCO
William H. Hooke	American Meteorological Society
Robert P. Hopkins	National Oceanic and Atmospheric Administration
Eli W. Hopson	House Science Committee
Christopher C. Horner	Competitive Enterprise Institute
Fiona M. C. Horsfall	National Weather Service
Linda L. Horton	Oak Ridge National Laboratory

Andrew H. Horvitz	NOAA/National Weather Service
Wayne Hosking	Horne Engineering
John C. Houghton	U.S. Department of Energy
Richard A. Houghton	Woods Hole Research Center
Thomas F. Hourigan	NOAA Fisheries
Paul R. Houser	NASA Goddard Space Flight Center
Bob Howarth	Cornell University
Douglas Hoyt	Retired
James M. Hrubovcak	U.S. Department of Agriculture
Jin Huang	National Oceanic and Atmospheric Administration
Ho Chun Huang	Illinois State Water Survey
Robert E. Huie	National Institute of Standards and Technology
Allen G. Hunt	National Science Foundation
Thomas G. Huntington	U.S. Geological Survey
James Hurrell	University Corporation Atmospheric Research
George C. Hurtt	University of New Hampshire
Robert C. Hyman	Cambridge Systematics
Helen Ingram	University of California, Irvine
Ron Isaacs	Atmospheric and Environmental Research, Inc. (AER)
Eric C. Itsweire	National Science Foundation
R. Cesar Izaurralde	Joint Global Change Research Institute
Clifford A. Jacobs	National Science Foundation
Daniel J. Jacob	Harvard University
Katharine L. Jacobs	Arizona State Department of Water Resources
Henry D. Jacoby	Massachusetts Institute of Technology
L. Douglas James	National Science Foundation
Anthony C. Janetos	H. John Heinz, III Center for Science, Economics, and the Environment
Jack E. Janisch	Washington Department of Ecology
Steven Japar	Ford Motor Company
Emily A. Jarvis	Senator Snowe's Office
Julie D. Jastrow	Argonne National Laboratory
Pincas Jawetz	New York University
Michael D. Jawson	USDA/Agricultural Research Service
David A. Jay	Oregon Health & Science University
Tara Jay	University Corporation for Atmospheric Research (UCAR)
Steven Jayne	Woods Hole Oceanographic Institution
Ming Ji	National Oceanic and Atmospheric Administration
Gensuo J. Jia	University of Virginia
J. Scott Jiusto	George Perkins March Institute
Anne K. Johnson	U.S. General Accounting Office
B. Carol Johnson	National Institute of Standards and Technology
Mark Johnson	University of Alaska Fairbanks
Michael R. Johnson	NOAA Office of Global Programs
Gary H. Johnston	National Park Service
Amber L. Jones	National Science Foundation
Carol A. Jones	U.S. Department of Agriculture
James Jones	University of Florida
Philip W. Jones	Los Alamos National Laboratory
Russell O. Jones	American Petroleum Institute
Sonja B. Jones	Oak Ridge National Laboratory
Linda A. Joyce	Rocky Mountain Research Station
Glenn Patrick Juday	University of Alaska Fairbanks
Lesley Julian	Retired National Weather Service
Susan Julius	USEPA/Office of Research and Development
Chris O. Justice	University of Maryland
John R. Justus	Library of Congress - Congressional Research Service
Tina M. Kaarsberg	United States Congress Science Committee
Menas Kafatos	George Mason University
Ramesh K. Kakar	National Aeronautics and Space Administration
Eugenia Kalnay	University of Maryland
Sally M. Kane	National Science Foundation
Ben J. Kangasniemi	British Columbia Ministry of Water, Land and Air Protection
Zoe H. Kant	The Nature Conservancy
Marlene Kaplan	National Oceanic and Atmospheric Administration
Jeffrey S. Kargel	U.S. Geological Survey
Dave Karl	University of Hawaii
Thomas R. Karl	NOAA/National Climatic Data Center
Miriam Kastner	Scripps Institution of Oceanography, UCSD
Seiji Kato	Hampton University
Kelli L. Kay	American Legislative Exchange Council
Leslie Kay	National Aeronautics and Space Administration
Jack A. Kaye	National Aeronautics and Space Administration
Terry J. Keating	U.S. Environmental Protection Agency Office of Air and Radiation
Alvin E. Keaton	Senior Scientists and Engineers
Michael J. Keeler	Ecologic Solution
Ralph Keeling	Scripps Institution of Oceanography
Timothy R.E. Keeney	National Oceanic and Atmospheric Administration
Stephen E. Kelleher	World Wildlife Fund-WWF
Bill Keller	Ontario Ministry of the Environment
John J. Kelly, Jr.	NOAA/National Weather Service
Walter B. Kelly	U.S. Department of State
Willett Kempton	University of Delaware
Charles F. Kennel	Scripps Institution of Oceanography, UCSD

John L. Kermond	University Corporation for Atmospheric Research/NOAA
William Kessler	NOAA Pacific Marine Environmental Laboratory
Ehsan U. Khan	U.S. Department of Energy
Reza M. Khanbilvardi	City University of New York
Haroon S. Kheshgi	Exxon Mobil Research and Engineering Company
Margaret Kieffer	National Aeronautics and Space Administration
Jeffrey Kiehl	National Center for Atmospheric Research
Timothy Killeen	National Center for Atmospheric Research
Edward J. Kim	National Aeronautics and Space Administration
John Kimble	U.S. Department of Agriculture
Anthony W. King	Oak Ridge National Laboratory
Michael D. King	NASA Goddard Space Flight Center
Roger King	National Aeronautics and Space Administration
James L. Kinter	Institute of Global Environment and Society
William Kirby	U.S. Geological Survey
S.M. Kirschner	Union of Concerned Scientists
David Kirtland	U.S. Geological Survey
David E. Kissel	University of Georgia
Michael J. Klassen	Science Applications Int'l Corporation
Steven L. Kleespie	Rio Tinto
Axel Kleidon	University of Maryland
George A. Klouda	National Institute of Standards and Technology
Robert G. Knox	NASA Goddard Space Flight Center
Malcolm K. Ko	NASA Langley Research Center
Chester J. Koblinsky	NASA and CCSPO
Kristen Koch	DOC/NOAA/OAR/Office of Scientific Support
Louisa Koch	National Oceanic and Atmospheric Research (NOAA)
Drew K. Kodjak	National Commission on Energy Policy
Russell Koffler	NOAA (Retired); Consultant
Felix Kogan	National Oceanic and Atmospheric Administration/NESDIS
Christine R Kojac	House Appropriations Committee
Charles Kolstad	Massachusetts Institute of Technology
Karen H. Koltes	U.S. Department of the Interior
George J. Komar	NASA/Goddard Space Flight Center
Burton H. Koske	Idaho National Engineering and Environmental Laboratory (INEEL)
Thomas Krafft	German National Committee on Global Change Research
Richard J. Kramer	DynCorp
Kristen L. Krapf	National Research Council
Jamie M. Krauk	National Oceanic and Atmospheric Administration
David A.R. Kristovich	Illinois State Water Survey
Herbert W. Kroehl	National Oceanic and Atmospheric Administration
Dina Kruger	U.S. Environmental Protection Agency
Edward W. Kruse	National Oceanic and Atmospheric Administration
Jeff Kueter	George Marshall Institute
Jeff R. Kuhn	University of Hawaii
Keelin S. Kuipers	National Oceanic and Atmospheric Administration
George Kukla	Lamont Doherty Earth Observatory of Columbia University
Betsy A. Kulle	Maryland Department of Natural Resources
Arun Kumar	National Oceanic and Atmospheric Administration
Sunil Kumar	Metropolitan Washington Council of Governments
Michael J. Kurylo	National Aeronautics and Space Administration; NIST
Won-Tae Kwon	Meteorological Research Institute
Kent P. Laborde	National Oceanic and Atmospheric Administration
Klaus S. Lackner	Columbia University
Murari Lal	Indian Institute of Technology Delhi
Rattan Lal	The Ohio State University
Peter J. Lamb	The University of Oklahoma
H. Richard Lane	National Science Foundation
Nadine T. Laporte	University of Maryland
Daniel A. Lashof	Natural Resources Defense Council
John A. Lasley	SAIC
Istvan Laszlo	National Oceanic and Atmospheric Administration
Jim Latimer	U.S. Environmental Protection Agency
Thomas Laughlin	NOAA/International Affairs
Conrad C. Lautenbacher	Under Secretary of Commerce for Oceans and Atmosphere
Peggy A. Lautenschlager	State of Wisconsin Department of Justice
James D. Laver	NOAA/NWS/NCEP
Richard G. Lawford	National Oceanic and Atmospheric Administration
Edward A. Laws	University of Hawaii
Linda L. Lawson	U.S. Department of Transportation
David Layzell	Queens University, Canada
Judith L. Lean	Naval Research Laboratory
Neil A. Leary	START
Stephen Leatherman	Florida International University
Sharon K. LeDuc	NOAA/National Environmental Satellite, Data, and Information Service
Albert Lee	National Institute of Standards and Technology
Arthur Lee	ChevronTexaco Corporation
Cindy Lee	State University of New York at Stony Brook
Tsengdar J. Lee	NASA Earth Science Enterprise
Virginia Lee	RI Sea Grant/ URI CRC
Ants Leetmaa	NOAA/Geophysical Fluid Dynamics Laboratory
Robert J. Leffler	NOAA National Weather Service
David M. Legler	U.S. CLIVAR Office

Margaret Leinen	National Science Foundation
Richard Lempert	National Science Foundation
Barry M. Lesht	Argonne National Laboratory
Dennis Lettenmaier	University of Washington
Daniel W. Leubecker	Maritime Administration, U.S. Department of Transportation
L. Ruby Leung	Pacific Northwest National Laboratory
Sydney Levitus	NODC/NOAA
Joel M. Levy	National Oceanic and Atmospheric Administration
Gad Levy	NorthWest Research Associates (NWRA)
Orest Lewinter	New York State Department of Environmental Conservation (retired)
Craig A. Lewis	ChevronTexaco
Marlo Lewis, Jr.	Competitive Enterprise Institute
Clement D. Lewsey	NOAA/NOS
Frances C. Li	National Science Foundation
Tim Li	University of Hawaii
Zin-Zhong Liang	Illinois State Water Survey
Sanjay S. Limaye	Office of Space Science Education, Space Science and Engineering Center
Allison F. Linden	Department of Health and Human Services
John J. Lipsey	Tetra Tech, Inc.
Michael M. Little	NASA Langley Research Center
Chung-Ming Liu	National Taiwan University
Diana M. Liverman	University of Arizona
Robert E. Livezey	National Weather Service
Steven A. Lloyd	The Johns Hopkins University
Maryalice Locke	U.S. Federal Aviation Administration
Eric R. Locklear	National Oceanic and Atmospheric Administration
Brent M. Lofgren	NOAA Great Lakes Environmental Research Laboratory
William Logan	National Research Council
Susan Long	Columbia University
Geoff B. Love	Intergovernmental Panel on Climate Change
Thomas R. Loveland	U.S. Geological Survey
Alan A. Lucier	National Council for Air and Stream Improvement, Inc.
Roger Lukas	University of Hawaii
Peter Lunn	U.S. Department of Energy
Le N. Ly	NPS/NOAA
Sheila Lynch	Northeast Advanced Vehicle Consortium
Lingfang Ma	University of Virginia
Michael MacCracken	International Commission on Climate
Sandy MacCracken	U.S. Global Change Research Program/CCSPO
Alexander MacDonald	NOAA/Oceanic and Atmospheric Research
Anthony MacDonald	Coastal States Organization
Tahamoana Macpherson	Embassy of New Zealand
Caren W. Madsen	NOAA Research
Lorenz Magaard	University of Hawaii
Judith Maguire	Environmental Protection Agency
Obed A. Mahena	Vlijana Vision of Tanzania
Jerry D. Mahlman	National Center for Atmospheric Research
Bernard J. Mahoney	Independent
James R. Mahoney	NOAA/U.S. Climate Change Science Program
Martha E. Maiden	National Aeronautics and Space Administration
Gyorgy Major	Hungarian Meteorological Society
Gene Makar	Forte Associates, Inc.
Jonathan T. Malay	Ball Aerospace and Technologies Corp.
Kate Maliga	NASA Office of External Relations
Elizabeth L. Malone	Joint Global Change Research Institute
Robert C. Malone	Los Alamos National Laboratory
Deborah R. Mangis	U.S. Environmental Protection Agency
Martin R. Manning	Intergovernmental Panel on Climate Change/WG I
Antonio Mannino	NASA Goddard Space Flight Center
Nathan Mantua	University of Washington
John H. Marburger III	Office of Science and Technology Policy
Kristin Marcell	NYS DEC Hudson River Estuary Program
Suzanne K. M. Marcy	U.S. Environmental Protection Agency
Jan W. Mares	EOP Group
Hank Margolis	Universite Laval, Canada
Hal Maring	National Aeronautics and Space Administration
Helaine W. Markewich	Department of the Interior/U.S. Geological Survey
Gregg Marland	Oak Ridge National Laboratory
Robert Marlay	U.S. Department of Energy
Todd Marse	University of New Orleans
Robert R.R. Marston	National Oceanic and Atmospheric Administration
R. Gene Martin	University Corporation for Atmospheric Research
Judy Martz	Western Governors' Association
Paul L. Marx	Federal Transit Administration
Byron Mason	National Academies
Cheryl L. Mason	National Science Foundation
Katie Mastriani	Columbia University
Edward J. Masuoka	NASA Goddard Space Flight Center
Roser Matamala	Argonne National Laboratory
David Mathis	USACE
Gary C. Matlock	NOAA/National Ocean Science
Patty Matrai	Bigelow Laboratory
David A. Matthews	DOI/U.S. Bureau of Reclamation
Mike A. Matthews	National Oceanic and Atmospheric Administration
Lourdes Q. Maurice	Federal Aviation Administration
Nancy G. Maynard	NASA Goddard Space Flight Center
Wilfred E. Mazur	NOAA/National Environmental Satellite, Data, and Information Service
Michael Mazzella	Science Applications International Corp.
Margaret R. McCalla	NOAA/Office of the Federal Coordinator
Eileen McCauley	California Air Resources Board
Bonnie McCay	Rutgers University
Douglas A. McChesney	Washington Department of Ecology
Charles R. McClain	NASA Goddard Space Flight Center
Lindsay R. McClelland	National Park Service
John L. McCormick	Citizens Coal Council
John McCoy	
Julian McCreary	University of Hawaii
Kenneth R. McDonald	NASA Goddard Space Flight Center
Robert H. McFadden	GHG Associates
Mack McFarland	DuPont Fluoroproducts
Jim McGrath	Retired
James Mcguire	National Aeronautics and Space Administration
Edward L. McIntosh	Citizen
Douglas G. McKinney	National Risk Management Research Laboratory
Diane McKnight	University of Colorado
Andrea McMakin	Pacific Northwest National Laboratory
Larry M. McMillin	NOAA/National Environmental Satellite, Data, and Information Service
John A. McMorris	AgCert International, LLC
Catherine P. McMullen	Allophilia
James M. McNeal	U.S. Geological Survey
Richard T. McNider	National Space Science and Technology Center
Diane McNight	University of Colorado
Ronald D. McPherson	American Meteorological Society
Steve Meacham	National Science Foundation
Linda O. Mearns	National Center for Atmospheric Research
Esther Mecking	Burke Mountain Academy
Tom Medaglia	RWE
Michelle Medeiros	Greenpeace
Jerry Meehl	University Corporation Atmospheric Research
J. Patrick Megonigal	Smithsonian Environmental Research Center
Vikram M. Mehta	The Center for Research on the Changing Earth System
Eric A. Meindl	National Data Buoy Center
Jill R. Meldon	National Oceanic and Atmospheric Administration
Jie Meng	NOAA/National Environmental Satellite, Data, and Information Service
Phil Merilees	National Center for Atmospheric Research (NCAR)
Karen L. Metchis	U.S. Environmental Protection Agency
Alden M. Meyer	Union of Concerned Scientists
Jim Meyer	Shannon Lab
Frederick Meyerson	AAAS/NSF Science and Technology Policy Fellow
Horst Meyrahn	RWE Rheinbraun AG
Patrick J. Michaels	University of Virginia
Aurelia J. Micko	UCAR/NOAA
Elizabeth M. Middleton	NASA/Goddard Space Flight Center
Paulette B. Middleton	Panorama Pathways
Edward Miles	University of Washington
Constance I. Millar	U.S. Forest Service, PSW Research Station
Alan Miller	The World Bank
Charles E. Miller	Haverford College
Christopher D. Miller	National Oceanic and Atmospheric Administration
David L. Miller	Idaho National Engineering and Environmental Laboratory
Douglas A. Miller	Pennsylvania State University
Mark Miller	Brookhaven National Laboratory
Norman L. Miller	Lawrence Berkeley National Laboratory
Ralph Milli	NorthWest Research Associates (NWRA)
John L. Mimikakis	United States House of Representatives Committee on Science
Laura Miner-Nordstrom	U.S. Environmental Protection Agency
Jean-Bernard Minster	University of California, San Diego
Irving Mintzer	Pacific Institute
Sara E. Mirabilio	National Oceanic and Atmospheric Administration
Ron Mitchell	University of Oregon
Humio Mitsudera	University of Hawaii
Alan C. Mix	Oregon State University
John R. Moisan	National Aeronautics and Space Administration
Bruce F. Molnia	U.S. Geological Survey
Craig F. Montesano	National Oceanic and Atmospheric Administration
Linda V. Moodie	National Oceanic and Atmospheric Administration
Harold A. Mooney	Stanford University
Kenneth A. Mooney	National Oceanic and Atmospheric Administration
Berrien Moore III	University of New Hampshire
Dennis W. Moore	Pacific Marine Environmental Laboratory, NOAA
Melinda Moore	U.S. Department of Health and Human Services
Thomas G. Moore	Hoover Institution, Stanford University
Joseph M. Moran	American Meteorological Society
Stephen G. Moran	Raytheon Company
Barbara J. Morehouse	University of Arizona
Jack A. Morgan	USDA Agricultural Research Service

Jennifer Morgan	World Wildlife Fund	Julie M. Palais	OPP/NSF
Richard E. Moritz	University of Washington	Michael A. Palecki	Illinois State Water Survey
John H. Morrill	Arlington County	Anna C. Palmisano	U.S. Department of Energy
John M. Morrison	North Carolina State University	Theresa Paluszkiewicz	National Science Foundation
Wayne A. Morrissey	Library of Congress	Robert J. Pankiewicz	Product Application Services
Susanne Moser	Union of Concerned Scientists	Kathryn L. Parker	U.S. Environmental Protection Agency
John F. Moses	National Aeronautics and Space Administration	Claire L. Parkinson	NASA Goddard Space Flight Center
Richard Moss	U.S. Climate Change Science Program	Martin I. Parry	School of Environmental Scientists
Philip W. Mote	University of Washington	Edward A. Parson	Harvard University
Jarvis L. Moyers	National Science Foundation	Mark Parsons	University of Colorado, Boulder (CIRES)
Glenn D. Mroz	Michigan Tech University	David F. Paskausky	Wind, Water and Waves
Hussin Muhamed	Department of Geography, Damascus	Toral Patel-Weynand	U.S. Department of State
Patrick J. Mulholland	Oak Ridge National Laboratory	Ari A. Patrinos	U.S. Department of Energy
Duane M. Muller	U.S. Agency for International Development	James W. Patten	Battelle Pacific Northwest National Laboratory
Charles H. Murphy	Unidata/UCAR	Jonathan A. Patz	Johns Hopkins Bloomberg School of Public Health
William L. Murray	NOAA/Office of Global Programs	Eldor A. Paul	Colorado State University
John N. Musinsky	Conservation International	James H. Paul	House Science Committee
John C. Mutter	Columbia University	Milan J. Pavich	U.S. Geological Survey
Carolyn Z. Mutter	International Research Institute for Climate Prediction, Columbia University	Steven Pawson	UMBC/GEST & NASA/DAO
Charles E. Myers	National Science Foundation	D. S. Peacock	National Science Foundation
Gerald Myers	Ophir Corporation	Donald H. Pearlman	The Climate Council
Knute J. Nadelhoffer	National Science Foundation	Sam Pearsall	The Nature Conservancy - North Carolina Chapter
Hironori Nakanishi	NEDO	William T. Pennell	Pacific Northwest National Laboratory
Don Nangira	World Meteorological Organization	Diana L. Perfect	NOAA National Weather Service
Arthur F. Napolitano	Raytheon - Space and Airborne Systems	Alan D. Perrin	U.S. Environmental Protection Agency
Syed Mahmood Nasir	Ministry of Environment, Local Government & Rural Development	Veronika Pesinova	U.S. Environmental Protection Agency
Patrick J. Neale	Smithsonian Environmental Research Center	Christa Peters-Lidard	NASA Goddard Space Flight Center
Thomas R. Neff	Mitretek Systems	William Peterson	NOAA Fisheries
William D. Neff	National Oceanic and Atmospheric Administration/OAR	Annie Petsonk	Environmental Defense Fund
Juniper M. Neill	National Oceanic and Atmospheric Administration	Lawrence Pettinger	U.S. Geological Survey
Ronald P. Neilson	U.S. Department of Agriculture Forest Service	Rickey C. Petty	U.S. Department of Energy
Loren Nelson	Ophir Corporation	Jonathan T. Phinney	National Oceanic and Atmospheric Administration
Paul Nelson	Southwest Research Institute	Roger A. Pielke, Sr.	American Association of State Climatologists
Patrick J. Neuman	National Oceanic and Atmospheric Administration	Crispin H. Pierce	University of Washington
Sandra G. Neuzil	U.S. Geological Survey	Henry J. Pierce	SAIC
Richard E. Newell	U.S. Department of Commerce (retired)	Raymond Pierrehumbert	The University of Chicago
George B. Newton	U.S. Arctic Research Commission	Leonard J. Pietrafesa	North Carolina State University
Mary D. Nichols	California Resources Agency	Paul R. Pike	Ameren
Elizabeth E. Nicholson	National Oceanic and Atmospheric Administration	Rick S. Piltz	Climate Change Science Program Office/USGCRP
Ala Vladimir Nicolenco	Moldavian Academy of Sciences	Nicholas Pinter	Southern Illinois University
Claudia Nierenberg	National Oceanic and Atmospheric Administration	Rodrigo G. Pinto	Brazilian Embassy
Ian R Noble	The World Bank	Stephen R. Piotrowicz	NOPP/Ocean.US
Juan C. Nogales	JRM Inc.	Hugh M. Pitcher	Pacific Northwest National Laboratory
Vivian Pardo Nolan	U.S. Geological Survey	Warren Piver	National Institutes of Health
Richard J. Norby	Oak Ridge National Laboratory	William A. Pizer	Resources for the Future
John J. Novak	Electric Power Research Institute	Robert J. Plante	Raytheon Information Systems
Albert Nunez	City of Takoma Park, Cities for Climate Protection Campaign	Gayther L. Plummer	GEORGIA Climatology Association
Frank Nutter	Reinsurance Association	Mitchell A. Plummer	Idaho National Engineering and Environmental Laboratory
James J. O'Brien	Florida State University	Ethan J. Podell	Consultant
Maureen R O'Brien	Office of Science and Technology Policy	John Drew Polisar	U.S. Department of State
Robert E. O'Connor	National Science Foundation	Igor Polyakov	University of Alaska Fairbanks
Sean O'Keefe	National Aeronautics and Space Administration	David M. Pomerantz	House Committee on Appropriations
William F. O'Keefe	George C. Marshall Institute	William M. Porch	Los Alamos National Laboratory
James M. O'Sullivan	National Oceanic and Atmospheric Administration	Dianne L. Poster	National Institute of Standards and Technology
Godwin O.P. Obasi	World Meteorological Organization	Krishna Rao Potarazu	TRW
Walter C. Oechel	San Diego State University	Joanne R. Potter	Cambridge Systematics
Robert J. Oglesby	National Aeronautics and Space Administration/MSFC	Murray Poulter	National Institute of Water and Atmospheric Research, New Zealand
John Ogren	NOAA Climate Monitoring and Diagnostics Laboratory	Kendall S. Powell	Nature
Eiji Ohira	Ministry of Economy, Trade and Industry	Daniel Power	Climate Institute
Richard L Ohlemacher	National Oceanic and Atmospheric Administration	Michael J. Prather	University of California, Irvine
Peter E. Ohonsi	York University	Cary Presser	National Institute of Standards and Technology
George Ohring	National Oceanic and Atmospheric Administration	Benjamin L. Preston	Pew Center on Global Climate Change
Dennis S. Ojima	Colorado State University	Stuart V. Price	RSVP Communications
Charles Chinedu Okeke	NUSA International ltd.	Joseph Prospero	University of Miami
Kirsten U. Oldenburg	U.S. Department of Transportation	Alex Pszenny	University of New Hampshire
Lembulung O. Kosyando	Naadutaro (Pastoralists' Survival Options)	Roger Pulwarty	NOAA/CIRES
Curtis R. Olsen	University of Massachusetts Boston	Mary Quillian	Nuclear Energy Institute
John O. Olsen	Heatlog Industries Inc.	John Quinn	Constellation Energy Group
Kathie L. Olsen	Office of Science and Technology Policy	Nancy Rabalais	Louisiana Universities Marine Consortium
Lola M. Olsen	NASA/Goddard Space Flight Center	Frank R. Rack	Joint Oceanographic Institutions
Joan Oltman-Shay	NorthWest Research Associates (NWRA)	Alexander Radichevich	National Oceanic and Atmospheric Administration
James Opaluch	University of Rhode Island	Patricia A. Rainey	The Boeing Company
Michael Oppenheimer	Princeton University	Michael T. Rains	USDA Forest Service
William C. Orr	National Alternative Fuels Foundation	Don R. Raish	California Casualty Insurance
Jessica P. Orrego	Climate Change Science Program Office	Sudhir Chella Rajan	Tellus Institute
Charles B. Osmond	Columbia University	V. Ramanathan	Scripps Institution of Oceanography
Doug Osugi	State of California, Department of Water Resources	Bruce H. Ramsay	NOAA/National Environmental Satellite, Data, and Information Service
Ralph P. Overend	National Renewable Energy Laboratory	Robert L. Randall	The RainForest ReGeneration Institute
James E. Overland	PMEL/NOAA	Randy Randol	Exxon Mobil
Jonathan T. Overpeck	University of Arizona	R. Keith Raney	Johns Hopkins University Applied Physics Laboratory
Stphen W Pacala	Princeton University	Robert H. Rank	NOAA/National Environmental Satellite, Data, and Information Service
Rajendraa K. Pachauri	TATA Energy Research Institute. TERI, The Habitat Place	Kenneth J. Ranson	NASA Goddard Space Flight Center
Pamela E. Padgett	USDA Forest Service	S. Trivikrama Rao	National Oceanic and Atmospheric Administration

Aaron G. Rappaport	Union of Concerned Scientists
Beth G. Raps	Independent Scholar
A. R. Ravishankara	National Oceanic and Atmospheric Administration
Walter J. Rawls	USDA/Agricultural Research Service
K.B. N. Rayana	IAMMA
Scott C. Rayder	Department of Commerce/NOAA
Donald G. Rea	Senior Scientists and Engineers, AAAS
Anthony L. Reale	NOAA/National Environmental Satellite, Data, and Information Service
Ruth A. Reck	University of California
Kambham Raja Reddy	Mississippi State University
Michelle Redlin	Hart Downstream Energy Services
Kelly T. Redmond	Western Regional Climate Center
Denise Reed	University of New Orleans
Michelle A. Reed	National Oceanic and Atmospheric Administration
Eileen Regan	Strategic Environmental Research and Development
David E. Reichle	Environmental Consultant
Stephen J. Reid	National Science Foundation
Daniel A. Reifsnyder	U.S. Department of State
Chris Reilly	Argonne National Laboratory
John M. Reilly	Massachusetts Institute of Technology
Thomas Reilly	Commonwealth of Massachusetts
David S. Renné	National Renewable Energy Laboratory
Nilton O. Renno	The University of Michigan
Andrew C. Revkin	The New York Times
Jose A. Rial	University of North Carolina at Chapel Hill
Charles W. Rice	Kansas State University
Don Rice	National Science Foundation
Joseph P. Rice	National Institute of Standards and Technology
Richard C. Rich	Virginia Tech
Kelvin Richards	University of Hawaii
Richard Richels	Electrical Power Research Institute
Michele M. Rienecker	NASA/Goddard Space Flight Center
Ray A. Rigert	U.S. Army
David Rind	National Aeronautics and Space Administration
Kathleen E. Ritzman	Scripps Institution of Oceanography
Alberto L.R.-Rentas	Metropolitan University
John O. Roads	Scripps Institution of Oceanography, UCSD
Franklin R. Robertson	NASA/MSFC
David A. Robinson	Rutgers University
Richard Robinson	Coastal and Watershed Resources Advisory Committee
Margaret Robnett	Chesapeake Climate Action Network
Alan Robock	Rutgers University
Barrett N. Rock	University of New Hampshire
Arturo E. Rodriguez	Miami-Dade Community College
Phillip J. Roehrs	American Coastal Coalition
Dean Roemmich	Scripps Institution of Oceanography, UCSD
Catriona E. Rogers	U.S. Environmental Protection Agency
Richard B. Rood	NASA Goddard Space Flight Center
Peter W. Rooney	House Science Committee
Maurice D. Roos	California Department of Water Resources
Eugene Rosa	Washington State University
Richard D. Rosen	Atmospheric and Environmental Research, Inc.
Norman J. Rosenberg	Pacific Northwest National Laboratory
Jeffrey O. Rosenfeld	American Meteorological Society
Joshua P. Rosenthal	National Institutes of Health - Fogarty International Center
Joyce E. Rosenthal	Columbia University
Cynthia Rosenzweig	NASA Goddard Institute for Space Studies
Richard Rosenzwieg	Natsource LLC
Kevin T. Rosseel	U.S. Environmental Protection Agency
William B. Rossow	NASA Goddard Institute for Space Studies
Doug Rotman	NASA Earth System Modeling Framework Advisory Board
Steven Rowe	Office of the Attorney General of Maine
Amy E. Royden	STAPPA/ALAPCO
Howard Runge	Orbital Sciences Corporation
George B. Runion	USDA/Agricultural Research Service
Steven W. Running	University of Montana
Leslie F. Ruppert	U.S. Geological Survey
David P. Russ	U.S. Geological Survey
Thomas A. Russell	ITT Industries
William (Bill) E. Russo	U.S. Environmental Protection Agency
Gene Russo	THE SCIENTIST News Journal
Glenn K. Rutledge	NOAA National Climatic Data Center
Robert Ryan	NBC4
Jeffrey D. Sachs	The Earth Institute at Columbia University
Rafe D. Sagarin	Office of Congresswoman Hilda Solis
Raphael Sagarin	University of California, Santa Barbara
William A. Salas	Applied GeoSolutions, LLC
Ross J. Salawitch	Jet Propulsion Laboratory, Caltech
Michael J. Sale	Oak Ridge National Laboratory
Vincent V. Salomonson	NASA Goddard Space Flight Center
Jason P. Samenow	U.S. Environmental Protection Agency
Alan H. Sanstad	Lawrence Berkeley National Laboratory
Jessica A. Santacruz	

Benjamin D. Santer	Lawrence Livermore National Laboratory
Edward S. Sarachik	NASA Earth System Modeling Framework Advisory Board/University of Washington
Myriam Sarachik	American Physical Society
Jorge Sarmiento	Princeton University
Jayant A. Sathaye	Lawrence Berkeley National Laboratory
Tetsuya Sato	Japan Maritime, Science and Technology Center
Janet L. Sawin	Worldwatch Institute
Karyn Sawyer	University Corporation for Atmospheric Research
Earl C. Saxon	The Nature Conservancy
Donald Scavia	NOAA National Ocean Service
John C. Schaake	NOAA/National Weather Service
Jacqueline Schafer	U.S. Agency for International Development
Joel D. Scheraga	U.S. Environmental Protection Agency
Kenneth Schere	U.S. Environmental Protection Agency
Craig Schiffries	National Council for Science and the Environment
David Schimel	National Center for Atmospheric Research
Michael E. Schlesinger	University of Illinois at Urbana-Champaign
William H. Schlesinger	Duke University
Adam Schlosser	GEST/UMBC/NASA
David Schmalzer	Argonne National Laboratory
Cynthia E. Schmidt	University Corporation for Atmospheric Research
Gavin Schmidt	NASA Earth System Modeling Framework Advisory Board
John A. Schmidt	NASA Earth System Modeling Framework Advisory Board
Raymond W. Schmitt	Woods Hole Oceanographic Institution
Anne-Marie Schmoltner	National Science Foundation
Marcus Schneider	U.S. Clean Energy Program
Niklas Schneider	University of Hawaii
Stephen Schneider	Stanford University
Russ Schnell	NOAA Climate Monitoring and Diagnostics Laboratory
Mark Schoeberl	NASA Goddard Space Flight Center
Paul S. Schopf	NASA Earth System Modeling Framework Advisory Board
Siegfried D. Schubert	NASA Goddard Space Flight Center
Laurel J. Schultz	U.S. Environmental Protection Agency
Heidi Z. Schuttenberg	NOAA Coral Team
Richard A. Schwabacher	The Cousteau Society
Stephen E. Schwartz	Brookhaven National Laboratory
Franklin B. Schwing	NOAA Pacific Fisheries Environmental Laboratory
Karen M. Scott	U.S. Environmental Protection Agency
Roger A. Sedjo	Resources for the Future
Dian J. Seidel	NOAA Air Resources Laboratory
Piers Sellers	NASA/JSC
Cynthia E. Sellinger	NOAA/GLERL
Hratch G. Semerjian	National Institute of Standards and Technology
Jeff Severinghaus	Scripps Institution of Oceanography, UCSD
Craig Shafer	National Park Service
Steven R. Shafer	U.S. Department of Agriculture
Awdhesh K. Sharma	NOAA/National Environmental Satellite, Data, and Information Service
Gary D. Sharp	Center for Climate/Ocean Resources Study
Walter Shaub	U.S. Chamber of Commerce
Eileen L. Shea	East-West Center
Edwin J. Sheffner	National Aeronautics and Space Administration
J. Marshall Shepherd	NASA Goddard Space Flight Center
Noriko L. Shoji	Office of Senator Daniel K. Inouye
Steve E. Short	Short & Associates, Inc.
Randy Showstack	American Geophysical Union
Jagadish Shukla	George Mason University
Martha Shulski	University of Alaska Fairbanks
David Shultz	U.S. Geological Survey
C. K. Shum	Ohio State University
Smita Siddhanti	Endyna Inc.
Hilary H. Sills	Sills Associates
Katherine Silverthorne	World Wildlife Fund
Lynn Simarski	National Science Foundation
Emmy B. Simmons	U.S. Agency for International Development
Michael E. Simmons	North Carolina A&T State University
Robert J. Simon	Chlorine Chemistry Council
Fiorella V. Simoni	George Mason University
Caitlin F. Simpson	National Oceanic and Atmospheric Administration
Michael M. Simpson	U.S. Congressional Research Service
N. Michael Simpson	NOAA/National Environmental Satellite, Data, and Information Service
Naaman M. Simpson	Department of Commerce/NOAA/NESDIS
Michael J. Singer	Soil Science Society of America
Michael W. Singer	Harris Corp
S. Fred Singer	Science and Environmental Policy Project
Ashbindu Singh	United Nations Environment Programme
Vijai Pratap Singh	Institute of Environment Management & Plant Sciences, Vikram University
Douglas L. Sisterson	Argonne National Laboratory
Sreeshankar S. Nair	SES, NOAA (contractor)
Howard Skinner	USDA-Agricultural Research Service
William Skirving	NOAA/National Environmental Satellite, Data, and Information Service
David Skole	Michigan State University
Francis P. Slakey	Public Affairs

Michael W. Slimak	U.S. Environmental Protection Agency	John Taylor	Argonne National Laboratory
Julia M. Slingo	NERC Centre for Global Atmospheric Modelling	Phillip R. Taylor	National Science Foundation
Richard L. Slonaker	Raytheon	Kathy A. Tedesco	NOAA Office of Global Programs
James R. Slusser	Colorado State University	Miguel J. C. A. Tenreiro	U.L.H.T.
John A. Small	National Institute of Standards and Technology	Angel R. Terán	National Weather Service
D. Brent Smith	NOAA/National Environmental Satellite, Data, and Information Service	Maxime J. Thibon	Consulate General of France
Ellen S. Smith	Columbia University	Mark Thiemens	University of California, San Diego
Lowell F. Smith	Retired	Alan R. Thomas	World Meteorological Organization (WMO)
Neville R Smith	Australian Bureau of Meteorology	Julie E. Thomas	DOI/National Park Service
Robert E. Smith	Compass International Development	Elvia H. Thompson	National Aeronautics and Space Administration
Steven J. Smith	Joint Global Change Research Institute	Louisa R. Thompson	Maryland Native Plant Society
Wade B. Smith	Mitretek Systems	Peter Thompson	NOAA/SEFSC
Anthony D. Socci	U.S. Environmental Protection Agency	Elizabeth Thorndike	Center for Environmental Information
Robert H. Socolow	Princeton University	Carolyn Thoroughgood	University of Delaware
Brian Soden	NOAA Geophysical Fluid Dynamics Laboratory	Alan J. Thorpe	University of Reading
Youngsinn Sohn	University of Maryland Baltimore County	Sidney Thurston	National Oceanic and Atmospheric Administration
Allen M. Solomon	U.S. Environmental Protection Agency	James G. Titus	U.S. Environmental Protection Agency
Susan Solomon	NOAA Aeronomy Laboratory	Andreas Tjernshaugen	CICERO
Andrew Solow	Woods Hole Oceanographic Institution	James F. Todd	National Oceanic and Atmospheric Administration
Shiv Someshwar	International Research Institute for Climate Prediction, Columbia University	Sean P. Todd	Fox Potomac Resources LLC
W. Somplatsky-Jarman	Presbyterian Church (USA)	Christine Todd-Whitman	Administrator, US Environmental Protection Agency
J. Richard Soulen	Retired	Pamela L. Tomski	EES Group, Ltd.
Thomas W. Spence	National Science Foundation	John C. Topping	Climate Institute
Peter L. Spencer	U.S. House of Representatives	Margaret S. Torn	Lawrence Berkeley Lab/UCB
Elliott C. Spiker	Retired	Michael P. Totten	Conservation International
Eliot Spitzer	State of New York Office of the Attorney General	John R. Townshend	University of Maryland
Martin A. Spitzer	House Science Committee	Charles C. Trees	National Aeronautics and Space Administration
Michael S. Spranger	University of Florida	Kevin E. Trenberth	National Center for Atmospheric Research
Bill Sprigg	University of Arizona	Thomas Triller	German Embassy
Margaret F. Spring	U.S. Senate Commerce Committee	Dennis A. Trout	Retired
Everett P. Springer	Los Alamos National Laboratory	Juli M. Trtanj	University Corporation for Atmospheric Research
Jordan P. St. John	National Oceanic and Atmospheric Administration	Paul D. Try	International GEWEX Project Office
Judson E. Stailey	NASA Goddard Space Flight Center	Lucia S. Tsaoussi	National Aeronautics and Space Administration
Detlef Stammer	Scripps Institution of Oceanography, UCSD	Nancy C. Tuchman	National Science Foundation
John M. Stamos	U.S. Department of Energy	Diane M. Turchetta	U.S. Department of Transportation
Carroll L. Stang	National Weather Service Office of Science and Technology	Karl K. Turekian	Yale University
Diane Stanitski-Martin	Shippensburg University, PA	Vaughan C. Turekian	U.S. State Department
Carl P. Staton	National Oceanic and Atmospheric Administration/CIO	Kenneth W. Turgeon	U.S. Commission on Ocean Policy
Amanda C. Staudt	The National Academies	William T. Turnbull	NOAA/CIO
Eugene Steadman, Jr.	CELANESE	Kelly M. Turner	National Oceanic and Atmospheric Administration
Konstantinos Stefanidis	NASA/Goddard Space Flight Center	William W. Turner	National Aeronautics and Space Administration
Will Steffen	IGBP: International Geosphere-Biosphere Programme	Alexander Tuyahov	National Aeronautics and Space Administration
David E. Steitz	National Aeronautics and Space Administration	Paul F. Twitchell	International GEWEX Project Office
Sue P. Stendebach	National Science Foundation	Henry Tyrrell	U.S. Department of Agriculture
Pamela L. Stephens	National Science Foundation	Michael Uhart	National Oceanic and Atmospheric Administration
Louis T. Steyaert	U.S. Geological Survey	Deborah P. Underwood	North Carolina A&T State University
Raphaela Stimmelmayr	Tanana Chiefs Conference, Inc.	Sushel Unninayar	National Aeronautics and Space Administration
Alexander R. Stine	Massachusetts Institute of Technology	Ana L. Unruh	Office of Rep. Edward Markey
Bryce J. Stokes	USDA Forest Service	Lall Upmanu	Columbia University
Carrie Stokes	U.S. Agency for International Development	Deanne R. Upson	Teletrips
Gerald M. Stokes	Joint Global Change Research Institute	David P. Urbanski	Air Force Combat Climatology Center
Douglas H. Stone	American Meteorological Society	Susan Ustin	Department of Land, Air, and Water Resource
John M.R. Stone	Environment Canada	Arthur C. Vailas	University of Houston
William E. Stoney	Mitretek Systems	Francisco P.J. Valero	Scripps Institution of Oceanography University of California, San Diego
Ronald J. Stouffer	NOAA/Geophysical Fluid Dynamics Laboratory	Azita Valinia	NASA/Goddard Space Flight Center
Greg Stover	NASA Langley Research Center	Deborah G. Vane	Jet Propulsion Laboratory
Tana L. Stratton	Department of Foreign Affairs and International Trade	Lelia B. Vann	NASA Langley Research Center
Alan E. Strong	NOAA/National Environmental Satellite, Data, and Information Service	Dalia E. Varanka	U.S. Geological Survey
Chris Struble	Software Engineer	Lisa Farrow Vaughan	National Oceanic and Atmospheric Administration
Jeffrey A. Stuart	National Weather Service	Robert Venezia	National Aeronautics and Space Administration
Max J. Suarez	NASA/Goddard Space Flight Center	David J. Verardo	National Science Foundation
Susan E. Subak	University of East Anglia	Michael Verkouteren	National Institute of Standards and Technology
Carla C. Sullivan	National Oceanic and Atmospheric Administration	Hans J.H. Verolme	British Embassy
Charles Sun	National Oceanic and Atmospheric Administration	Konstantin Vinnikov	University of Maryland
Shan Sun	NASA Goddard Institute for Space Studies	Hassan Virji	START Secretariat
S.S. Sundarvel	Pondicherry University	Martin Visbeck	Lamont-Doherty Earth Observatory
Eric T. Sundquist	U.S. Geological Survey	Stephen M. Volz	National Aeronautics and Space Administration
Nicholas A. Sundt	Climate Change Science Program Office	Thomas H. Vonder Haar	Cooperative Institute for Research in the Atmosphere
Terry Surles	California Energy Commission	Richard Wagener	Brookhaven National Laboratory
Jon Sutinen	University of Rhode Island	David L. Wagger	Independent Researcher
Jan A. Suurland	Royal Netherlands Embassy	Frederic H. Wagner	Utah State University
Freyr Sverrisson	Independent Energy Consultant	Thomas W. Wagner	University of Michigan
Lea E. Swanson	Global Environment & Technology Foundation	Cameron Wake	University of New Hampshire
William Sydeman	PRBO Conservation Science	Shelby E. Walker	National Science Foundation
Gregory H. Symmes	National Academies	Margaret K. Walsh	ICF Consulting
Keiko Takahashi	Earth Simulator Center	John Walsh	University of Alaska Fairbanks
Taro Takahashi	Lamont-Doherty Earth Observatory, Columbia University	Chris Walvoord	National Environmental Trust
Eugene S. Takle	Iowa State University	Bin Wang	University of Hawaii
Pieter Tans	NOAA Climate Monitoring and Diagnostics Laboratory	Julian X.L. Wang	Air Resources Laboratory
Jay J. Tanski	State University of New York	Rik Wanninkhof	NOAA/AOML
Leland Tarnay	NASA Earth System Modeling Framework Advisory Board/National Park Service	James T. Waples	University of Wisconsin-Milwaukee/Great Lakes WATER Institute
		M. Neil Ward	International Research Institute for Climate Prediction (IRI)
Dan Tarpley	NOAA/National Environmental Satellite, Data, and Information Service	Morris (Bud) A. Ward	Moris A. Ward, Inc.

David A. Warrilow	UK Department for Environment, Food and Rural Affairs
Warren M. Washington	National Center for Atmospheric Research
Lewis T. Waters	Consultant
Harlan L. Watson	U.S. Department of State
Kathy L. Watson	NOAA, Office of Global Programs
Thomas B. Watson	National Oceanic and Atmospheric Administration
Mark P. Weadon	U.S. Air Force
John W. Weatherly	Cold Regions Research Lab, ERDC, US Army Corps of Engineers
Ronald L. S. Weaver	National Snow and Ice Data Center
Robert S. Webb	NOAA/OAR/Climate Diagnostics Center
Elke U. Weber	Columbia University
Larry H. Weber	National Science Foundation
Ted Weber	Chesapeake and Coastal Watershed Service, Maryland Department of Natural Resources
Eric Webster	United States House of Representatives
Michael F. Wehner	Lawrence Berkeley National Laboratory
Rodney Weiher	National Oceanic and Atmospheric Administration
Allan M. Weiner	Harris Corporation
Joseph M. Welch	NOAA/National Ocean Service
Gunter Weller	University of Alaska Fairbanks
Robert A. Weller	Woods Hole Oceanographic Institution
Mark Weltz	USDA/Agricultural Research Service
Gerd Wendler	University of Alaska Fairbanks
Frank J. Wentz	Remote Sensing Systems
Carol A. Werner	Environmental and Energy Study Institute
Jon Werner	U.S. Department of Agriculture
Jason West	AAAS Environmental Fellow
Jordan M. West	U.S. Environmental Protection Agency
June M. Whelan	National Petrochemical and Refiners Association
Elizabeth A. White	National Oceanic and Atmospheric Administration
James C. White	Center for Environmental Information
Robert M. White	Washington Advisory Group
Jeff Whitlow	U.S. Environmental Protection Agency
Carol E. Whitman	National Rural Electric Cooperative Association
Gene Whitney	Office of Science and Technology Policy
Diane E. Wickland	National Aeronautics and Space Administration
Bruce A. Wielicki	NASA Langley Research Center
John Wiener	Independent Reviewer
Tom M.L. Wigley	UCAR, WORI, Acacia Project
Mark Wigmosta	Pacific Northwest National Laboratory
Thomas J. Wilbanks	Oak Ridge National Laboratory
George T. Wilcox	American Meteorological Society
Harold Wilhite	University of Oslo
Richard R. Wilk	Indiana University
Diana Wilkins	UK Department for Environment, Food and Rural Affairs
Debra A. Willard	U.S. Geological Survey
Sean R. Willard	NOAA Climate Office
Ted Z. Willard	NASA Programs/CSC
Allen Williams	Illinois State Water Survey
Darrel L. Williams	NASA Goddard Space Flight Center
Elizabeth O. Williams	National Aeronautics and Space Administration
Gregory J. Williams	National Aeronautics and Space Administration
Larry J. Williams	Electric Power Research Institute
Richard S. Williams, Jr.	U.S. Geological Survey
Samuel P. Williamson	NOAA/Office of the Federal Coordinator for Meteorology
Stanley Wilson	Department of Commerce
Doug Wilson	National Oceanic and Atmospheric Administration
Gary P. Wilson	Ball Aerospace and Technologies Corp.
Jim Wilson	University of Dublin
Tyrone P. Wilson	U.S. Geological Survey
David M. Winker	NASA Langley Research Center
Julie A. Winkler	Michigan State University
Darrell A. Winner	U.S. Environmental Protection Agency
Robert S. Winokur	Earth Satellite Corporation (EarthSat)
Derek Winstanley	Illinois State Water Survey
Creighton D. Wirick	Brookhaven National Laboratory
Warren Wiscombe	NASA Goddard Space Flight Center
Gregory W. Withee	DOC/NOAA/NESDIS
James M. Witting	Unaffiliated
Steve Wittrig	BP Amoco Chemicals Company
Steven C. Wofsy	Harvard University
David E. Wojick	Climatechangedebate.org
Henry Wolf	George Mason University School of Computational Science
David W. Wolfe	Cornell University
George Wolff	General Motors Corporation
Kevin R. Wood	National Oceanic and Atmospheric Administration
George Wooddell	University of Louisiana at Lafayette
Ginny L. Worrest	United States Senate
Robert C. Worrest	U.S. Global Change Research Information Office (GCRIO)
Darlene Wright	Tanana Chiefs Conference, Inc.
Xiangqian Wu	DOC/NESDIS/Office of Research and Application
Stan D. Wullschleger	Oak Ridge National Laboratory
Annie Wyndham	Independent Reviewer
David Yap	Ontario Ministry of the Environment
Brent Yarnal	Pennsylvania State University
Jim Yoder	National Science Foundation
Shira Yoffe	U.S. Department of State
Abby M. Young	International Council for Local Environmental Initiatives
Michael T. Young	Short and Associates
Barbara G. Yuhas	International City/County Management Association
Cheryl L. Yuhas	National Aeronautics and Space Administration
Scott A. Yuknis	Meteorlogix
Bernard Zak	Sandia National Laboratory
Rahul A. Zaveri	Pacific Northwest National Laboratory
Stephen E. Zebiak	NASA Earth System Modeling Framework Advisory Board/Int'l Research Institute for Climate Prediction
Steve Zebiak	NASA Earth System Modeling Framework Advisory Board
Richard Zepp	U.S. Environmental Protection Agency
Chris Zganjar	The Nature Conservancy
Renyi Zhang	Texas A&M University
Jiayu Zhou	NOAA/National Weather Service
Herman B. Zimmerman	National Science Foundation
Patrick R. Zimmerman	South Dakota School of Mines and Technology
Rae Zimmerman	New York University
Mark A. Zito	U.S. Department of Agriculture
Francis Zwiers	Canadian Centre for Climate Modelling and Analysis

B | References

CCOIIG, 2003: *An Ocean Implementation Plan for the U.S. Carbon Cycle Science Program* [Doney, S (ed.)]. Carbon Cycle Ocean Interim Implementation Group, University Corporation for Atmospheric Research, Washington, DC, USA (manuscript).

CCWG, 1999: *A U.S. Carbon Cycle Science Plan* [Sarmiento, J.L. and S.C. Wofsy (eds.)]. Carbon and Climate Working Group, University Corporation for Atmospheric Research, Washington, DC, USA, 69 pp.

EHP, 2001: *Environmental Health Perspectives: Human Health Consequences of Climate Variability and Change for the United States* [Bernard, S.M., M.G. McGeehin, and J.A. Patz (eds.)]. EHP Supplement 109(2), Public Health Service, U.S. Department of Health and Human Services, Washington, DC, USA, 64 pp.

FAO, 2002: *Terrestrial Carbon Observation: The Rio de Janeiro Recommendations for Terrestrial and Atmospheric Measurements* [Cihlar, J., S. Denning, F. Ahern, F. Bretherton, J. Chen, C. Dobson, C. Gerbig, R. Gibson, R. Gommes, T. Gower, K. Hibbard, T. Igarashi, R. Olson, C. Potter, M. Raupach, S. Running, J. Townshend, D. Wickland, and Y. Yasuoka (eds.)]. FAO Environment and Natural Resources Series No. 2. United Nations Food and Agriculture Organization, Rome, Italy, 42 pp.

Global Carbon Project, 2003: *Science Framework and Implementation. Earth System Science Partnership (IGBP, IHDP, WCRP, Diversitas).* Global Carbon Project No. 1, Canberra, Australia (in press).

GCOS, 2003: *The Second Report on the Adequacy of the Global Observing Systems for Climate in Support of the UNFCCC.* WMO-IOC-UNEP-ICS, GCOS-82, Technical Document No. 1143, World Meteorological Organization, Geneva, Switzerland, 85 pp.

GOOS, 2002: *A Global Ocean Carbon Observation System - A Background Report* [Doney, S. and M. Hood (eds.)]. Global Ocean Observing System Report No. 118, UNESCO Intergovernmental Oceanographic Commission, IOC/INF-1173, 55 pp.

IPCC, 2000a: *Land Use, Land-Use Change, and Forestry. A Special Report of the IPCC* [Watson, R.T., I.R. Noble, B. Bolin, N.H. Ravindranath, D.J. Verardo, and D.J. Dokken (eds.)]. Cambridge University Press, Cambridge, United Kingdom and New York, NY, USA, 377 pp.

IPCC, 2000b: *Emissions Scenarios. A Special Report of IPCC Working Group III* [Nakicenovic, N., J. Alcamo, G. Davis, B. de Vries, J. Fenhann, S. Gaffin, K. Gregory, A. Grubler, T.Y. Jung, T. Kram, E.L. La Rovere, L. Michaelis, S. Mori, T. Morita, W. Pepper, H. Pitcher, L. Price, K. Riahi, Al. Roehrl, H.-H. Rogner, A. Sankovski, M. Schlesinger, P. Shukla, S. Smith, R. Swart, S. van Rooijen, N. Victor, and Z. Dadi (eds.)]. Cambridge University Press, Cambridge, United Kingdom and New York, NY, USA, 599 pp.

IPCC, 2001a: *Climate Change 2001: The Scientific Basis. A Contribution of Working Group I to the Third Assessment Report of the Intergovernmental Panel on Climate Change* [Houghton, J.T., Y. Ding, D.J. Griggs, M. Noguer, P.J. van der Linden, X. Dai, K. Maskell, and C.A. Johnson (eds.)]. Cambridge University Press, Cambridge, United Kingdom and New York, NY, USA, 881 pp.

IPCC, 2001b: *Climate Change 2001: Impacts, Adaptation, and Vulnerability. A Contribution of Working Group II to the Third Assessment Report of the Intergovernmental Panel on Climate Change* [McCarthy, J.J., O.F. Canziani, N.A. Leary, D.J. Dokken, and K.S. White (eds.)]. Cambridge University Press, Cambridge, United Kingdom and New York, NY, USA, 1032 pp.

IPCC, 2001c: *Climate Change 2001: Mitigation. A Contribution of Working Group III to the Third Assessment Report of the Intergovernmental Panel on Climate Change* [Metz, B., O. Davidson, R. Swart, and J. Pan (eds.)]. Cambridge University Press, Cambridge, United Kingdom and New York, NY, USA, 752 pp.

IPCC, 2001d: *Climate Change 2001: Synthesis Report. A Contribution of Working Groups I, II, and III to the Third Assessment Report of the Intergovernmental Panel on Climate Change* [Watson R.T. and the Core Writing Team (eds.)]. Cambridge University Press, Cambridge, United Kingdom and New York, NY, USA, 397 pp.

LSCOP, 2002: *A Large-Scale CO_2 Observing Plan: In Situ Oceans and Atmosphere* [Bender, M., S. Doney, R. Feely, I. Fung, N. Gruber, D. Harrison, R. Keeling, J. Moore, J. Sarmiento, E. Sarachik, B. Stephens, T. Takahashi, P. Tans, and R. Wanninkhof (eds.)]. National Technical Information Service, National Oceanic and Atmospheric Administration, Silver Spring, MD, USA, 201 pp.

MA Secretariat, 2002: *Millennium Ecosystem Assessment: Millennium Ecosystem Assessment Methods* [Reid, W., N. Ash, E. Bennett, P. Kumar, M. Lee, N. Lucas, H. Simons, V. Thompson, and M. Zurek (eds.)]. Millenium Assessment Secretariat, ICLARM Office, Penang, Malaysia, 81 pp.

NACP, 2002: *The North American Carbon Program (NACP). A Report of the NACP Committee of the U.S. Carbon Cycle Science Steering Group* [Wofsy, S. and R. Harriss (eds.)]. University Corporation for Atmospheric Research, Boulder, Colorado, USA, 56 pp.

NASA, 2002: *Air Pollution as a Climate Forcing: Proceedings of a Workshop (Honolulu, Hawaii • 29 April - 3 May 2002)* [Hansen, J.E. (ed.)]. National Aeronautics and Space Administration, Goddard Institute for Space Studies, New York, NY, USA, 169 pp.

NRC, 1999a: *Global Environmental Change: Research Pathways for the Next Decade.* Committee on Global Change Research, National Research Council, National Academy Press, Washington, DC, USA, 616 pp.

NRC, 1999b: *Capacity of U.S. Climate Modeling to Support Climate Change Assessment Activities.* Climate Research Committee, National Research Council, National Academy Press, Washington, DC, USA, 78 pp.

NRC, 1999c: *Our Common Journey: A Transition toward Sustainability*. Board on Sustainable Development, National Research Council, National Academy Press, Washington, DC, USA, 363 pp.

NRC, 1999d: *From Monsoons to Microbes: Understanding the Ocean's Role in Human Health*. Committee on the Ocean's Role in Human Health, National Research Council, National Academy Press, Washington, DC, USA, 144 pp.

NRC, 1999e: *Making Climate Forecasts Matter*. Committee on the Human Dimensions of Global Change, National Research Council, National Academy Press, Washington, DC, USA, 187 pp.

NRC, 1999f: *A Question of Balance: Private Rights and the Public Interest in Scientific and Technical Databases*. National Research Council, National Academy Press, Washington, DC, USA, 158 pp.

NRC, 1999g: *Adequacy of Climate Observing Systems*. Panel on Climate Observing Systems Status, Climate Research Committee, National Research Council, National Academy Press, Washington, DC, USA, 66 pp.

NRC, 2000: *Reconciling Observations of Global Temperature Change*. Commission on Geosciences, Environment, and Resources, National Academy Press, Washington, DC, USA, 104 pp.

NRC, 2001a: *Climate Change Science: An Analysis of Some Key Questions*. Committee on the Science of Climate Change, National Research Council, National Academy Press, Washington, DC, USA, 42 pp.

NRC, 2001b: *Grand Challenges in Environmental Sciences*. Committee on Grand Challenges in Environmental Sciences, National Research Council, National Academy Press, Washington, DC, USA, 106 pp.

NRC, 2001c: *Under the Weather: Climate, Ecosystems, and Infectious Disease*. Board on Atmospheric Sciences and Climate, National Research Council, National Academy Press, Washington, DC, USA, 160 pp.

NRC, 2001d: *Improving the Effectiveness of U.S. Climate Modeling*. Board on Atmospheric Sciences and Climate, National Research Council, National Academy Press, Washington, DC, USA, 142 pp.

NRC, 2001e: *The Science of Regional and Global Change: Putting Knowledge to Work*. Committee on Global Change Research, National Research Council, National Academy Press, Washington, DC, USA, 19 pp.

NRC, 2001f: *Global Air Quality: An Imperative for Long-Term Observational Strategies*. Committee on Atmospheric Chemistry, Board on Atmospheric Sciences and Climate, National Research Council, National Academy Press, Washington, DC, USA, 56 pp.

NRC, 2002: *Abrupt Climate Change: Inevitable Surprises*. Committee on Abrupt Climate Change, Ocean Studies Board, Polar Research Board, Board on Atmospheric Sciences and Climate, Division on Earth and Life Sciences, National Research Council, National Academy Press, Washington, DC, USA, 230 pp.

Ocean Theme Team, 2001: *An Ocean Theme for the IGOS Partnership* [Lindstrom, E., J. Fellous, M. Drinkwater, R. Navalgund, J. Marra, T. Tanaka, J. Johannessen, C. Summerhayes, and L. Charles (eds.)]. NP-2001-01-261-HQ. National Aeronautics and Space Administration [on behalf of the Integrated Global Observing Strategy (IGOS) Partnership], Washington, DC, USA, 36 pp.

USGCRP, 2000: *High-End Climate Science: Development of Modeling and Related Computing Capabilities*. Ad Hoc Working Group on Climate Modeling (report to the USGCRP, December 2000), U.S. Global Change Research Program, Washington, DC, USA, 44 pp.

USGCRP, 2001: *A Plan for a New Science Initiative on the Global Water Cycle* [Hornberger, G.M. J.D. Aber, J. Bahr, R.C. Bales, K. Beven, E. Foufoula-Georgiou, G. Katul, J.L. Kinter III, R.D. Koster, D.P. Lettenmaier, D. McKnight, K. Miller, K. Mitchell, J.O. Roads, B.R. Scanlon, and E. Smith (eds.)]. The USGCRP Water Cycle Study Group, U.S. Global Change Research Program, Washington, DC, USA, 118 pp.

WAG, 1997: *Climate Change and Vector-Borne and Other Infectious Diseases: A Research Agenda*. The Washington Advisory Group, Washington, DC, USA, 25 pp.

WCRP, 1995: *CLIVAR Science Plan - A Study of Climate Variability and Predictability*. World Climate Research Programme Report No. 89 (WMO/TD No. 690), World Meteorological Organization, Geneva, Switzerland, 157 pp.

WMO, 1999: *Scientific Assessment of Ozone Depletion: 1998*. Global Ozone Research and Monitoring Project - Report No. 44, World Meteorological Organization, Geneva, Switzerland, 732 pp.

WMO, 2003: *Scientific Assessment of Ozone Depletion: 2002*. Global Ozone Research and Monitoring Project - Report No. 47, World Meteorological Organization, Geneva, Switzerland, 498 pp.

Only Strategic Plan graphics with more detailed captions and/or source information have been included in this annex—hence the gaps in figure numbering.

Figure 3-2: Schematic illustrating a comparison among several factors that influence Earth's climate. These factors can be broadly compared using the concept of radiative forcing. The principal forcings arise from changes in atmospheric composition of gases, aerosols, clouds, land use, and solar output. The rectangular bars represent best estimates of the contributions of these forcings, some of which yield warming and some cooling. The forcing contributions are shown for different aerosol types and for those produced or emitted by aviation. The indirect effect shown for aerosols is their effect on the size and number of cloud droplets. A second indirect effect, namely the effect on cloud lifetime, which would lead to a negative forcing, is not shown. The vertical line about the rectangular bars indicates the range of estimates. A vertical line without a rectangular bar denotes a forcing for which no best estimate can be given owing to large uncertainties. Aerosol forcings are generally not uniform over the globe, unlike those of well-mixed greenhouse gases (e.g., CO_2). A more detailed description of this schematic and its interpretation is provided by the IPCC (IPCC, 2001a,d).

Figure 3-3: Schematic of the processes that cause the formation and transformation of aerosol particles in the atmosphere. Aerosols can be emitted directly into the atmosphere or be formed there from the emissions of gaseous precursors. Particles grow by condensation of gases and by coagulation with other particles. Sizes of important atmospheric particles vary over several orders of magnitude. Reactions can occur on the surfaces of particles that can alter the composition of the particle or the surrounding atmosphere. Particles can also grow to become cloud droplets or ice crystals. Particle number and composition can influence the formation and radiative characteristics of clouds. Particles scavenge a variety of gases from the atmosphere and are eventually removed from the atmosphere by wet or dry deposition to Earth's surface. Graphic adapted from: Heintzenberg, J., F. Raes, and S.E. Schwartz, 2003: Tropospheric aerosols. In: *Atmospheric Chemistry in a Changing World: An Integration and Synthesis of a Decade of Tropospheric Chemistry Research* [Brasseur, G.P., R.G. Prinn, and A.A.P. Pszenny (eds)]. Springer-Verlag, New York, NY, USA, pp. 125-156.

Figure 3-4: Schematic illustrating the estimated recovery of stratospheric ozone in the coming decades. Observations show the decline of global total ozone (top panel) and minimum values of total ozone over Antarctica (lower panel) beginning in 1980. In the future, the amounts of chlorine- and bromine-containing gases are expected to further diminish in the stratosphere as a result of international compliance with the Montreal Protocol and, in response,

ozone amounts are expected to recover significantly toward values observed in 1980 and before. The range of recovery times shown is based on predictions of atmospheric models. The model results differ because they have different assumptions about the composition and meteorology of the future stratosphere. The research needs as outlined in Section 3.4 address reducing the range of uncertainty in ozone recovery estimates. More detail on the recovery of stratospheric ozone is provided in the *Scientific Assessment of Ozone Depletion: 2002* (WMO, 2003).

Figure 4-1: The Earth's surface temperature has increased by about 0.6°C over the record of direct temperature measurements (1860-2000, top panel)—a rise that is unprecedented, at least based on proxy temperature data for the Northern Hemisphere, over the last millennium (bottom panel). In the top panel the departures from global mean surface temperature are shown year-by-year (red bars with *very likely* ranges as thin black line) and approximately decade-by-decade (continuous red line). Analyses take into account data gaps, random instrumental errors and uncertainties, uncertainties in bias corrections in the ocean surface temperature data, and also uncertainties in adjustments for urbanization over the land. The lower panel merges proxy data (year-by-year blue line with *very likely* ranges as gray band, 50-year-average purple line) and the direct temperature measurements (red line) for the Northern Hemisphere. The proxy data consist of tree rings, corals, ice cores, and historical records that have been calibrated against thermometer data. Insufficient data are available to assess such changes in the Southern Hemisphere. Source: IPCC (2001d).

Figure 4-2: Simulating the Earth's temperature variations and comparing the results to the measured changes can provide insight into the underlying causes of the major changes. A climate model can be used to simulate the temperature changes that occur from both natural and anthropogenic causes. The simulations represented by the band in (a) were done with only natural forcings: solar variation and volcanic activity. Those encompassed by the band in (b) were done with anthropogenic forcings: greenhouse gases and an estimate of sulfate aerosols. Those encompassed by the band in (c) were done with both natural and anthropogenic forcings included. From (b), it can be seen that the inclusion of anthropogenic forcings provides a plausible explanation for a substantial part of the observed temperature changes over the past century, but the best match with observations is obtained in (c) when both natural and anthropogenic factors are included. These results show that the forcings included are sufficient to explain the observed changes, but do not exclude the possibility that other forcings may also have contributed. Similar results to those in (b) are obtained with other models with anthropogenic forcing. Source: IPCC (2001d).

Figure 5-1: Conceptualization of the global water cycle and its interactions with all other components of the Earth-climate system. The figure illustrates the transport and transformation of water within the Earth system, and the distribution of freshwater over the Earth's surface. The cycling of water in the Earth system involves water in all three of its phases: solid (snow, ice), liquid (precipitation, the oceans, land water bodies, groundwater, rivers/lakes, etc.), and gaseous (atmospheric water vapor, fluxes between the atmosphere and the land surface in the form of evaporation, evapotranspiration from vegetation, and evaporation from the oceans). The water cycle operates on a continuum of time and space scales and exchanges large amounts of energy as water undergoes phase changes and is moved dynamically from one part of the Earth system to another. The energy exchanges involving water in all its phases include interactions with radiation and dynamics of the atmospheric circulation; together, they intertwine the water and energy cycles of the Earth system. Source: Paul Houser and Adam Schlosser, NASA GSFC.

Figure 5-3: Observational evidence showing the effect of aerosols on cloud formation and precipitation processes. TRMM Visible and Infrared Scanner (VIRS) paints yellow pollution tracks in the clouds over South Australia due to reduced droplet size (Area 2). The TRMM Precipitation Radar shows precipitation as white patches only outside the pollution tracks (Areas 1 and 3). TRMM Microwave Imager (TMI) data (not shown) indicated ample water in the polluted clouds. The effective droplet radius retrieved by TRMM VIRS does not exceed 14 μm in polluted clouds within Area 2, lending credence to the hypothesis that pollution causes effective cloud water droplet size to drop below this precipitation threshold. Source: NASA GSFC (research results from Daniel Rosenfield, The Hebrew University of Jerusalem).

Figure 5-5: Factors affecting the predictability of precipitation in summer (June, July, and August). The influence on the predictability of precipitation is shown for (a) sea surface temperature (SST) alone and (b) both SST and land surface moisture state. The values of the predictability index were determined through an analysis of an ensemble of multi-decade atmospheric general circulation model simulations. The plotted precipitation predictability index varies from 0 to 1. Areas with values close to 1 indicate that precipitation is strongly determined by SST (and/or soil moisture), and is thus "predictable" on seasonal time scales if the SSTs (and soil moisture) themselves are predictable; whereas values close to 0 indicate that a foreknowledge of these fields may not lead to useful seasonal precipitation predictions. In the latter areas (values close to 0), nonlinear turbulent atmospheric processes would appear to dominate over SST (or soil moisture) control. Source: R.D. Koster, M. Suarez, and M. Heiser, NASA.

Figure 6-2: Forest cover increase and abandonment of marginal agricultural lands in Grand Traverse County, Michigan. Analyses of the upper midwestern United States show an increase in forest cover since 1970. This increase results in greater carbon sequestration because forest cover is primarily replacing pastures and croplands. The left image illustrates land-use change by parcel interpreted from aerial photographs. Green colors are forest, beige/yellow agriculture, and pink color residential development. The right image is forest cover from Landsat MSS satellite images. Green is

forest and light yellow not forest. The images illustrate the forest regrowth that is occurring across the Upper Midwest concurrently with parcel fragmentation, agricultural abandonment, and rural residential development. Source: School of Natural Resources and Environment, University of Michigan.

Figure 6-3: Land-cover change in eastern U.S. ecosystems, 1973-2000. An analysis of land-use and land-cover change in eastern U.S. ecological regions provides evidence of distinctive regional variation in the rates and characteristics of changes. USGS, in cooperation with EPA and NASA, used Landsat images from 5 years (1973, 1980, 1986, 1992, and 2000) to map the rates of ecoregion change in each time interval (portrayed in ecoregion color), and the primary land-cover transformations (portrayed in the pie charts). Land cover of approximately 20 percent of the land in the Mid-Atlantic Coastal Plain and Southeastern Plain changed during the nearly 30-year period due to the rapid, cyclic harvesting and replanting of forests. The adjacent Piedmont region also showed substantial change in forest cover. Urbanization was the dominant conversion in the Northern Piedmont and Atlantic Coast Pine Barrens. The two Appalachian regions studied (Blue Ridge and North Central Appalachia) had comparatively low overall change, with the primary transformations being urban development and forest conversion, respectively. Source: USGS EROS Data Center.

Figure 6-4: The map on the left is a true reflection of land use types as of 1994 in the seven-county area of central Maryland and the map on the right is the predicted distribution of land use types based on a 'polycentric' city model. Comparing the two maps, one sees remarkable similarity in commercial, high, and medium densities, but disagreement in the low density residential category. This finding is important because it is this fragmented, low density residential development that is consuming a disproportionate amount of open space and is causing high public service costs and possibly large environmental costs. Models with better projections of this type of development will better support development-related decisionmaking. Work is being done with grants from NASA to develop models that can predict this type of development. Source: Nancy Bockstael, University of Maryland.

Figure 6-5: Increasing atmospheric carbon dioxide (CO_2), climatic variation, and fire disturbance play substantial roles in the historical carbon dynamics of Alaska. Analyses of stand-age distribution in Alaska indicate that fire has likely become less frequent compared to the first half of the 20th century. The contemporary stand-age distribution of forest in interior Alaska is reproduced by assuming that fire return interval (FRI) prior to 1950 was 55% of FRI since 1950. Logging associated with the gold rush appears to be responsible for the discrepancy between observed and simulated for stands that are approximately 100 years old. Application of the Terrestrial Ecosystem Model indicates that regrowth under a less frequent fire regime leads to substantial carbon storage in the state between 1980 and 1989. Source: David McGuire and Dave Verbyla, University of Alaska, Fairbanks.

Figure 7-1: The global carbon cycle. Storages [in petagrams carbon (PgC)] and fluxes (in PgC yr[-1]) of carbon for the major components of the Earth's carbon cycle are estimated for the 1990s. The carbon storage and flux figures are from the IPCC's Third Assessment

Report (IPCC, 2001a). Carbon is exchanged among the atmosphere, oceans, and land, and, more slowly, with sediments and sedimentary rocks. The natural cycling of carbon within the Earth system is fundamental to the capture and storage of the energy that fuels life and human societies and to regulating the climate of the planet. In this figure, the components are simplified and average values are presented. There is strong evidence of significant year-to-year fluctuations in many of the key fluxes (see, e.g., Figure 7-3), which this static view cannot portray. Source: Graphic courtesy of David Schimel and Lisa Dilling, NCAR; it is updated from UCAR, 2000: *The Carbon Cycle* [Wigley T. and D. Schimel (eds.)]. Cambridge University Press, New York, NY, 292 pp. Figure update courtesy of D. Schimel and L. Dilling, NCAR.

Figure 7-3: Rate of increase of atmospheric CO_2 and fossil-fuel emissions. The upper curve shows the annual global amount of carbon added to the atmosphere (in PgC yr^{-1}) in the form of CO_2 by burning coal, oil, and natural gas. The strongly varying curve shows the annual rate of increase of carbon in the atmosphere based on measurements, originally at Mauna Loa Observatory, and later at a global network of monitoring sites. The difference between the two curves represents the total net amount of CO_2 absorbed each year by the oceans and terrestrial ecosystems. Source: CMDL, 2002: *Climate Monitoring and Diagnostics Laboratory Summary Report 26, 2000-2001* [King, D.B., et al. (eds.)]. Climate Monitoring and Diagnostics Laboratory, Boulder, Colorado.

Figure 7-4: Map showing the climatological mean annual distribution of net sea-air CO_2 flux (moles CO_2 m^{-2} yr^{-1}) over the global oceans in a reference year 1995. The blue-magenta areas indicate that the ocean is a sink for atmospheric CO_2, and the yellow-red areas, a source. The annual CO_2 uptake by the oceans has been estimated to be in the range of 1.5 to 2.1 PgC. This map has been constructed on the basis of about 1.4 million measurements of partial pressure of CO_2 (pCO_2) in surface waters made during the past 40 years. The sea-air pCO_2 difference is computed using atmospheric CO_2 concentrations from the GLOBALVIEW-CO_2 data set compiled by CMDL/NOAA. The monthly net sea-air CO_2 flux in each 4° x 5° pixel area has been estimated using the NCEP 41-mean wind speed at 10 meters. Source: After Takahashi et al. (2002). Reference: Takahashi, T., Sutherland, S.C., Sweeney, C., Poison, A., Metzl, N., Tilbrook, B., Bates, N., Wanninkhof, R., Feely, R.A., Sabine, C., Olafsson, J., and Nojiri, Y., 2002: Global sea-air CO_2 flux based on climatological surface ocean pCO_2, and seasonal biological and temperature effects. *Deep Sea Research II*, **49**, 1601-1622.

Figure 7-5: Soil processes influence carbon sequestration and transport. The dynamics of carbon transformations and transport in soil are complex and can result in sequestration in the soil as organic matter or in groundwater as dissolved carbonates, increased emissions of CO_2 to the atmosphere, or export of carbon in various forms into aquatic systems. Source: DOE, 1999: *Carbon Sequestration Research and Development* [Reichle, D., J. Houghton, B. Kane, J. Ekmann (eds.), S. Benson, J. Clarke, R. Dahlman, G. Hendrey, H. Herzog, J. Hunter-Cevera, G. Jacobs, R. Judkins, J. Ogden, A. Palmisano, R. Socolow, J. Stringer, T. Surles, A. Wolsky, N. Woodward, and M. York]. Department of Energy, Oak Ridge, TN.

Figure 9-2: Possible pathways of public health impacts from climate change. Source: Adapted from Patz and Balbus, 1996: Methods for assessing public health vulnerability to climate change. *Climate Research*, **6**, 113-125.

Figure 12-1a: Three-dimensional representation of the latitudinal distribution of atmospheric carbon dioxide (CO_2) in the marine boundary layer. The data are from the NOAA Climate Monitoring and Diagnostics Laboratory (CMDL) cooperative air sampling network. The surface represents data smoothed in time and latitude. Source: Pieter Tans and Thomas Conway, NOAA CMDL Carbon Cycle Greenhouse Gases, Boulder, Colorado.

Figure 12-1b: The NOAA CMDL Carbon Cycle Greenhouse Gases group operates four measurement programs. *In situ* measurements are made at the CMDL baseline observatories: Barrow, Alaska; Mauna Loa, Hawaii; Tutuila, American Samoa; and South Pole, Antarctica. The cooperative air sampling network includes samples from fixed sites and commercial ships. Measurements from tall towers and aircraft began in 1992. Presently, atmospheric CO_2, methane, carbon monoxide, hydrogen, nitrous oxide, sulfur hexafluoride, and the stable isotopes of CO_2 and methane are measured. Source: Pieter Tans, NOAA CMDL Carbon Cycle Greenhouse Gases, Boulder, Colorado.

Figure 12-2: The U.S. Geological Survey (USGS) Hydro-Climatic Data Network (HCDN) was compiled in 1988 and included 854 gauges that were active and had at least 36 years of record. Stations that are still active are shown in green. Stations discontinued since 1988 are shown in red. These stations would have at least 50 years of record as of 2002 if they had been kept in operation. HCDN is a subset of the USGS stream gauge network that includes continuous long-record stations that have relatively accurate records and are relatively free of overt regulation or diversions. Source: William Kirby, Global Hydroclimatology Program, USGS.

Appendix 12.1: Adapted from Unninayar, S. and R.A. Schiffer, 2002: Earth Observing Systems. In: *Encyclopedia of Global Environmental Change, Volume 1: The Earth System: Physical and Chemical Dimensions of Global Environmental Change* [MacCracken, M.C. and J. S. Perry (eds.)]. John Wiley & Sons, Ltd, Chichester, United Kingdom, pp. 61-81.

Glossary of Terms

Acclimatization
The physiological adaptation to climatic variations. Biologically, acclimation is a physiological (phenotypic) adjustment by an organism to an environmental change (distinguished from *adaptation*, which is genotypic).

Adaptation
Adjustment in natural or *human systems* to a new or changing environment that exploits beneficial opportunities or moderates negative effects.

Adaptive capacity
The ability of a *system* to adjust to *climate change* (including *climate variability* and extremes) to moderate potential damages, to take advantage of opportunities, or to cope with the consequences.

Adaptive management
Operational decisions principally for managing entities that are influenced by *climate variability* and *change*.

Adaptive management decisions
Operational decisions, principally for managing entities that are influenced by *climate variability* and *change*. These decisions can apply to the management of infrastructure (e.g., a waste water treatment plant), the integrated management of a natural resource (e.g., a watershed), or the operation of societal response mechanisms (e.g., health alerts, water restrictions). *Adaptive management* operates within existing policy frameworks or uses existing infrastructure, and the decisions usually occur on time scales of a year or less. See *policy decisions*.

Aerosols
Airborne solid or liquid particles, with a typical size between 0.01 and 10 μm that reside in the atmosphere for at least several hours. Aerosols may be of either natural or anthropogenic origin. Aerosols may influence *climate* in two ways: directly through scattering and absorbing radiation, and indirectly through acting as condensation nuclei for cloud formation or modifying the optical properties and lifetime of clouds.

Albedo
The fraction of solar radiation reflected by a surface or object, often expressed as a percentage. Snow-covered surfaces have a high albedo; the albedo of soil ranges from high to low; vegetation-covered surfaces and oceans have a low albedo. The Earth's albedo varies mainly through varying cloudiness, snow, ice, leaf area, and *land-cover* changes.

Assessments
Processes that involve analyzing and evaluating the state of

---

*Italicized words or phrases within definitions cross-reference to other glossary terms.

scientific knowledge (and the associated degree of scientific certainty) and, in interaction with users, developing information applicable to a particular set of issues or decisions.

Atmosphere
The gaseous envelope surrounding the Earth. The dry atmosphere consists almost entirely of nitrogen (78.1% volume mixing ratio) and oxygen (20.9% volume mixing ratio), together with a number of trace gases such as argon (0.93% volume mixing ratio), helium, and radiatively active *greenhouse gases* such as carbon dioxide (0.035% volume mixing ratio) and *ozone*. In addition, the atmosphere contains water vapor, whose amount is highly variable but typically 1% volume mixing ratio. The atmosphere also contains clouds and *aerosols*.

Attribution
See *detection and attribution*.

Biosphere
The part of the Earth system comprising all *ecosystems* and living organisms in the atmosphere, on land (terrestrial biosphere), or in the oceans (marine biosphere), including derived dead organic matter such as litter, soil organic matter, and oceanic detritus.

Carbon cycle
The term used to describe the flow of carbon [in various forms such as carbon dioxide (CO_2), organic matter, and carbonates] through the atmosphere, ocean, terrestrial biosphere, and lithosphere.

Climate
Climate can be defined as the statistical description in terms of the mean and variability of relevant measures of the atmosphere-ocean system over periods of time ranging from weeks to thousands or millions of years.

Climate change
A statistically significant variation in either the mean state of the *climate* or in its variability, persisting for an extended period (typically decades or longer). Climate change may be due to natural internal processes or to external forcing, including changes in solar radiation and volcanic eruptions, or to persistent human-induced changes in atmospheric composition or in *land use*. See also *climate variability*.

Climate feedback
An interaction among processes in the *climate system* in which a change in one process triggers a secondary process that influences the first one. A positive feedback intensifies the change in the original process, and a negative feedback reduces it.

Climate model
A numerical representation of the *climate system* based on

the physical, chemical, and biological properties of its components, their interactions and *feedback* processes, and accounting for all or some of its known properties. The climate system can be represented by models of varying complexity—that is, for any one component or combination of components a "hierarchy" of models can be identified, differing in such aspects as the number of spatial dimensions, the extent to which physical, chemical or biological processes are explicitly represented, or the level at which empirical *parametrizations* are involved. Coupled atmosphere/ocean/sea-ice general circulation models provide a comprehensive representation of the climate system. There is an evolution towards more complex models with active chemistry and biology. Climate models are applied, as a research tool, to study and simulate the climate, but also for operational purposes, including monthly, seasonal, and interannual climate *predictions*.

Climate scenario

A plausible and often simplified representation of the future *climate*, based on an internally consistent set of climatological relationships, that has been constructed for explicit use in investigating the potential consequences of anthropogenic *climate change*, often serving as input to impact models. Climate *projections* often serve as the raw material for constructing climate scenarios, but climate scenarios usually require additional information such as about the observed current climate. A "climate change scenario" is the difference between a climate scenario and the current climate.

Climate sensitivity

In IPCC assessments, "equilibrium climate sensitivity" refers to the equilibrium change in global mean surface temperature following a doubling of the atmospheric (equivalent) CO_2 concentration. More generally, equilibrium climate sensitivity refers to the equilibrium change in surface air temperature following a unit change in radiative forcing ($°C/Wm^{-2}$). In practice, the evaluation of the equilibrium climate sensitivity requires very long simulations with coupled general circulation models. The "effective climate sensitivity" is a related measure that circumvents this requirement. It is evaluated from model output for evolving non-equilibrium conditions. It is a measure of the strengths of the *feedbacks* at a particular time and may vary with forcing history and climate state. See also *climate model*.

Climate system

The highly complex system consisting of five major components: the *atmosphere*, the *hydrosphere*, the *cryosphere*, the land surface, and the *biosphere*, and the interactions among them. The climate system evolves in time under the influence of its own internal dynamics and because of external forcings such as volcanic eruptions, solar variations, and human-induced forcings such as the changing composition of the atmosphere and *land-use change*.

Climate variability

Variations in the mean state and other statistics of climatic features on temporal and spatial scales beyond those of individual *weather* events. These often are due to internal processes within the *climate system*. Examples of cyclical forms of climate variability include *El Niño-Southern Oscillation*, the *North Atlantic Oscillation* (NAO), and Pacific Decadal Variability (PDV). See also *climate change*.

Critical dependencies

Topics within the Strategic Plan for which progress in one research element is only possible if related research is first completed in other areas.

Cryosphere

The component of the *climate system* consisting of all snow, ice, and permafrost on and beneath the surface of the Earth and ocean.

Decision support

The provision of timely and useful information that addresses specific questions. See also *decision support resources*.

Decision support resources

The set of *observations*, analyses, interdisciplinary research products, communication mechanisms, and operational services that provide timely and useful information to address questions confronting policymakers, resource managers, and other users. See also *decision support*.

Detection and attribution

Climate varies continually on all *time scales*. Detection of *climate change* is the process of demonstrating that climate has changed in some defined statistical sense, without providing a reason for that change. Attribution of causes of climate change is the process of establishing the most likely causes for the detected change with some defined level of confidence.

Ecosystem

A community (i.e., an assemblage of populations of plants, animals, fungi, and microorganisms that live in an environment and interact with one another, forming together a distinctive living system with its own composition, structure, environmental relations, development, and function) and its environment treated together as a functional system of complementary relationships and transfer and circulation of energy and matter.

Ecosystem goods and services

Through numerous biological, chemical, and physical processes, ecosystems provide both goods and services to humanity. Goods include food, feed, fiber, fuel, pharmaceutical products, and wildlife. Services include maintenance of hydrologic cycles, cleansing of water and air, regulation of climate and weather, storage and cycling of nutrients, and provision of beauty and inspiration. Many goods pass through markets, but services rarely do.

El Niño-Southern Oscillation (ENSO)

El Niño, in its original sense, is a warmwater current that periodically flows along the coast of Ecuador and Peru, disrupting the local fishery. This oceanic event is associated with a fluctuation of the intertropical surface pressure pattern and circulation in the Indian and Pacific Oceans, called the Southern Oscillation. This coupled atmosphere-ocean phenomenon is collectively known as El Niño-Southern Oscillation. During an El Niño event, the prevailing trade winds weaken and the equatorial countercurrent strengthens, causing warm surface waters in the Indonesian area to flow eastward to overlie the cold waters of the Peru current. This event has great impact on the wind, sea surface temperature, and precipitation patterns in the tropical Pacific. It has climatic effects throughout the

Pacific region and in many other parts of the world. The opposite of an El Niño event is called La Niña.

Emissions
In the *climate change* context, emissions refer to the release of *greenhouse gases* and/or their precursors and *aerosols* into the atmosphere over a specified area and period of time.

Emissions scenario
A plausible representation of the future development of *emissions* of substances that are potentially radiatively active (e.g., *greenhouse gases*, *aerosols*), based on a coherent and internally consistent set of assumptions about driving forces (such as demographic and socio-economic development, technological change) and their key relationships. Concentration scenarios, derived from emissions scenarios, are used as input into a *climate model* to compute *climate projections*.

Evapotranspiration
The combined process of evaporation from the Earth's surface and transpiration from vegetation.

External forcing
See *climate system*.

Extreme weather event
An extreme weather event is an event that is rare within its statistical reference distribution at a particular place. Definitions of "rare" vary, but an extreme weather event would normally be as rare as or rarer than the 10th or 90th percentile. By definition, the characteristics of what is called extreme weather may vary from place to place. An extreme *climate* event is an average of a number of weather events over a certain period of time, an average which is itself extreme (e.g., rainfall over a season).

Feedback
See *climate feedback*.

Full carbon accounting
Complete accounting of all carbon stocks and changes in them for all carbon pools related to a given spatial area in a given time period.

Global change
Changes in the global environment (including alterations in *climate*, land productivity, oceans or other water resources, atmospheric chemistry, and ecological systems) that may alter the capacity of the Earth to sustain life (from the Global Change Research Act of 1990, PL 101-606).

Global change research
Study, monitoring, *assessment*, prediction, and information management activities to describe and understand the interactive physical, chemical, and biological processes that regulate the total Earth system; the unique environment that the Earth provides for life; changes that are occurring in the Earth system; and the manner in which such system, environment, and changes are influenced by human actions.

Greenhouse gas
Greenhouse gases are those gaseous constituents of the atmosphere, both natural and anthropogenic, that absorb and emit radiation at specific wavelengths within the spectrum of infrared radiation emitted by the Earth's surface, the atmosphere, and clouds. This property causes the greenhouse effect. Water vapor (H_2O), carbon dioxide (CO_2), nitrous oxide (N_2O), methane (CH_4), and ozone (O_3) are the primary greenhouse gases in the Earth's atmosphere. Moreover there are a number of entirely human-made greenhouse gases in the atmosphere, such as the halocarbons and other chlorine- and bromine-containing substances, dealt with under the Montreal Protocol.

Human system
Any system in which human organizations play a major role. Often, but not always, the term is synonymous with "society" or "social system" (e.g., agricultural system, political system, technological system, economic system).

Hydrosphere
The component of the *climate system* composed of liquid surface and subterranean water, such as oceans, seas, rivers, freshwater lakes, underground water, etc.

(Climate) Impact assessment
The practice of identifying and evaluating the detrimental and beneficial consequences of *climate change* on natural and *human systems*.

(Climate) Impacts
Consequences of *climate change* on natural and *human systems*. Depending on the consideration of *adaptation*, one can distinguish between potential impacts and residual impacts.
- Potential impacts: All impacts that may occur given a projected change in *climate*, without considering adaptation.
- Residual impacts: The impacts of climate change that would occur after adaptation.

Information
Knowledge derived from study, experience, or instruction.

Integrated assessment
A method of analysis that combines results and models from the physical, biological, economic, and social sciences, and the interactions between these components, in a consistent framework, to evaluate the status and the consequences of environmental change and the policy responses to it.

Land cover
The vegetation and artificial built-up materials covering the land surface. This includes areas of vegetation (forests, shrublands, crops, deserts, lawns), bare soil, developed surfaces (paved land, buildings), and wet areas and bodies of water (watercourses, wetlands).

Land use
The total of arrangements, activities, and inputs undertaken in a certain land cover type (a set of human actions). The social and economic purposes for which land is managed (e.g., grazing, timber extraction, and conservation).

Land-use change
A change in the use or management of land by humans, which may

lead to a change in land cover. Land cover and land-use change may have an impact on the albedo, evapotranspiration, *sources*, and *sinks* of *greenhouse gases*, or other properties of the *climate system*, and may thus have an impact on *climate*, locally or globally.

Lifetime

Lifetime is a general term used for various *time scales* characterizing the rate of processes affecting the concentration of trace gases. In general, lifetime denotes the average length of time that an atom or molecule spends in a given reservoir, such as the atmosphere or oceans. The following lifetimes may be distinguished:

- "Turnover time" (T) or "atmospheric lifetime" is the ratio of the mass M of a reservoir (e.g., a gaseous compound in the atmosphere) and the total rate of removal S from the reservoir: $T = M/S$. For each removal process separate turnover times can be defined.
- "Adjustment time," "response time," or "perturbation lifetime" (T_a) is the time scale characterizing the decay of an instantaneous pulse input into the reservoir. The term adjustment time is also used to characterize the adjustment of the mass of a reservoir following a step change in the source strength. Half-life or decay constant is used to quantify a first-order exponential decay process. See *response time* for a different definition pertinent to *climate* variations. The term "lifetime" is sometimes used, for simplicity, as a surrogate for "adjustment time."

In simple cases, where the global removal of the compound is directly proportional to the total mass of the reservoir, the adjustment time equals the turnover time: $T = T_a$. An example is CFC-11 which is removed from the atmosphere only by photochemical processes in the *stratosphere*. In more complicated cases, where several reservoirs are involved or where the removal is not proportional to the total mass, the equality $T = T_a$ no longer holds. CO_2 is an extreme example. Its turnover time is only about 4 years because of the rapid exchange between atmosphere and the ocean and terrestrial biota. However, a large part of that CO_2 is returned to the atmosphere within a few years. Thus, the adjustment time of CO_2 in the atmosphere is actually determined by the rate of removal of carbon from the surface layer of the oceans into its deeper layers. Although an approximate value of 100 years may be given for the adjustment time of CO_2 in the atmosphere, the actual adjustment is faster initially and slower later on. In the case of CH_4, the adjustment time is different from the turnover time, because the removal is mainly through a chemical reaction with the hydroxyl radical OH, the concentration of which itself depends on the CH_4 concentration. Therefore the CH_4 removal S is not proportional to its total mass M.

Mitigation (climate change)

An intervention to reduce the causes of change in climate. This could include approaches devised to reduce emissions of greenhouse gases to the atmosphere; to enhance their removal from the atmosphere through storage in geological formations, soils, biomass, or the ocean; or to alter incoming solar radiation through several "geo-engineering" options.

Mitigative capacity

The social, political, and economic structures and conditions that are required for effective *mitigation*.

Monitoring

A scientifically designed system of continuing standardized measurements and observations and the evaluation thereof.

North Atlantic Oscillation (NAO)

The North Atlantic Oscillation consists of opposing variations of barometric pressure near Iceland and near the Azores. On average, a westerly current, between the Icelandic low pressure area and the Azores high pressure area, carries cyclones with their associated frontal systems towards Europe. However, the pressure difference between Iceland and the Azores fluctuates on *time scales* of days to decades, and can be reversed at times. It is the dominant mode of winter climate variability in the North Atlantic region, ranging from central North America to Europe.

Observations

Standardized measurements (either continuing or episodic) of variables in *climate* and related systems.

Parameterization

In *climate models*, this term refers to the technique of representing processes that cannot be explicitly resolved at the spatial or temporal resolution of the model (sub-grid-scale processes), by relationships between the area- or time-averaged effect of such sub-grid-scale processes and the larger scale flow.

Place-based

Related to the locus (regional, sectoral, cultural) of a particular object or action (e.g., place-based decisions).

Planning

A process inherently important for both *policy decisions* and *adaptive management*. It usually occurs in the framework of established or projected policy options.

Policy decisions

Decisions that result in laws, regulations, or other public actions. These decisions are typically made in government settings (federal, state, local) by elected or appointed officials. These decisions, which usually involve balancing competing value issues, can be assisted by—but not specified by—scientific analyses. See *adaptive management decisions*.

Prediction (climate)

A probabilistic description or forecast of a future *climate* outcome based on *observations* of past and current climatological conditions and quantitative models of climate processes (e.g., a prediction of an El Niño event).

Projection (climate)

A description of the response of the *climate system* to an assumed level of future radiative forcing. Changes in radiative forcing may be due to either natural sources (e.g., volcanic emissions) or human-induced causes (e.g., emissions of *greenhouse gases* and *aerosols*, or changes in *land use* and *land cover*). Climate "projections" are distinguished from climate "predictions" in order to emphasize that climate projections depend on *scenarios* of future socioeconomic, technological, and policy developments that may or may not be realized.

Radiative forcing

A process that directly changes the average energy balance of the Earth-atmosphere *system* by affecting the balance between incoming solar radiation and outgoing or "back" radiation. A positive forcing tends to warm the surface of the Earth and a negative forcing tends to cool the surface.

Rapid climate change

The non-linearity of the *climate system* may lead to rapid *climate change*, sometimes called abrupt events or even surprises. Some such abrupt events may be imaginable, such as a dramatic reorganization of the thermohaline circulation, rapid deglaciation, or massive melting of permafrost leading to fast changes in the *carbon cycle*. Others may be truly unexpected, as a consequence of a strong, rapidly changing, forcing of a non-linear system.

Regional reanalysis

The process of "freezing" or holding constant a recent version of a regional climate or weather model with the latest process representations and assimilation capabilities, and rerunning that model with historical satellite and *in situ* data sets to generate products for the period covered by the historic records. This process allows climatological analyses to be carried out using the best consistent data products possible.

Resilience

The ability of an organism or other entity to recover from or to adjust easily to change or other stress.

Response time

The response time or adjustment time is the time needed for the *climate system* or its components to re-equilibrate to a new state, following a forcing resulting from external and internal processes or *feedbacks*. It is very different for various components of the climate system. The response time of the *troposphere* is relatively short, from days to weeks, whereas the *stratosphere* comes into equilibrium on a *time scale* of typically a few months. Due to their large heat capacity, the oceans have a much longer response time, typically decades, but up to centuries or millennia. The response time of the strongly coupled surface-troposphere system is, therefore, slow compared to that of the stratosphere, and mainly determined by the oceans. The biosphere may respond fast (e.g., to droughts), but also very slowly to imposed changes. See *lifetime* for a different definition of response time pertinent to the rate of processes affecting the concentration of trace gases.

Scenario

A coherent description of a potential future situation that serves as input to more detailed analyses or modeling. Scenarios are tools to explore, "if ..., then..." statements, and are not *predictions* of or prescriptions for the future. See *climate scenario* and *emissions scenario*.

Sensitivity

Sensitivity is the degree to which a system is affected, either adversely or beneficially, by climate-related stimuli. The effect may be direct (e.g., a change in crop yield in response to a change in the mean, range, or variability of temperature) or indirect (e.g., damages caused by an increase in the frequency of coastal flooding due to sea-level rise). See also *climate sensitivity*.

Sequential decisionmaking

Stepwise decisionmaking aiming to identify short-term strategies in the face of long-term uncertainties, by incorporating additional information over time and making mid-course corrections.

Sequestration

The process of increasing the carbon content of a carbon reservoir other than the atmosphere. Biological approaches to sequestration include direct removal of CO_2 from the atmosphere through *land-use change*, afforestation, reforestation, and practices that enhance soil carbon in agriculture. Physical approaches include separation and disposal of CO_2 from flue gases or from processing fossil fuels to produce hydrogen- and CO_2-rich fractions and long-term storage in underground depleted oil and gas reservoirs, coal seams, and saline aquifers.

Sink

Any process, activity, or mechanism that removes a *greenhouse gas*, an *aerosol*, or a precursor of a greenhouse gas or aerosol from the atmosphere.

Socio-economic potential

The socio-economic potential represents the level of greenhouse gas *mitigation* that would be approached by overcoming social and cultural obstacles to the use of technologies that are cost-effective.

Source

Any process, activity, or mechanism that releases a *greenhouse gas*, an *aerosol*, or a precursor of a greenhouse gas or aerosol into the atmosphere.

Spatial and temporal scales

Climate may vary on a large range of spatial and temporal scales. Spatial scales may range from local (less than 100,000 km²), through regional (100,000 to 10 million km²) to continental (10 to 100 million km²). Temporal scales may range from seasonal to geological (up to hundreds of millions of years).

Stabilization

The achievement of stabilization of atmospheric concentrations of one or more *greenhouse gases*.

Stakeholders

Individuals or groups whose interests (financial, cultural, value-based, or other) are affected by *climate variability*, *climate change*, or options for adapting to or mitigating these phenomena. Stakeholders are important partners with the research community for development of *decision support resources*.

Stratosphere

The highly stratified region of the atmosphere above the *troposphere* extending from about 10 km (ranging from 9 km in high latitudes to 16 km in the tropics on average) to about 50 km.

System

Integration of interrelated, interacting, or interdependent components into a complex whole.

Technology

A piece of equipment or a technique for performing a particular activity.

Technology transfer

The broad set of processes that cover the exchange of knowledge, money, and goods among different *stakeholders* that lead to the spreading of *technology* for adapting to or mitigating *climate change*. As a generic concept, the term is used to encompass both diffusion of technologies and technological cooperation across and within countries.

Time scale

Characteristic time for a process to be expressed. Since many processes exhibit most of their effects early, and then have a long period during which they gradually approach full expression, for the purpose of this report the time scale is numerically defined as the time required for a perturbation in a process to show at least half of its final effect.

Troposphere

The lowest part of the atmosphere from the surface to about 10 km in altitude in mid-latitudes (ranging from 9 km in high latitudes to 16 km in the tropics on average) where clouds and "weather" phenomena occur. In the troposphere, temperatures generally decrease with height.

Uncertainty

An expression of the degree to which a value (e.g., the future state of the *climate system*) is unknown. Uncertainty can result from lack of information or from disagreement about what is known or even knowable. It may have many types of sources, from quantifiable errors in the data to ambiguously defined concepts or terminology, or uncertain projections of human behavior. Uncertainty can therefore be represented by quantitative measures (e.g., a range of values calculated by various models) or by qualitative statements (e.g., reflecting the judgment of a team of experts).

Vulnerability

The degree to which a system is susceptible to, or unable to cope with, adverse effects of *climate change*, including *climate variability* and extremes. Vulnerability is a function of the character, magnitude, and rate of climate variation to which a system is exposed, its *sensitivity*, and its *adaptive capacity*.

Weather

The specific condition of the atmosphere at a particular place and time. It is measured in terms of parameters such as wind, temperature, humidity, atmospheric pressure, cloudiness, and precipitation.

ACIA	Arctic Climate Impact Assessment
ACRIM	Active Cavity Radiometer Irradiance Monitor
ACSYS	Arctic Climate System Study
AMIP	Atmospheric Model Intercomparison Project
AMSR	Advanced Microwave Scanning Radiometer
AOSB	Arctic Ocean Sciences Board
APN	Asia-Pacific Network for Global Change Research
AQRS	Air Quality Research Subcommittee
ARM	Atmospheric Radiation Measurement
ASOF	Arctic and Subarctic Ocean Fluxes Study
ATBD	Algorithm Theoretical Basis Document
AVHRR	Advanced Very High Resolution Radiometer
BAHC	Biospheric Aspects of the Hydrological Cycle (IGBP core project)
BER	Biological and Environmental Research
CAP	Climate Action Partnership
CART	Cloud Atmospheric Radiation Testbed
CCCSTI	Committee on Climate Change Science Science and Technology Integration
CCGG	Carbon Cycle Greenhouse Gases
CCRI	Climate Change Research Initiative
CCSM	Community Climate System Model
CCSP	Climate Change Science Program
CCSPO	Climate Change Science Program Office
CCTP	Climate Change Technology Program
CDC	Centers for Disease Control and Prevention
CDR	Climate Data Record
CDST	Climate Data Science Team
CEOP	Coordinated Enhanced Observing Period
CEOS	Committee on Earth Observation Satellites
CEQ	Council on Environmental Quality
CERES	Clouds and the Earth's Radiant Energy System
CFC	chlorofluorocarbon
CGIAR	Consultative Group on International Agricultural Research
CH_3SCH_3	dimethyl sulfide
CH_4	methane
CliC	Climate and Cryosphere
CLIVAR	Climate Variability and Predictability
CMDL	Climate Monitoring and Diagnostics Laboratory
CMI	Common Modeling Infrastructure
CMIP	Coupled Model Intercomparison Project
CO	carbon monoxide
CO_2	carbon dioxide
COLA	Center for Ocean-Land-Atmosphere Studies
CONCAUSA	Central American-United States of America Joint Accord
CPT	Climate Process and Modeling Team
DIF	Directory Interchange Format
DIS	Data and Information System
DL	Digital Library
DMWG	Data Management Working Group
DOC	Department of Commerce
DOD	Department of Defense
DOE	Department of Energy
DOI	Department of the Interior
DOS	Department of State
DOT	Department of Transportation
DPC	Domestic Policy Council
EMF	Energy Modeling Forum
ENSO	El Niño-Southern Oscillation
EOS	Earth Observing System
EPA	Environmental Protection Agency
ESA	European Space Agency
ESG	Earth System Grid
ESMF	Earth System Modeling Framework
ESS-P	Earth System Science Partnership
EU	European Union
FAO	Food and Agriculture Organization
FAQ	Frequently Asked Question
FCS	Fire-Climate-Society
FEWS NET	Famine Early Warning System Network
FGDC	Federal Geographic Data Clearinghouse
FRI	Fire Return Interval
GACC	Geographic Area Coordination Center
GAW	Global Atmosphere Watch Programme
GCM	general circulation model
GCMD	Global Change Master Directory
GCP	Global Carbon Project
GCOS	Global Climate Observing System
GCTE	Global Climate and Terrestrial Ecosystems
GECAFS	Global Environmental Change and Food Systems
GEF	Global Environment Facility
GEOHAB	Global Ecology and Oceanography of Harmful Algal Bloom
GEWEX	Global Energy and Water Cycle Experiment
GFDL	Geophysical Fluid Dynamics Laboratory
GHG	greenhouse gas
GIS	geographic information system
GLOBEC	Global Ocean Ecosystem Dynamics
GOFC-GOLD	Global Observation of Forest Cover and Global Observations of Land Cover Dynamics
GOIN	Global Observation Information Network
GOOS	Global Ocean Observing System
GOS	Global Observing System
GPM	Global Precipitation Measurement
GSFC	Goddard Space Flight Center
GSN	GCOS Surface Network
GTOS	Global Terrestrial Observing System
GUAN	GCOS Upper-Air Network

GWS	Global Water System		NCAR	National Center for Atmospheric Research
HAB	harmful algal bloom		NCEP	National Centers for Environmental Prediction
HCDN	Hydro-Climatic Data Network		NEC	National Economic Council
HCFC	hydrochlorofluorocarbon		NEESPI	Northern Eurasia Earth Science Partnership Initiative
HECRTF	High-End Computing Revitalization Task Force		NERSC	National Energy Research Scientific Computing
HELP	Hydrology for Environment, Life, and Policy		NGO	Non-governmental organization
HFC	hydrofluorocarbon		NH_3	ammonia
HHS	Department of Health and Human Services		NIEHS	National Institute of Environmental Health Sciences
HLC	High-Level Consultations on Climate Change		NIH	National Institutes of Health
HPS	hantavirus pulmonary syndrome		NIST	National Institute of Standards and Technology
HRDLS	High-Resolution Dynamics Limb Sounder		NOAA	National Oceanic and Atmospheric Administration
HSB	Humidity Sounder for Brazil		NOPP	National Oceanographic Partnership Program
IAI	Inter-American Institute for Global Change Research		NO_x	nitrogen oxides
IASC	International Arctic Sciences Committee		NPEO	North Pole Environmental Observatory
ICSU	International Council of Scientific Unions		NPOESS	National Polar-Orbiting Operational Environmental Satellite System
IDN	International Directory Network		NRC	National Research Council
IGAC	International Global Atmospheric Chemistry		NSC	National Security Council
IGBP	International Geosphere-Biosphere Programme		NSF	National Science Foundation
IGFA	International Group of Funding Agencies for Global Change Research		O_2	oxygen
IGOS	International Global Observing Strategy		O_3	ozone
IGOS-P	International Global Observing Strategy Partnership		ODP	Ocean Drilling Program
IHDP	International Human Dimensions Programme		OH	hydroxyl radical
ILTER	International Long-Term Ecological Research		OMB	Office of Management and Budget
IOC	Intergovernmental Oceanographic Commission		ONR	Office of Naval Research
IODP	Integrated Ocean Drilling Program		OSMEC	Observing System Management Executive Council
IPCC	Intergovernmental Panel on Climate Change		OSOEC	Observing System Operations Executive Council
IRC	International Research and Cooperation		OSSE	Observation System Simulation Experiment
IRI	International Research Institute for Climate Prediction		OSSEC	Observing System Science Executive Council
ITCZ	Intertropical Convergence Zone		OSTP	Office of Science and Technology Policy
IWG	Interagency Working Group		PAGES	Past Global Changes
IWGCCST	Interagency Working Group on Climate Change Science and Technology Integration		PDSI	Palmer Drought Severity Index
			PDV	Pacific Decadal Variability
JGOFS	Joint Global Ocean Flux Study		PFC	perfluorocarbon
LBA	Large-Scale Biosphere-Atmosphere Experiment in Amazonia		PMIP	Paleoclimate Model Intercomparison Project
			PNNL	Pacific Northwest National Laboratory
LOICZ	Land-Ocean Interactions in the Coastal Zone project		RCM	regional climate model
			RESACs	Regional Earth Science Applications Centers
LUCC	Land-Use and Land-Cover Change		RISA	Regional Integrated Sciences and Assessments
MA	Millennium Ecosystem Assessment		SAGE	Stratospheric Aerosol and Gas Experiment
MEXT	Ministry of Education, Culture, Sport, Science and Technology		SAM	southern hemisphere annular mode
			SBSTA	Subsidiary Body for Scientific and Technical Advice
MMV	Measurement, Monitoring, and Verification		SCAR	Scientific Committee on Antarctic Research
MODIS	Moderate-Resolution Imaging Spectroradiometer		SCOR	Scientific Committee on Oceanic Research
MOPITT	Measurement of Pollution in the Troposphere		SeaWiFS	Sea-Viewing Wide Field-of-View Sensor
MSS	Multispectral Scanner		SF_6	sulfur hexafluoride
MSU	Microwave Sounding Unit		SGCR	Subcommittee on Global Change Research
N	nitrogen		SGP	Southern Great Plains
N_2O	nitrous oxide		SO_2	sulfur dioxide
NACIP	National Aerosol–Climate Interactions Program		SOLAS	Surface Ocean–Lower Atmosphere Study
NACP	North American Carbon Program		SPARC	Stratospheric Processes and their Role in Climate
NAM	northern hemisphere annular mode		SSS	sea surface salinity
NAO	North Atlantic Oscillation		SST	sea surface temperature
NAS	National Academy of Sciences		START	Global Change SysTem for Analysis, Research, and Training
NASA	National Aeronautics and Space Administration		SWAQ	Sub-committee on Water Availability and Quality
NASDA	Japan's National Space Development Agency		TAV	Tropical Atlantic Variability

TISO	Tropical Intra-Seasonal Oscillation	**USGCRP**	U.S. Global Change Research Program
TM	Thematic Mapper	**USGS**	U.S. Geological Survey
TMI	TRMM Microwave Imager	**UV-B**	ultraviolet-B
TOGA	Tropical Ocean and Global Atmosphere project	**VIRS**	Visible and Infrared Scanner
TOMS	Total Ozone Mapping Spectrometer	**VOC**	volatile organic carbon
TRMM	Tropical Rainfall Measuring Mission	**VAMOS**	Variability of the American Monsoon Systems
UNDP	United Nations Development Programme	**WCRP**	World Climate Research Programme
UNEP	United Nations Environment Programme	**WDCS**	World Data Center System
UNESCO	United Nations Educational, Scientific, and Cultural Organization	**WGCM**	Working Group on Coupled Modeling
		WHO	World Health Organization
UNFCCC	United Nations Framework Convention on Climate Change	**WMO**	World Meteorological Organization
		WOCE	World Ocean Circulation Experiment
USAID	U.S. Agency for International Development	**WRAP**	Water Resources Applications Project
USDA	U.S. Department of Agriculture	**WWW**	World Weather Watch

UNITS

SI (Systéme Internationale) Units

Physical Quantity	Name of Unit	Symbol
length	meter	m
mass	kilogram	kg
time	second	s
thermodynamic temperature	kelvin	K
amount of substance	mole	mol

Special Names and Symbols for Certain SI-Derived Units

Physical Quantity	Name of SI Unit	Symbol for SI Unit	Definition of Unit
force	newton	N	$kg\ m\ s^{-2}$
pressure	pascal	Pa	$kg\ m^{-1}\ s^{-2}\ (= Nm^{-2})$
energy	joule	J	$kg\ m^2\ s^{-2}$
power	watt	W	$kg\ m^2\ s^{-3}\ (= Js^{-1})$
frequency	hertz	Hz	s^{-1} (cycle per second)

Decimal Fractions and Multiples of SI Units Having Special Names

Physical Quantity	Name of Unit	Symbol for Unit	Definition of Unit
length	ångstrom	Å	$10^{-10}\ m = 10^{-8} cm$
length	micrometer	μm	$10^{-6} m = μm$
area	hectare	ha	$10^4\ m^2$
force	dyne	dyn	$10^{-5}\ N$
pressure	bar	bar	$10^5\ N\ m^{-2}$
pressure	millibar	mb	$1hPa$
weight	ton	t	$10^3\ kg$

CLIMATE CHANGE SCIENCE PROGRAM OFFICE

James R. Mahoney, CCSP Director

Richard H. Moss, CCSPO Director

David M. Allen

Jeff Amthor

Susan K. Avery

James H. Butler

Margarita Conkright Gregg

David J. Dokken [TECHNICAL EDITOR]

Susanna Eden

Genene Fisher

Stephanie A. Harrington

Chester J. Koblinsky

David M. Legler

Sandy MacCracken

Jessica Orrego

Rick S. Piltz

Nicholas A. Sundt

Ahsha N. Tribble

Bud Ward

Robert C. Worrest

To obtain a copy of this document, contact:

Climate Change Science Program Office
1717 Pennsylvania Avenue, NW
Suite 250
Washington, DC 20006
202-223-6262 (voice)
202-223-3065 (fax)
information@climatescience.gov
http://www.climatescience.gov/
http://www.usgcrp.gov/

The Climate Change Science Program
incorporates the U.S. Global Change
Research Program and the Climate Change
Research Initiative.

www.ingramcontent.com/pod-product-compliance
Lightning Source LLC
Chambersburg PA
CBHW08063718052 6
45168CB00008B/3204